Birkhäuser

Ellina Grigorieva

Methods of Solving Sequence and Series Problems

 Birkhäuser

Ellina Grigorieva
Department of Mathematics and Computer Science
Texas Woman's University
Denton, TX, USA

ISBN 978-3-319-83348-4   ISBN 978-3-319-45686-7   (eBook)
DOI 10.1007/978-3-319-45686-7

Mathematics Subject Classification (2010): 97I30, 97U40, 97I50, 97I40

Printed on acid-free paper

This book is published under the trade name Birkhäuser, www.birkhauser-science.com
The registered company is Springer International Publishing AG
The registered company address is: Gewerbestrasse 11, 6330 Cham, Switzerland

To my Beautiful Daughter,
  Sasha
And to my Wonderful Parents,
  Natali** and **Valery Grigoriev
Your Encouragement Made this Book
Possible
And to my University Mentor and Scientific
Advisor Academician,
  Stepanov Nikolay Fedorovich
Without your Help and Brilliant Mind my
Career as a Scientist Would not be
Successful!

Preface

Many contest problems concern series and sequences. Each year contest problems become more and more challenging and even an "A" math student without special preparation for such problems could feel frustrated and lost. Based on my own experience, the first time I had to consider a series, different from an arithmetic or geometric progression, it was at my city math olympiad when I was in 9th grade. I remember that one of the problems looked like this.

> Evaluate the sum $\frac{1}{1+\sqrt{2}} + \frac{1}{\sqrt{2}+\sqrt{3}} + \frac{1}{\sqrt{3}+\sqrt{4}} + \ldots + \frac{1}{\sqrt{1977}+\sqrt{1978}}$.

I knew that this sum exists because they would not ask to evaluate it otherwise. I noticed that there are 1977 terms to add, and of course it would not be math contest if there was not some interesting approach to find this sum without putting 1977 over the common denominator. The first thing I tried worked. I multiplied the numerator and denominator of one fraction by the quantity that differed from its denominator by only the sign. Then I applied the formula of the difference of two squares, which made the denominator of the fraction one.

For example, $\frac{1}{\sqrt{2}+\sqrt{3}} = \frac{\sqrt{3}-\sqrt{2}}{(\sqrt{3}+\sqrt{2})(\sqrt{3}-\sqrt{2})} = \frac{\sqrt{3}-\sqrt{2}}{(\sqrt{3})^2-(\sqrt{2})^2} = \frac{\sqrt{3}-\sqrt{2}}{1} = \sqrt{3} - \sqrt{2}$.

Replacing each fraction as above by "rationalizing its denominator," I noticed that all radicals were canceled, except the first term and last terms, $\sqrt{2} - 1 + \sqrt{3} - \sqrt{2} + \sqrt{4} - \sqrt{3} + \sqrt{5} - \sqrt{4} + \ldots + \sqrt{1978} - \sqrt{1977} = \sqrt{1978} - 1$.

The answer was obtained!

After winning the City Olympiad, I was sent to the Regional Math Olympiad and was again surprised that two or three problems there were on topics that were not yet covered in our classes at school. One of the problems was on sequences, but again, it was different from the arithmetic and geometric progressions that we learned in algebra class. Here it is.

> Find the formula for the nth term of the sequence of numbers: 1, 1, 2, 3, 5, 8, 13, 21, 34, ...

I remember that I looked at those numbers, noticed that each term starting from the third one is the sum of the two preceding terms. That allowed me to create the formula for the sequence: $a_{n+2} = a_n + a_{n+1}$, $a_1 = a_2 = 1$.

Knowing nothing about the Fibonacci sequence and what this sequence actually described, I started thinking this way, "This sequence is not an arithmetic progression because the difference of any consecutive terms is not the same." I asked myself a question, "What if the terms of this sequence belong to a geometric progression?" Then they must satisfy the formula above. The idea appeared to be good and after manipulations, I found the answer. I solved this problem without any preliminary knowledge about the Fibonacci sequence and derived "my method" of dealing with the sequences given by recursion.

I show how I solved that problem from 10th grade in detail by demonstrating it as Problem 24, Chapter 1 of this book. Why do I write about these two examples from my own Olympiad experience and emphasize my lack of the knowledge about special sequences? There are several reasons for this but the first is to understand that nobody knows everything. We learn by organizing information and thoughts, not by simply storing them. I do not ask you to reinvent the wheel each time. However, I ask you to understand rather than simply memorize.

My method of teaching mathematics is constructed on four simple premises:

1. It is my opinion that creativity can be developed by considering some interesting approaches while also gaining routine background knowledge. For example, a difference of squares formula that I used to solve the first problem is not boring if considered in conjunction with an example of use such as one from my other book *Methods of Solving Nonstandard Problems* for the solution of $39999 \cdot 40001$ without a calculator.

2. Math education is now mostly oriented on teaching mathematics by "having fun." But "fun" should not be skin deep. I noticed that many math educators show their students the amazing Fibonacci sequence, generate it, and show its properties using videos or slides. Yes, students probably would recognize that the sequence given at the Math Olympiad in 1978 was Fibonacci. Many would be able to find some of its terms either by hand or by using a graphing calculator. I am not sure that many would derive the formula for its nth term. A deeper understanding of mathematical concepts can be fun, and it is far more rewarding in the long term.

3. Concepts should not stand in isolation. For example, there is a connection between the golden ratio and the Fibonacci sequence which generates it. The golden ration of nature is the result of a simple recurrent relationship! Connections generate new insight and further enhance concepts.

4. The learning of mathematics should have a human purpose. Perhaps that purpose is merely to compete. That is good enough! Many modern contest problems have sequences and series either directly or as a part of a problem. Hence, as an instructor, it is a good idea to help those who want to participate in math contests to learn more about sequences and series by exploring the topics in order that one would create their own beautiful solutions to a problem.

If you are struggling with math, this book is for you. Most math books start from theoretical facts and give one or two examples and then a set of problems. In this book almost every statement is followed by problems. You are not just memorizing a theorem—you apply the knowledge immediately. Upon seeing a similar problem in the homework section you will be able to recognize and solve it. While each section of the book can be studied independently, the book is constructed to reinforce patterns developed at stages throughout the book. This helps you see how math topics are connected.

What Is This Book About?

This book is not a textbook. It is a learning and teaching tool that helps the reader to develop a creative learning experience. It gives many examples of series, partial, or infinite sums of which can be evaluated using methods taught in this book. Let us consider the problem to evaluate the series (Problem 50),

$$\frac{1}{1\cdot2} + \frac{1}{2\cdot3} + \frac{1}{3\cdot4} + \ldots + \frac{1}{2016\cdot2017}.$$

I want to share my experience with my Calculus 2 class when learning series and sequences. Although some of my students answered correctly that there are 2016 terms and recognize the formula of the nth term as $\frac{1}{n\cdot(n+1)}$, usually nobody in class can find this finite sum. I tell them a story how I evaluated this sum in 9th grade by noticing that each fraction can be written as a difference of two unit fractions

$$\frac{1}{2\cdot3} = \frac{1}{2} - \frac{1}{3}, \quad \ldots \quad \frac{1}{n(n+1)} = \frac{1}{n} - \frac{1}{n+1}.$$

They quickly replace each term by the difference and evaluate the sum as $1 - \frac{1}{2017} = \frac{2016}{2017}$. This would do nothing for most of my students and they would not remember this "trick" as nobody remembers telephone numbers anymore unless I asked them next class to evaluate the following infinite sum:

$$\frac{1}{2} + \frac{1}{6} + \frac{1}{12} + \frac{1}{20} + \frac{1}{30} + \ldots. \tag{P.1}$$

Those who recognized that this series is an infinite form of the finite series given before would evaluate its partial sum as $S_n = 1 - \frac{1}{n+1}$ and hence using the limit as n approaches infinity would state that the series is convergent to 1.

Further, we will discuss the so-called Leibniz harmonic triangle, related to Pascal's triangle, but that has only unit fractions recorded in the form of a triangle. The sum of the series represents the sum of all elements of the second diagonal of the infinite Leibniz triangle and my students learn that the sum is one using a different approach (Problem 67). Next, I ask my students to modify the Leibniz triangle so that it has only denominators of each fraction instead of fractions themselves (Figure 2.2). Let us construct a sequence of the denominators: 2, 6, 12, 20, 30, 42, $\ldots, n(n+1), \ldots$. Students see that the same numbers belong to the second diagonal of the modified Leibniz triangle! I tell my students that these numbers are special and that each of them is a double so-called triangular number, known by the Ancient Greeks, 2000 BC.

Consider the sequence 1, 3, 6, 10, 15, 21, $\ldots \frac{n(n+1)}{2}$, $\ldots$. Greeks visualized each such number placed in a triangle of side 1, 2, 3, 4, etc. The total number of the balls that could fit a triangle of side n would represent the nth triangular number. We construct by hand several triangular numbers and learn their properties (Figure P.1).

Figure P.1 Triangular
Numbers

Many properties are formulated and solved as problems in this book. For example,

Can you explain why the formula for the nth triangular number is $\frac{n(n+1)}{2}$?

This is where we recall how each was constructed and where my students see that it is the sum of all natural numbers between 1 and n. Thus,

$$1 = 1$$
$$3 = 1 + 2$$
$$6 = 1 + 2 + 3$$
$$\cdots \quad \cdot$$

It is useful to be reminded how the famous mathematician Gauss evaluated such a sum at the age of 10. The story is told frequently in algebra and calculus books, but here students can actually use the idea in deriving $T_n = 1 + 2 + 3 + \ldots + n = \frac{n(n+1)}{2}$. For example, can we evaluate the sum of the first n triangular numbers, $1 + 3 + 6 + 10 + 15 + \ldots + \frac{n(n+1)}{2}$? Different methods of finding this and other sums are taught in this book. Younger students would probably enjoy a geometric approach, and calculus students would really benefit from applying sigma notation and well-known summation formulas (Problem 38).

What actually impresses all my present and former students is the connection between the sequence of natural numbers, sequence of triangular numbers, sequence of triangular numbers and tetrahedral numbers, etc. I demonstrate that the nth partial sum of triangular numbers is the corresponding tetrahedral number (Problems 36 and 39). Denote a tetrahedral number by $Tr(n)$. Then $Tr(n) = \sum_{1}^{n} T_n$

$$= \sum_{1}^{n} \frac{n(n+1)}{2} = \frac{n(n+1)(n+2)}{6}$$ (Figure P.2). Here we can briefly discuss that although formulas for the nth terms of either triangular or tetrahedral numbers look like fractions, the numbers are always integers, because the product of two consecutive natural numbers is always a multiple of two and a product of three consecutive

Figure P.2 Tetrahedral numbers

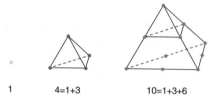

1   4=1+3   10=1+3+6

integers is always divisible by 6.

Let us return to the sum of the numbers in the second diagonal of the Leibniz-modified triangle (Figure 2.2),

$$2 + 6 + 12 + 20 + 30 + 42 + \ldots + n(n+1) = 2\sum_{k=1}^{n} \left(\frac{k(k+1)}{2} \right)$$
$$= \sum_{k=1}^{n} k(k+1)$$
$$= \frac{n(n+1)(n+2)}{3}.$$

We can see that with n increasing, this sum will increase without bound and that the corresponding infinite sum of the reciprocals of each number $\left(\frac{1}{k(k+1)} \right)$ converges to unity. The series of Eq. P.1 is called a "telescopic series" and plays a very important role in the convergence of infinite series. Jacob Bernoulli used a slight modification of this series for the comparison test and found the upper boundary for the infinite sum of the Dirichlet series, $\frac{1}{1^2} + \frac{1}{2^2} + \frac{1}{3^2} + \ldots$. This problem is named Basel's problem and was solved by Leonhard Euler 40 years after being proposed (Chapters 2 and 3). Euler found that the series converges to $\frac{\pi^2}{6}$.

Finding the sum for infinite series for which it is impossible to evaluate the partial sum is often a challenging problem. Many mathematicians of all ages at some point of their life tried to find a number associated with certain infinite series.

The first step is to establish whether the series is convergent or not. This is why I describe the famous convergence theorems for numerical and functional series in Chapter 3. Chapter 3 might not look like competition material, but it does have many unique methods for finding partial and infinite sums. Let us consider one of the problems from Chapter 3.

> Find the sum of an infinite series $\frac{1}{4} + \frac{1}{36} + \frac{1}{144} + \frac{1}{400} + \frac{1}{900} + \frac{1}{1764} + \cdots$.

This series can be rewritten in terms of the Dirichlet and telescopic series and converges to $\frac{\pi^2}{3} - 3$.

I start from an exploration of the properties of well-known arithmetic and geometric sequences that are familiar to high school students. By giving my students many problems during the 25 years of my teaching experience, I noticed that they are very adept in pattern recognition. They might recognize that this is a Fibonacci sequence and determine succeeding terms by the two preceding terms. However, as I mentioned above, it is usually hard for them to analytically find the formula for the nth term or even to add the numbers: $2 + 9 + 16 + 23 + \ldots + 352$.

Yes, they find that the terms differ by 7 and that the first term is 2. But many students panic because they do not know how many terms there are, in order to apply the Gauss counting approach. This is why students need to study arithmetic and geometric progressions. For example, the nth term of the series, 352, can be written as $a_n = 2 + (n-1) \cdot 7 = 7n - 5 = 352 \Rightarrow n = 51$. Next, we can use Gauss's formula and evaluate the sum as

$$S_{51} = \left(\frac{2 + 352}{2} \right) \cdot 51 = 9027.$$

Many challenging problems of arithmetic, geometric, and other sequences can be found in the book. For example, knowledge of geometric series will allow you to solve interesting problems such as,

> Find the sum of 2016 numbers $3 + 33 + 333 + 3333 + \ldots + \underbrace{333...3}_{2016}$.

Other methods will be used to evaluate a sum like,

> Evaluate the sum, $S = \frac{1}{2!} + \frac{2}{3!} + \frac{3}{4!} + \ldots + \frac{2015}{2016!}$.

An important feature of this book is that most Statements, Lemmas, and Theorems have detailed proofs. I remember how one graduate student who was teaching geometry in a private high school rushed to report to me about finding "the formula for a prime number." He stated that it is $2^n - 1$. On my question "Why?," he replied that $2^2 - 1 = 3$, $2^3 - 1 = 7$ are primes. When I asked what about $2^4 - 1 = 15$ that is not prime? He was confused and said "I did not go that far. ..." This story sounds like a joke, but it really happened and demonstrates that any statement must be proven. His "formula" was wrong and it was proven wrong by contradiction. Particular cases must be generalized and proven, for example, by mathematical induction, directly, or by contradiction.

This book is a collection of simple and complex problems on series and sequences that are selected to motivate the reader to start solving challenging problems. For example, the following problem requires similar ideas to Problem 50 and also generalizes the method and develops proof skills.

The numbers $a_1, a_2, \ldots, a_n, a_{n+1}$ are terms of an arithmetic sequence. Prove that $\frac{1}{a_1 \cdot a_2} + \frac{1}{a_2 \cdot a_3} + \ldots + \frac{1}{a_n \cdot a_{n+1}} = \frac{n}{a_1 \cdot a_{n+1}}$.

After recognizing different sequences, one might like the following problem,

Find the nth term of a sequence 3, 13, 30, 54, 85, 123,....

The given sequence is not an arithmetic sequence; however, the differences of two consecutive terms are 10, 17, 24, 31, 38, ... and are in an arithmetic progression with common difference $d = 7$. This means that the given sequence of numbers 3, 13, 30, 54, ... is the sequence of partial sums of this arithmetic progression and that its nth term can be evaluated as $a_n = S_n(d = 7) = \frac{2 \cdot 3 + (n-1) \cdot 7}{2} \cdot n = \frac{(7n-1) \cdot n}{2}$.

The techniques used in this book are basic to understanding series and sequences. As early as 2000 BC, the Babylonians created tables of cubes and squares of the natural numbers and proved the summation of natural numbers, their squares, and cubes by a geometric approach. These formulas are in nearly every textbook still today and are used in finding other sums of finite series. While remembering these formulas by heart is a very good idea, it is better to be able to prove each formula by at least one of the methods demonstrated in this book. Memorization cannot replace understanding. Read the book with a pen and paper and be ready to derive a forgotten mathematical identity if it is needed.

Versions of problems solved by the Ancient Babylonians and Greeks often reappear in modern math contests. Their importance to modern mathematics is fundamental and unavoidable. For example, here is one of the problems of Chapter 1 of the book that was known to ancient mathematicians.

Prove that a cube of a natural number n can be uniquely written as a sum of precisely n odd consecutive numbers.

We prove this statement and find formulas for the first and last odd numbers for any given n. For example, the cube of 7 is uniquely represented by the sum of seven odd numbers, $7^3 = 43 + 45 + 47 + 49 + 51 + 53 + 55 = 343$, 1000 by 10 consecutive odd numbers, etc. Would you like to know how? The answer is in the book.

This book is not a textbook. Some knowledge of algebra and geometry such as what is introduced in secondary school is necessary to make full use of the material of Chapters 1 and 2. Knowledge of calculus is needed for better understanding of Chapter 3. However, a mastery of these subjects is not a prerequisite. You will use your knowledge of secondary school mathematics in order to better delve into the analysis of sequence and series and their properties as you develop problem-solving skills and your overall mathematical abilities.

The book is divided into four chapters: Introduction to Sequences and Series, Further Study of Sequences and Series, Series Convergence Theorems and Applications, Real-Life Applications of Arithmetic and Geometric Sequences. One hundred twenty homework problems with hints and detailed solutions are given at the end of the book. There are overlaps in knowledge and concepts between chapters. These overlaps are unavoidable since the threads of deduction we follow from the central ideas of the chapters are intertwined well within our scope of interest. For example, we will on occasion use the results of a particular lemma or theorem in a solution but wait to prove that lemma or theorem until it becomes essential to the thread at hand. If you know that property you can follow along right away and, if not, then you may find it in the following sections or in the suggested references.

Many figures are prepared with MAPLE, Excel, and Geometer's Sketchpad. Additionally, Chapters 1 and 4 have a number of screenshots produced by a popular graphing calculator by Texas Instruments. These graphs are shown especially for the benefit of students accustomed to using calculators in order to introduce them to analytical methods. Sometimes by comparing solutions obtained numerically and analytically, we can more readily see the advantages of analytical methods while referring to the numerically calculated graphs to give us confidence in our results. Following the new rules of the US Mathematics Olympiad, I suggest that you prepare all sketches by hand and urge you not to rely on a calculator or computer to solve the homework problems.

This book covers geometric, arithmetic and other sequences and their applications, sigma notation, and series. You will learn how to evaluate a limit in calculus analytically using arithmetic and geometric sequences and how to take an integral in just one step by recognizing a similarity with a sum like $\frac{1}{1 \cdot 5} + \frac{1}{5 \cdot 9} + \frac{1}{9 \cdot 13} + \cdots$.

Additionally, we will teach you how to find any term of a sequence given by a recursion formula and will introduce the so-called generating functions. You will have fun learning about figurate numbers and their properties and the application of mathematical induction to sequences and series. This book will also assist the reader in how to prove lemmas and theorems using different methods. Working on projects that Chapter 4 offers, you will see a connection of arithmetic and geometric series and sequences with real-life problems (radioactive decay, mortgage, loan, debts, etc.) and the wise use of technology for mathematics. This book will be very useful for beginners and for those who are looking challenge themselves. The book can be helpful for self-education, for people who want to do well in math classes, or for those preparing for competitions. It is also meant for math teachers and college professors who would like to use it as an extra resource in their classroom.

How Should This Book Be Used?

Here are my suggestions about how to use the book. Read the corresponding section and try to solve the problem without looking at my solution. If a problem is not easy, then sometimes it is important to find an auxiliary condition that is not a part of the problem, but that will help you to find a solution in a couple of additional steps. I will point out ideas we used in the auxiliary constructions so that you can develop your own experience and hopefully become an expert soon. If you find any question or section too difficult, skip it and go to another one. Later you may come back and try to master it. Different people respond differently to the same question. Return to difficult sections later and then solve all the problems. Read my solution when you have found your own solution or when you think you are just absolutely stuck. Think about related problems that you could solve using the same or similar approach and compare that to corresponding problems in the Homework section. Create your own problem and write it down along with your original solution. Now it is your powerful method. You will use it when it is needed.

I promise that this book will make you successful in problem solving. If you do not understand how a problem was solved or if you feel that you do not understand my approach, please remember that there are always other ways to do the same problem. Maybe your method is better than one proposed in this book. If a problem requires knowledge of trigonometry or number theory or another field of mathematics that you have not learned yet, then skip it and do other problems that you are able to understand and solve. This will give you a positive record of success in problem solving and will help you to attack the harder problem later. Do not ever give up!

I hope that upon finishing this book you will love math and its language as I do. Good luck and my best wishes to you!

Denton, TX, USA   Ellina Grigorieva, Ph.D.

Acknowledgments

During the years working on this book, I received feedback from my friends and colleagues at Lomonosov Moscow State University for which I will always be indebted.

I am especially grateful for the patient and conscientious work of Dr. Paul Deignan (University of Texas at Dallas) and for his contributions in the final formatting and preparation of the book and for the multitude of useful and insightful comments on its style and substance.

I appreciate the extremely helpful suggestions from the reviewers: your feedback made this book better! I would also like to acknowledge the editor Benjamin Levitt at Birkhäuser who has always been encouraging, helpful, and positive.

Denton, TX, USA   Ellina Grigorieva

Contents

Chapter 1
Introduction to Sequences and Series

I think everyone has met a problem like this at least once in life, "What is the next number of $1, 4, 9, 16, \ldots$" or "Continue the sequence: $1, 1, 2, 3, 5, 8, \ldots$." Perhaps, "Predict the 49th term of the sequence, $\frac{1}{3}, \frac{1}{10}, \frac{1}{17}, \frac{1}{24}, \ldots$." These three problems have something in common. They are sequences of numbers (numerical sequences) and can be written in the form $a_1, a_2, \ldots, a_k \ldots, a_n, \ldots$, where a_1, a_2, and a_n are the first, second, and n^{th} terms of the sequence.

> **Definition.** A **finite sequence** of numbers $\{a_n\}$ is a list for which there is a rule that associates each natural number $n = 1, 2, 3, \ldots, N$ $(n \in \mathbb{N})$ with only one member of the list, a_n. An **infinite sequence** associates all natural numbers to a unique element of the list.

For example, $a_n = n^2$, $a_n = \sin(n)$, $a_n = (-1)^n \cdot (2n + 1)$, etc.

In the aforementioned problems, you want to find a rule relating numbers in the sequence—a pattern. You might notice that the first sequence is a sequence of the squares of all natural numbers starting from 1, therefore, the next term after 16 would be 25, then 36 and so on. The second sequence is a Fibonacci sequence. It would take a while to find the relationship between its terms, but having some experience, you would come up with the idea,

$$2 = 1 + 1$$
$$3 = 1 + 2$$
$$5 = 2 + 3$$
$$8 = 3 + 5.$$

Every succeeding term of the sequence is the sum of the two preceding terms, so we can say that the term after 8 will be $5 + 8 = 13$, then $8 + 13 = 21$ and so on. This can be written in a recursive form as

© Springer International Publishing Switzerland 2016
E. Grigorieva, *Methods of Solving Sequence and Series Problems*,
DOI 10.1007/978-3-319-45686-7_1

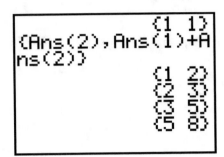

Figure 1.1 {1,1}   [ENTER]
{Ans(2), Ans(1) + Ans(2)}   [ENTER]
{1, 2}   [ENTER]
{2, 3}   [ENTER]
{3, 5}   [ENTER]
{5, 8}   [ENTER]
{8, 13} and so on

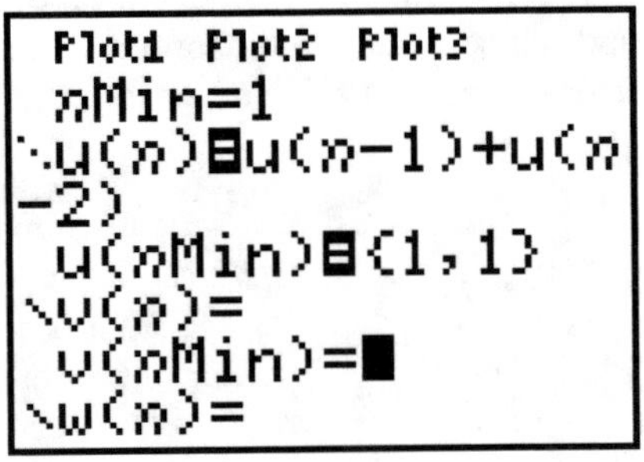

Figure 1.2 Using sequence
mode we can plot or make a
table of the sequence $u(n)$ as
a function of n and enter the
function

$$a_1 = 1$$
$$a_2 = 1$$
$$a_{n+2} = a_n + a_{n+1}, \quad n \geq 1.$$

So if you know the preceding terms, you can find the following term. However, this does not give you the answer right away for the value of the 100th or 2017th member of the sequence. For many sequences given recursively, finding an explicit formula for the nth term can be a challenging task. You will learn some analytical methods and get experience by reading this chapter.

In order to quickly explore the Fibonacci and other sequences, we can also use a graphing calculator (Figures 1.1, 1.2, and 1.3). A TI 83/84 graphing calculator can generate the Fibonacci sequence recursively on the home screen in FUNCTION MODE.

A TI 83/84 calculator treats a sequence as a function $u(n)$ whose domain is the set of positive integers. The functional value $u(n)$ represents the nth term, a_n, of the sequence.

Figure 1.3 Create a table

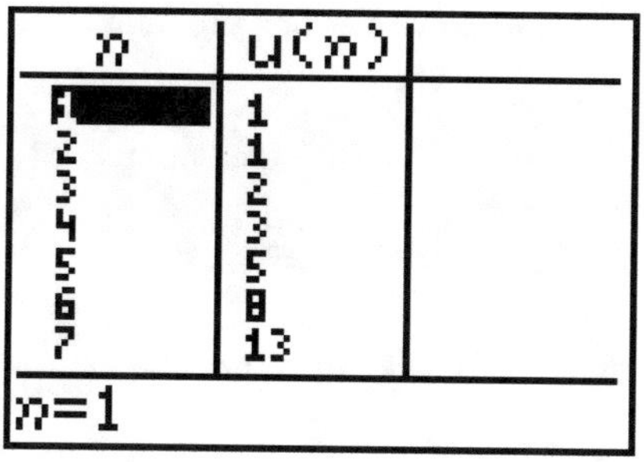

Returning to our original problems, the last sequence appears easier, because each fraction has one as the numerator. Considering only the sequence of numbers with denominators: 3, 10, 17, 24, ..., we notice that the difference between two consecutive terms is constant and equals 7. Such a sequence is called an arithmetic sequence with the first term 3 and common difference 7. So the numbers 3, 10, 17, 24, 31, 38, 45, ... can be described by the formula $u_n = 7n - 4$ and the given by a sequence of reciprocals, $a_n = \frac{1}{u_n} = \frac{1}{7n-4}$, $n \in N$.

At this point, we have introduced some practical sequences. You have recognized a pattern and even used technology to explore the sequences. Now it is time to look at the sequences from a different angle.

1.1 Sequences and Series

Although for all three sequences mentioned in the introduction in which any term can be evaluated, these numerical sequences are different. Thus, while the terms of the first and the second sequences are nondecreasing as n is increasing, the terms of the last sequence are decreasing.

Definition. A numerical sequence $\{a_n\}$ is called **nondecreasing** (**nonincreasing**) if for any natural number $n \in \mathbb{N}$, it is true that $a_n \leq a_{n+1}$ ($a_n \geq a_{n+1}$). The sequence, $\{a_n\}$ is *increasing* (*decreasing*) if for any natural number $n \in \mathbb{N}$, $a_n < a_{n+1}$ ($a_n > a_{n+1}$).

Corollary 1.1 describes sequences with positive terms.

Corollary 1.1 Consider the ratio of two consecutive terms of the sequence. If $\frac{a_{n+1}}{a_n} > 1$ then the sequence is strictly increasing. If $\frac{a_{n+1}}{a_n} < 1$, then it is a strictly decreasing sequence.

Let us solve our first problem.

Problem 1 Prove that a sequence $\{a_n\}$, $a_n = \frac{2^n}{n!}$, $n \in \mathbb{N}$ is strictly increasing starting from the second term.

Proof. Consider the ratio $\frac{a_{n+1}}{a_n} = \frac{2^{n+1}}{(n+1)!} \cdot \frac{n!}{2^n} = \frac{2}{n+1}$ and set it less than 1,

$$\frac{2}{n+1} < 1$$
$$n+1 > 2$$
$$n > 1.$$

The least natural number greater than 1 is $n = 2$. The proof is complete.

A series is the sum of the terms of a sequence. Finite sequences and series have defined first and last terms, whereas infinite sequences and series continue indefinitely. For example, the following series is a finite arithmetic series of 26 terms with the first term 3 and common difference $d = 8$, $3 + 11 + 19 + \ldots + 203$.

In mathematics, given an infinite sequence of numbers $\{a_n\}$ a series is informally the result of adding all those terms together, $a_1 + a_2 + a_3 + \ldots$. These can be written more compactly using the summation symbol $\sum$, (sigma notation). Thus, $\sum_{k=m}^{n} a_k$ means a sum of all terms, a_k, from the term with number $k = m$ to the term with number $k = n$. We say that $k = m$ is the lower index and $k = n$ is the upper index of summation. For example, $1 + 2 + 3 + \cdots + 100$ is the sum of all natural numbers from 1 to 100 and can be written as $\sum_{n=1}^{100} n$.

Let us consider the sum $1 + 4 + 9 + 16 + 25 + \cdots + 10,000$. Each term in this series is a square of a natural number, $a_k = k^2$, where $1 \le k \le 100$, so that the series can be written as $\sum_{n=1}^{100} n^2$. Also, the series $\frac{1}{8} + \frac{1}{27} + \frac{1}{81} + \ldots + \frac{1}{1,000,000} = \sum_{n=2}^{100} \frac{1}{n^3}$ represents summation from $n = 2$ to $n = 100$.

The terms of the series are often produced according to a certain rule, such as by a formula, or by an algorithm. As there are an infinite number of terms, this notion is often called an infinite series. Unlike finite summations, infinite series need tools from mathematical analysis to be fully understood and manipulated. In addition to their ubiquity in mathematics, infinite series are also widely used in other quantitative disciplines such as physics and computer science.

Definition. Expression $a_1 + a_2 + .. + a_n + \dots$, where $\{a_n\} = a_1, a_2, \dots, a_n, \dots$ is a numerical sequence called a **series**, and a_n is the common term of the series. A series is considered to be defined if the common term of the series is known as a function of number n, $a_n = f(n)$.

Problem 2 Find a series using the formula for u_n for:

a) $u_n = \frac{1}{n(n+1)}$; b) $u_n = 3 \cdot 2^{n-1}$; c) $u_n = (-1)^n$.

Solution.

a. If $u_n = \frac{1}{n(n+1)}$, then $\frac{1}{1\cdot2} + \frac{1}{2\cdot3} + \frac{1}{3\cdot4} + \dots + \frac{1}{n(n+1)} + \dots = \sum\limits_{n=1}^{\infty} \frac{1}{n(n+1)}$.

b. If $u_n = 3 \cdot 2^{n-1}$, then $3 + 3\cdot2 + 3\cdot2^2 + 3\cdot2^3 + \dots + 3\cdot2^{n-1} + \dots = \sum\limits_{n=1}^{\infty} 3\cdot2^{n-1}$

c. If $u_n = (-1)^n$, then $-1 + 1 - 1 + 1 + \dots + (-1)^n + \dots = \sum\limits_{n=1}^{\infty} (-1)^n$.

Problem 3 Given the common term of the series $u_n = \frac{n}{n^2+1}$, a) Evaluate u_5; b) Find the ratio $\frac{u_{n+1}}{u_n}$.

Solution.

a. $u_5 = \frac{5}{5^2+1} = \frac{5}{26}$

b. $\frac{u_{n+1}}{u_n} = \frac{(n+1)(n^2+1)}{((n+1)^2+1)n} = \frac{n^3+n^2+n+1}{n^3+2n^2+2n}$

Problem 4 Find the formula for the common term if the first five terms of the series are given by $\frac{1}{1\cdot3} + \frac{1}{2\cdot4} + \frac{1}{3\cdot5} + \frac{1}{4\cdot6} + \frac{1}{5\cdot7} + \dots$

Solution. It follows from the type of the denominators that $u_n = \frac{1}{n(n+2)}$.

Problem 5 Find the first three terms for each series:

a) $\sum\limits_{n=1}^{\infty} \frac{n^2}{2\cdot4\cdot6\cdot...\cdot2n}$

b) $\sum\limits_{n=1}^{\infty} \frac{\cos n\pi}{n^3}$

c) $\sum\limits_{n=1}^{\infty} \frac{2^n}{n!}.$

Solution.

a. $u_1 = 1/2, u_2 = \frac{2^2}{2\cdot4} = \frac{1}{2}, u_3 = \frac{3^2}{2\cdot4\cdot6} = \frac{9}{48} = \frac{3}{16}$

b. $u_1 = \frac{\cos\pi}{1^3} = -1, u_2 = \frac{\cos 2\pi}{2^3} = \frac{1}{8}, u_3 = \frac{\cos 3\pi}{3^3} = -\frac{1}{27}$

c. $u_1 = \frac{2}{1!} = 2, u_2 = \frac{2^2}{2!} = 2, u_3 = \frac{2^3}{3!} = \frac{8}{6} = \frac{4}{3}$

A partial sum S_n is the sum of n terms of a sequence. Let us consider the sequence $\{a_i\}$ and its partial sums,

$$S_1 = a_1$$
$$S_2 = a_1 + a_2$$
$$S_3 = a_1 + a_2 + a_3$$
$$\ldots$$
$$S_n = a_1 + a_2 + a_3 + \ldots + a_n.$$

Definition. The **partial sum** S_n of the series $u_1 + u_2 + .. + u_n + \ldots = \sum\limits_{n=1}^{\infty} u_n$ is the sum of the first n terms of the series, i.e., $S_n = u_1 + u_2 + .. + u_n = \sum\limits_{k=1}^{n} u_k.$

For the series $\sum\limits_{n=1}^{\infty} \frac{1}{2^n}$, $S_1 = u_1 = \frac{1}{2}$, $S_2 = u_1 + u_2 = \frac{1}{2} + \frac{1}{2^2} = \frac{3}{4}$, $S_3 = u_1 + u_2 + u_3 = S_2 + \frac{1}{2^3} = \frac{3}{4} + \frac{1}{8} = \frac{7}{8}$. Each series $\sum\limits_{n=1}^{\infty} u_n$ is associated with the sequence of its partial sums $\{S_n\}$. If the limit of the partial sums $(S_1, S_2, S_3, \ldots, S_k, \ldots)$ converges as $n \to \infty$, then we say the infinite series converges.

- If $\lim\limits_{n\to\infty} S_n = S$, the S is called the sum of the infinite series $a_1 + a_2 + \ldots + a_n + \ldots = S.$
- If $\lim\limits_{n\to\infty} S_n$ does not exist, then we say the infinite series diverges.

Although the following are both examples of infinite series, the first series are divergent and second are convergent. The first series is an infinite arithmetic series (a sum of all natural numbers is obviously divergent because it increases without bound) and the second is a geometric decreasing series, which soon will be shown to be always convergent:

$$1 + 2 + 3 + \ldots + n + \ldots = \sum_{n=1}^{\infty} n$$

$$1 + \frac{1}{3} + \frac{1}{9} + \ldots + \frac{1}{3^{n-1}} + \ldots = \sum_{n=1}^{\infty} \frac{1}{3^{n-1}}.$$

Problem 6 is not so straightforward because it asks to find the terms of the series given the formula for its partial sum.

Problem 6 Given a sequence $\left\{ \frac{n}{n+2} \right\}$. Find a series for which this sequence represents a sequence of its partial sums.

Solution. By the condition $S_n = \frac{n}{n+2}$, we can evaluate the first term as $u_1 = S_1 = \frac{1}{1+2} = \frac{1}{3}$ and the second term as the difference between the second and first partial sums, $u_2 = S_2 - S_1 = \frac{2}{2+2} - \frac{1}{1+2} = \frac{1}{6}$. The nth term formula is $u_n = S_n - S_{n-1} = \frac{n}{n+2} - \frac{n-1}{n-1+2} = \frac{n}{n+2} - \frac{n-1}{n+1} = \frac{2}{(n+1)(n+2)}$.

The series can be expressed as $\frac{1}{3} + \frac{1}{6} + \frac{1}{10} + \ldots + = \sum_{n=1}^{\infty} \frac{2}{(n+1)(n+2)}$. Note, that this series is an infinite series of reciprocals of triangular numbers. More information about which you can find in this and the following chapters.

Answer. $\frac{1}{3} + \frac{1}{6} + \frac{1}{10} + \ldots + = \sum_{n=1}^{\infty} \frac{2}{(n+1)(n+2)}$.

Definition. The series $\sum_{n=1}^{\infty} u_n$ is **convergent** if the sequence of its partial sums $\{S_n\}$ is convergent, i.e., there exists the limit $\lim_{n \to \infty} S_n = S$. Then we write $\sum_{n=1}^{\infty} u_n = S$.

Definition. The series $\sum_{n=1}^{\infty} u_n$ is **divergent** if the sequence of its partial sums $\{S_n\}$ is divergent, i.e., the limit of the sequence of partial sums does not exist or it is infinite.

In Problem 7 an infinite series is introduced using the limiting transition.

Problem   7 Investigate   convergence   (divergence)   of   the   series,
$$\frac{1}{1\cdot 2} + \frac{1}{2\cdot 3} + \ldots + \frac{1}{n(n+1)} + \ldots = \sum_{n=1}^{\infty} \frac{1}{n(n+1)}, \text{ if } S_n = 1 - \frac{1}{n+1}.$$

Solution. Because $\lim_{n\to\infty} S_n = 1$, the series is convergent and $\sum_{n=1}^{\infty} u_n = 1$.

Remark. To find S_n usually is a creativity problem. You may find it using methods explained further in Chapter 2 of this book.

1.2   Arithmetic Progression

Definition. If a sequence of numbers is such that the difference between any two consecutive numbers in the sequence is the same, the numbers are said to be in an **arithmetic sequence (progression)**.

Thus, 5, 10, 15, 20, ... forms an arithmetic progression in which any two consecutive numbers differ by 5. A number within an arithmetic sequence is called a term and a difference between any two consecutive terms is called the "common difference." Let us show that if the first term of an arithmetic progression and the common difference are given, the entire sequence is defined. Let a_1 be the first term of an arithmetic sequence and d be its common difference. Then

$$a_2 = a_1 + d$$
$$a_3 = a_2 + d = a_1 + 2d$$
$$a_4 = a_3 + d = a_1 + 3d$$
$$\ldots$$
$$a_n = a_{n-1} + d = a_1 + (n-1)d.$$

We notice that any particular term of the sequence can be found as a sum of the first, a_1, and the common difference d multiplied by the number of the term minu 1.

(This can be easily proven by mathematical induction, please try it as a homework exercise.) So if we wanted to find the 12th term of the previous sequence 5, 10, 15,..., we would use the formula,

$$a_n = a_1 + (n - 1)d \tag{1.1}$$

with $a_1 = 5$, $d = 5$, $a_{12} = a_1 + 11 \cdot d = 60$.

Maybe it is not always obvious but arithmetic progression has many applications in different branches of mathematics, for example in number theory. If a number n is divisible by m (n is a multiple of m), then it can be written as $n = m \cdot k$, where n, k, and m are integers. If m is an arbitrary integer, the multiples of m are all numbers: 0, $\pm m$, $\pm 2m$, $\pm 3m$, $\pm 4m$, ..., $\pm km$. If a natural number n is not divisible by a number m we say that n divided by m gives a remainder r and can be written in the form, $n = m \cdot k + r$, where $1 \leq r \leq m - 1$.

Suppose that you have 20 apricots and must divide them equally between three children. Everyone can get 6 apricots and two apricots will stay in the basket. If originally you had 19 apricots, then again every kid would get 6 with one remaining in the basket. Three children with 0 remaining would equally divide 18 apricots.

If a natural number N is a multiple of 3, it can be represented in the form $N = 3k$, where $k = 1, 2, 3,....$ For example, $18 = 3 \cdot 6$, $9 = 3 \cdot 3$, $300 = 3 \cdot 100$ etc. All natural numbers that are not multiples of 3 when divided by 3 give a remainder of 1 or 2. It can be written in the form $N = 3k + 1$ or $N = 3k + 2$. For example, $20 = 3 \cdot 6 + 2$ (2 is a remainder), $301 = 3 \cdot 100 + 1$ (1 is a remainder), etc. Therefore, all natural numbers and multiples of 3 can be written as

$$3, 6, 9, 12, 15, 18, \ldots, 3k. \tag{1.2}$$

All natural numbers divided by 3 with remainder of 1 can be listed as

$$1, 4, 7, 10, 13, 16, \ldots, 3k + 1, \ldots, \text{ where } k = 0, 1, 2, 3, \ldots \tag{1.3}$$

Finally, all natural numbers divided by 3 with remainder of 2 can be listed as

$$2, 5, 8, 11, 14, 17, \ldots, 3k + 2, \ldots, \text{ where } k = 0, 1, 2, 3, \ldots \tag{1.4}$$

We can visualize all existing natural numbers as those that are multiples of 3 ($3k$), those that divided by 3 leave a remainder of 1 ($3k + 1$) and those that when divided by 3 leave a remainder of 2 ($3k + 2$). It is like dividing a "big pie" of all natural numbers into three pieces. Moreover, the numbers that belong to one such "piece of the pie" given by Eqs. 1.2–1.4 are corresponding terms of the arithmetic progressions with the same common difference ($d = 3$) and different first term, $b_1 = 3$, $b_1 = 1$, $b_1 = 2$, respectively.

Problem 8 Find all numbers that are simultaneously the terms of the both arithmetic sequences: 3, 7, 11,..., 407 and 2, 9, 16,..., 709.

Solution. The nth term of the first sequence is

$$a_n = 3 + 4(n-1). \tag{1.5}$$

The kth term of the second sequence can be written as

$$b_k = 2 + 7(k-1). \tag{1.6}$$

Therefore, we have to find such numbers n and k that $a_n = b_k$, $1 \leq n \leq 102$, $1 \leq k \leq 102$. Equating Eqs. 1.5 and 1.6, we obtain $4n + 4 = 7k$. This equation has integer solutions if and only if its right hand side is divisible by 4, i.e., $k = 4s$. It is clear that s can be 1, 2, ..., 25 because $k = 1, 2,..., 102$,

$$4(n+1) = 7 \cdot 4s$$
$$n+1 = 7s$$
$$n = 7s - 1$$

since $1 \leq n \leq 102$, then $1 \leq s \leq 14$. Therefore, there are exactly 14 numbers that are terms of both arithmetic sequences. We can find all of them either from Eq. 1.5 using the substitution $n = 7s - 1$ or from Eq. 1.6 using $k = 4s$, $s = 1, 2, 3, \ldots, 14$.

Answer. 23, 51, 79, ..., 387.

Problem 9 Prove that the numbers $\sqrt{2}, \sqrt{3}, \sqrt{5}$ cannot be the terms of an arithmetic sequence.

Proof. When I give this problem to my students, they very often say that irrational numbers $\sqrt{2}, \sqrt{3}, \sqrt{5}$ cannot differ from each other by the same number. Others take calculators and try to validate this statement by estimation: $\sqrt{3} - \sqrt{2} \approx 0.318$, $\sqrt{5} - \sqrt{3} \approx 0.514$. However, estimation on a calculator cannot be considered as a rigorous proof. Moreover, even if we accept the fact that the consecutive differences are not the same, we still have to prove that the numbers cannot be just three non-neighboring terms of the arithmetic sequence.

Let us provide a correct proof by contradiction. Assume that $\sqrt{2}, \sqrt{3}, \sqrt{5}$ are the kth, the mth and the nth terms of the arithmetic sequence with the first term of a_1 and common difference of d:

$$\sqrt{2} = a_k = a_1 + (k-1)d$$
$$\sqrt{3} = a_m = a_1 + (m-1)d$$
$$\sqrt{5} = a_n = a_1 + (n-1)d.$$

Subtract the second and the first equations, and then the third and the second. Finally, dividing the results of subtractions we obtain $\frac{\sqrt{3}-\sqrt{2}}{\sqrt{5}-\sqrt{3}} = \frac{m-k}{n-m}$. The right side of this is a rational number, because m, k, n are natural numbers. Denote this number by r,

$$r = \frac{\sqrt{3} - \sqrt{2}}{\sqrt{5} - \sqrt{3}}.$$

which can be written as $r(\sqrt{5} - \sqrt{3}) = \sqrt{3} - \sqrt{2}$. Squaring both sides of the equation above, we obtain: $r^2\sqrt{15} - \sqrt{6} = \frac{8r^2-5}{2}$. The right side of this equation is again a rational number and we can denote it by s: $r^2\sqrt{15} - \sqrt{6} = s$. Squaring both sides again, after simplification we have $\sqrt{10} = \frac{15r^4-s^2+6}{6r^2}$. This relationship indicates that $\sqrt{10}$ is a rational number. However, $\sqrt{10}$ is irrational. Therefore, we obtained the contradiction. Our original assumption was wrong and $\sqrt{2}, \sqrt{3}, \sqrt{5}$ cannot be terms of the same arithmetic sequence. The proof is complete.

Suppose we want to find the sum of the first n terms of an arithmetic sequence, called a partial sum. Denote such a sum as S_n, then

$$S_1 = a_1$$
$$S_2 = a_1 + a_2$$
$$S_3 = a_1 + a_2 + a_3$$
$$\dots$$
$$S_k = a_1 + a_2 + a_3 + \dots + a_k$$
$$\dots$$
$$S_n = a_1 + a_2 + \dots + a_k + \dots + a_n.$$

How do we evaluate it? Some of you may remember the story of how 10-year-old Carl Friedrich Gauss (German mathematician, 1777–1855) added all natural numbers from 1 to 100 in his math class. He wrote them in two rows,

$$1 + 2 + 3 + \dots + 98 + 99 + 100 \quad \text{(ascending order) and}$$
$$100 + 99 + 98 + \dots + 3 + 2 + 1 \quad \text{(descending order).}$$

He noticed that the sum in each column is 101. There are 100 columns and the sum of in row is the same, it is the unknown. Therefore, he multiplied 101 by 100 then divided by 2 because we need only a single sum, not a double,

$$S_{100} = \frac{100(1 + 100)}{2} = 50 \cdot 101 = 5050.$$

Let us use the same idea to obtain S_n.

$$S_n = a_1 + (a_1 + d) + (a_1 + 2d) + \ldots + (a_1 + (n-2)d) + (a_1 + (n-1)d)$$
$$S_n = a_n + (a_n - d) + (a_n - 2d) + \ldots + (a_n - (n-2)d) + (a_n - (n-1)d).$$

Again if we add elements by columns, in each column we will get $(a_1 + a_n)$. There are n such columns, so $2S_n = n(a_1 + a_n)$ or dividing both sides by 2,

$$S_n = \frac{(a_1 + a_n)}{2} \cdot n. \tag{1.7}$$

Replacing the nth term, a_n, in terms of a_1 and d we obtain another form for S_n,

$$S_n = \frac{(2a_1 + (n-1)d)}{2} \cdot n. \tag{1.8}$$

We leave it to the reader to obtain this form on her own. Notice that Eq. 1.8 can be more useful, because a_n is usually unknown. Let us solve Problem 10 now.

Problem 10 Find the sum of all natural numbers between 1 and 1000 that are not divisible by 13.

Solution. Let S be the unknown sum, then S can be written in the form: $S = S_{1000} - M$, where S_{1000} is the sum of all natural numbers between 1 and 1000, and M is the sum of all multiples of 13 less than 1000. It is clear that S_{1000} is the sum of the first 1000 terms of an arithmetic sequence, $1 + 2 + 3 + \cdots + 999 + 1000$ with $a_1 = 1$ and $a_{1000} = 1000$. Using Eq. 1.7 we have $S_{1000} = \frac{1+1000}{2} \cdot 1000 = 500,500$. All multiples of 13 can be written in the form $13k$, and their sum as $M = 13 + 13 \cdot 2 + 13 \cdot 3 + \ldots + 13 \cdot k$,

$$13k \leq 1000 \tag{1.9}$$

$$M = 13 \cdot (1 + 2 + 3 + \ldots + k) \tag{1.10}$$

M can be evaluated as soon as we know k. From Eq. 1.9 we have $k \leq \frac{1000}{13} = 76.923$, then the greatest natural number k satisfying the inequality is 76, i.e.,

$1000 = 13 \cdot 76 + 12$, $k = 76$. Now we can derive M from Eq. 1.10.
$M = 13(1 + 2 + 3 + \cdots + 76)$ and using Eq. 1.7 again, we get
$M = 13 \cdot \frac{(1+76)}{2} \cdot 76 = 13 \cdot 77 \cdot 38 = 38038$. Subtracting M from S_{1000} we obtain
$S = 500,500 - 38,038 = 462,462$.

Answer. The sum is equal to 462,462.

Sometimes a problem can be solved in one or two steps even if it seems to have too many unknowns.

> **Problem 11** $\{a_n\}$ is an arithmetic progression with a non-zero common difference. The sum of all terms between the fourth and the fourteenth is 77. Find the number of the term of the sequence that is equal to 7.

Solution. From the premise of the problem, we can write

$$a_4 + a_5 + a_6 + \ldots + a_{14} = 77$$

or

$$\frac{a_4 + a_{14}}{2} \cdot 11 = \frac{a_1 + 3d + a_1 + 13d}{2} \cdot 11$$

$$(a_1 + 8d) \cdot 11 = 77$$

Dividing both sides of the last equation by 11 we obtain $a_1 + 8d = 7$. But we have to find which term of the sequence equals 7. Because for an arithmetic sequence $a_n = a_1 + (n-1)d$, then $a_1 + 8d = a_9 = 7$.

Answer. The 9th term is 7.

> **Problem 12** Find the sum of the first 19 terms of the arithmetic sequence $\{a_n\}$ such that $a_4 + a_8 + a_{12} + a_{16} = 224$.

Solution. 1) Because $a_n = a_1 + (n-1)d$ we can express each term of the given equality in terms of a_1 and d. Thus,

$$a_4 + a_8 + a_{12} + a_{16} = a_1 + 3d + a_1 + 7d + a_1 + 11d + a_1 + 15d = 224$$
$$4(a_1 + 9d) = 56 \cdot 4$$

so $a_1 + 9d = 56$. 2) Let us find S_{19} using Eq. 1.8: $S_{19} = \frac{(2a_1 + 18d)}{2} \cdot 19$
$= (a_1 + 9d) \cdot 19 = 56 \cdot 19 = 1064$.

Answer. The sum of the first 19 terms is 1064.

Remark. Sometimes even if you have too many unknowns you can solve a problem. In part 1, we obtained an expression that connects the first term of the sequence and its common difference. In part 2 we just replaced $a_1 + 9d$ by 56. Notice that we solved this problem without knowing the values of a_1 and d.

Problem 13 The sum of all even 2-digit numbers was divided by one such number, a multiple of 9. There was no remainder. The quotient differs from the divisor by only the order of the digits. What is the divisor?

Solution. All even two-digit numbers are terms of the arithmetic sequence with $a_1 = 10$ and $d = 2$. The last term is known as well, that is, 98. Using the formula of the nth term of an arithmetic sequence we can find the number of terms in the series.

$$a_n = a_1 + (n-1)d$$
$$98 = 10 + (n-1) \cdot 2$$
$$n = 45.$$

Now let us find the sum of 45 terms of the sequence, $S_{45} = \frac{a_1 + a_{45}}{2} \cdot 45 = \frac{10 + 98}{2} \cdot 45 = 54 \cdot 45 = 2430$, from which we see that $2430 = 54 \cdot 45$. 54 and 45 are multiples of 9, two-digit numbers, and also 54 is even. Because $2430 \div 54 = 45$, then 54 is the divisor, a multiple of 9. Forty-five is the quotient—a multiple of 9. Fifty-four and 45, like mirror images differ by only the order of digits. We find that the divisor is 54.

Let us show that there is only a single solution, 54, of this problem. By the condition of the problem, 2430 divided by some even two-digit number has no remainder, and the quotient differs from the divisor only by the order of its digits. Using a decimal representation of a two digit number xy, we can write it as $\overline{xy} = 10x + y$, and we know $x + y = 9$. (Remember? If a number is divisible by 9, then the sum of its digits must be a multiple of 9. In this case the sum of the digits must be nine only. Do you know why other multiples of 9, such as 18, 27, 36, etc. would not work?). Now we obtain a system of two equations in two variables:

$$\begin{cases} 2430 = (10x + y)(10y + x) \\ x + y = 9 \end{cases} \Leftrightarrow \begin{cases} x = 9 - y \\ y^2 - 9y + 20 = 0 \end{cases} \Leftrightarrow \begin{bmatrix} x = 5, \ y = 4 \\ x = 4, \ y = 5 \end{bmatrix}$$

We found two numbers 54 and 45 satisfying the given system. But 54 is the only answer because 54 is an even number.

Answer. 54.

Problem 14 The product of the third and the sixth terms of an arithmetic sequence is 406. The ninth term of the sequence divided by the fourth term gives a quotient of 2 and a remainder of 6. Find the first term and the common difference of the arithmetic sequence.

Solution. Let $\{a_n\}$ be an arithmetic sequence. We are going to rewrite the condition and then express each term using a_1 and d:

$$\begin{cases} a_3 \cdot a_6 = 406 \\ a_9 = a_4 \cdot 2 + 6 \end{cases} \quad \begin{cases} (a_1 + 2d)(a_1 + 5d) = 406 \\ a_1 + 8d = (a_1 + 3d) \cdot 2 + 6. \end{cases}$$

Remember the formula,

$$\text{Dividend} = \text{Divisor} \cdot \text{Quotient} + \text{Remainder}?$$

Let us first simplify the second equation of the system,

$$a_1 + 8d = 2a_1 + 6d + 6$$
$$a_1 = 2d - 6.$$

Substituting a_1 from here into the first equation of the system above, we have $(4d - 6)(7d - 6) = 406$ so $14d^2 - 33d - 185 = 0$. This quadratic equation has two roots, $d = \frac{33 \pm \sqrt{11449}}{28} = \frac{33 \pm 107}{28}$; so $d = 5$; $\frac{-37}{14}$.

Two different values for a common difference will give us two different arithmetic sequences:
1. $\{a_n\}$, $a_1 = 4$, $d = 5$ and
2. $\{a_n\}$, $a_1 = -\frac{79}{7}$, $d = -\frac{37}{14}$.

Answer. 1. $\{a_n\}$, $a_1 = 4$, $d = 5$ and 2. $\{a_n\}$, $a_1 = -\frac{79}{7}$, $d = -\frac{37}{14}$.

Problem 15 The second term of some arithmetic progression containing only whole numbers is 2 and the sum of squares of the third and the fourth terms is less than 4. Find the first term of the progression.

Solution.

$$\begin{cases} a_2 = 2 \\ a_3{}^2 + a_4{}^2 < 4 \end{cases}$$

Using properties of an arithmetic progression, we can rewrite a_3 and a_4 in terms of the second term, a_2, and a common difference, d,

$$a_3 = a_2 + d = 2 + d$$
$$a_4 = a_2 + 2d = 2 + 2d.$$

Now we can simplify the system

$$(2 + d)^2 + (2 + 2d)^2 < 4$$
$$4 + 4d + d^2 + 4 + 8d + 4d^2 < 4$$
$$5d^2 + 12d + 4 < 0$$
$$d = \frac{-6 \pm \sqrt{36 - 20}}{5} = \frac{-6 \pm 4}{5}.$$

In order to satisfy the inequality, d must be in the range $-2 < d < -0.4$, but because the arithmetic progression contains only whole numbers and the second term equals 2, its common difference, d, can be only a whole number. From the inequality above, we obtain $d = -1$. Then $a_1 = a_2 - d = 2 - (-1) = 3$.

Answer. 3.

Problem 16 Prove that if 25, 43, and 70 are terms of infinite arithmetic progression, then 2005 also belongs to this progression.

Proof. By the condition of the problem,

$$\begin{cases} 25 = a_1 + kd \\ 43 = a_1 + nd \\ 70 = a_1 + md \end{cases} \Rightarrow \begin{cases} 18 = (n - k)d \\ 27 = (m - n)d \end{cases} \Rightarrow 9 = (m - 2n + k)d.$$

Because $2005 = 1935 + 70$, and 1935 is a multiple of 9, we have the chain of true relationships:

$$2005 = 70 + 9 \cdot 215 \Rightarrow 2005 = a_1 + md + 215 \cdot (m - 2n + k) \cdot d$$
$$= a_1 + d(216m - 430n + 215k) = a_1 + d \cdot l.$$

The statement is proven.

1.3   Geometric Progression

Definition.

> A **geometric progression** is a sequence of numbers in which the ratio of any two consecutive terms is the same.

Thus, $3, 21, 147, 1029, 7203, \ldots$ is a geometric progression with the first term 3 and the common ratio $\dfrac{21}{3} = \dfrac{147}{21} = \cdots = 7$. Let us denote the first term of our sequence as b_1 and the common ratio as r. Then

$$
\begin{aligned}
& b_1 \\
& b_2 = b_1 \cdot r \\
& b_3 = b_2 \cdot r = b_1 \cdot r^2 \\
& \ldots \\
& b_n = b_1 \cdot r^{n-1} \\
& b_1, \; b_1 r, \; b_1 r^2, \; \ldots, \; b_1 r^{n-1}, \; \ldots
\end{aligned}
\tag{1.11}
$$

The geometric progression of Eq. 1.11 is said to be increasing or decreasing according to whether r is greater or less than one. If $r > 1$, each term is greater than the preceding term, while if $r > 0$ but is less than 1 each term is less than the preceding term.

Let us find the sum of the first n terms of a geometric sequence,

$$
S_n = b_1 + b_1 \cdot r + b_1 \cdot r^2 + \ldots + b_1 \cdot r^{n-1}.
\tag{1.12}
$$

Multiplying both sides of Eq. 1.12 by r, we obtain a new equation,

$$
r \cdot S_n = b_1 \cdot r + b_1 \cdot r^2 + \ldots + b_1 \cdot r^{n-1} + b_1 \cdot r^n.
\tag{1.13}
$$

Subtracting Eq. 1.13 from Eq. 1.12,

$$
\begin{aligned}
S_n - r S_n &= b_1 - b_1 r^n \\
S_n(1 - r) &= b_1(1 - r^n).
\end{aligned}
$$

Dividing both sides by $(1 - r)$ we obtain

$$
S_n = \frac{b_1(1 - r^n)}{1 - r}.
\tag{1.14}
$$

If $r > 1$ then it is better to use S_n in a different form,

$$S_n = \frac{b_1(r^n - 1)}{r - 1}. \tag{1.15}$$

Let us return to a geometric series with infinitely many terms,

$$S = b_1 + b_1 \cdot r + b_1 \cdot r^2 + \ldots + b_1 \cdot r^{n-1} + \ldots \tag{1.16}$$

If the number of terms is infinitely great (n approaches infinity), then the expression on the right of Eq. 1.16 is called an infinite geometric series. Consider an infinite decreasing geometric series. The partial sum is $S_n = \frac{b_1(1-r^n)}{1-r}$. If $|r| < 1$ and $n \to \infty$, then $r^n \to 0$ and

$$\lim_{n \to \infty} S_n = \frac{b_1}{1 - r}. \tag{1.17}$$

We notice that the sequence of partial sums has a limit.

However, if $|r| > 1$ by Eq. 1.15 the sequence of the partial sums will dramatically increase without bound. In order to illustrate this phenomenon, let us solve Problem 17.

Problem 17 Suppose a family has twin boys, Brian and Paul, and each is asked to choose a gift for their birthday that will be in three weeks. Brian said that he would like to get \$10 each day for three weeks and Paul asked for 2 cents on day one, 4 cents on day two and so on, doubling the amount each day for three weeks. Which gift would you choose?

Solution. What I usually hear right away is, "Of course 10 bucks is the better choice! I will end up with \$210!" When I say, "Let us check the second opportunity," some students think there might be a trick and ask for time to think.

Let us write down the sequence: $2, 4, 8, 16, 32, 64, 128, 256, \ldots, 1024, \ldots$ or 2^1, $2^2, 2^3, 2^4, 2^5, 2^6, \ldots, 2^{10}, \ldots$ Each term is the amount of money on a particular day n. For example, 8 cents $= 2^3$ is how much the parents would give Paul on day 3, and 2^n cents on day n. This sequence is an increasing geometric sequence with the first term equal to 2 and the common ratio, $r = 2 > 1$ and it grows very fast. On day 5 choosing the first gift, Brian would have \$50 and Paul, choosing the second, only $2(2^5 - 1) = 62$ cents. On day 21, however, Paul would get 2^{21} cents, and adding all amounts from the preceding 20 days using the formula for a sum of the first 21 terms of a geometric sequence, we have

$$2 + 2^2 + 2^3 + \ldots + 2^{21} = \frac{2(2^{21} - 1)}{2 - 1} = 2^{22} - 2 = \$41,943.02$$

which is much more than \$210! Maybe in offering two gifts these parents had no idea how much they could lose if they agreed with Paul's plan. You do not want to be that ignorant parent in the future!

Remark. On a TI 83/84 graphing calculator you can see how much money you would get after 2 days, 3 days and so on, up to 21 days. Type CUM SUM (Seq $(2^{\wedge}(n+1)-2, n, 1, 21)$ and press [ENTER] to get $\{2, 6, 14, \ldots, 4, 194, 300\}$.

Here is another simple example for you. You know that any rational number can be represented as a fraction $\frac{n}{m}$, such that the numerator n is an integer and the denominator, m is a natural number. It is easy to see that for example $2.3 = 23/10$ or $5 = 20/4$, but when asked how to rewrite $0.313131\ldots$ as a fraction, some students have a hard time. We will teach you how to use infinite geometric series to rewrite any repeated decimal as a fraction.

Problem 18 Rewrite these repeating decimals as fractions:
a) $0.333\ldots$ b) $0.777\ldots$ c) $0.454545\ldots$ d) $1.227027027\ldots$.

Solution. Every repeating decimal has one or more digits after the decimal point repeating infinitely many times and can be written as a geometric series. For example, many of us know that $0.333\ldots = 1/3$. Let us prove this using Eq. 1.17,

$$0.33\ldots3\cdots = \frac{3}{10} + \frac{3}{100} + \frac{3}{1000} + \cdots + \frac{3}{10^n} + \cdots = \frac{3}{10} \cdot \frac{1}{1 - \frac{1}{10}} = \frac{1}{3}.$$

By analogy

$$0.777\ldots7\ldots = \frac{7}{10} + \frac{7}{100} + \frac{7}{1000} + \cdots = \frac{7}{9} \text{ and}$$

$$0.4545\ldots45\ldots = \frac{45}{100} + \frac{45}{10000} + \cdots = \frac{45}{100} \cdot \frac{1}{1 - 1/100} = \frac{45}{99} = \frac{5}{11}.$$

In the number $1.227027\ldots27\ldots$, we have only 027 repeating infinitely many times. This number can be represented as

$$1.227027\ldots027\ldots = \frac{12}{10} + \frac{27}{1000} + \frac{27}{1000000} + \cdots$$
$$= \frac{6}{5} + \frac{27}{1000} \cdot \frac{1}{1 - \dfrac{1}{1000}} = \frac{6}{5} + \frac{3}{111} = \frac{681}{555}.$$

Answers a) 1/3, b) 7/9, c) 5/11, d) 681/555.

Problem 19 Let $\{a_n\}$ be an arithmetic sequence such that its 1st, 20th, and 58th terms are consecutive terms of some geometric sequence. Find the common ratio of the geometric sequence.

Solution. Because a_1, a_{20}, and a_{58} are terms of some arithmetic sequence; let us express them in terms of a_1 and d, where a_1 is the first term and d is a common difference,

$$a_1 = a_1$$
$$a_{20} = a_1 + 19d$$
$$a_{58} = a_1 + 57d,$$

but a_1, a_{20}, and a_{58} are simultaneously the first, second, and third terms of some geometric sequence $\{b_n\}$. Thus,

$$b_1 = a_1$$
$$b_2 = a_1 + 19d$$
$$b_3 = a_1 + 57d.$$

The common ratio is

$$r = \frac{b_2}{b_1} = \frac{b_3}{b_2} = \frac{a_1 + 19d}{a_1} = \frac{a_1 + 57d}{a_1 + 19d}$$

from which we have

$$(a_1 + 19d)^2 = a_1(a_1 + 57d)$$
$$a_1{}^2 + 38da_1 + 361d^2 = a_1{}^2 + 57da_1$$
$$19a_1 d - 361d^2 = 0.$$

Factoring the last equality we have $19d(a_1 - 19d) = 0$ because $a_1 \neq a_{20} \neq a_{58}$, $d \neq 0$, and $a_1 = 19d$. Replacing a_1 by $19d$ in the formula for the common ratio, we obtain $r = \frac{19d+57d}{19d+19d} = \frac{76}{38} = 2$.

Answer. The common ratio is 2.

Problem 20 (ASHME) Define a sequence of real numbers $a_1, a_2, a_3, \ldots$ by $a_1 = 1$ and $a_{n+1}^3 = 99a_n^3$ for all $n \geq 1$. Find a_{100}.

Solution. If $a_{n+1}^3 = 99a_n^3$, then taking the cubic root of both sides we obtain $\frac{a_{n+1}}{a_n} = \sqrt[3]{99}$. This expression means that the given sequence is geometric with common ratio $r = \sqrt[3]{99}$ and the first term $a_1 = 1$. To find a_{100} we apply the formula for the nth term, $a_n = a_1 r^{n-1}$ so $a_{100} = 1 \cdot \left(\sqrt[3]{99}\right)^{99} = 99^{33}$.

Answer. 99^{33}.

Problem 21 (Rivkin) The roots of the equation $x^3 - 7x^2 + 14x + a = 0$ are terms of an increasing geometric progression. Solve it.

Solution. Let x_1, x_2, and x_3 be the roots of the equation. Applying Vieta's Theorem for a cubic equation or factoring the given cubic equation as $(x - x_1)(x - x_2)(x - x_3) = 0$ and after equating corresponding coefficients,

$$x_1 x_2 x_3 = -a$$
$$x_1 + x_2 + x_3 = 7$$
$$x_1 x_2 + x_1 x_3 + x_2 x_3 = 14.$$

Since the roots are consecutive terms of a geometric sequence, then we have

$$x_1 + x_1 r + x_1 r^2 = 7$$
$$x_1^2 r + x_1^2 r^2 + x_1^2 r^3 = 14$$
$$x_1^3 r^3 = -a.$$

Dividing the second equation by the first, $x_1 = \frac{2}{r}$. Substituting it into the second equation we have $2r^2 - 5r + 2 = 0$ which has two roots, $r_1 = \frac{1}{2}$ and $r_2 = 2$. Since the progression would be increasing only for $r = 2$, then $x_1 = 1$, $x_2 = 2$, $x_3 = 4$. Moreover, $x_1^3 r^3 = 8 = -a$, $a = -8$.

Answer. $1, 2, 4$.

Problem 22 Find the sum of 2016 numbers: $3 + 33 + 333 + 3333 + \cdots + \underbrace{333\ldots3}_{\text{2016 times}}$.

Solution. Denote this sum by S. Because

$$3 \cdot 3 = 9 = 10 - 1$$
$$3 \cdot 33 = 99 = 100 - 1$$
$$3 \cdot 333 = 999 = 1000 - 1, \ldots$$

we can evaluate the given sum as

$$3S = (10 - 1) + (100 - 1) + (1000 - 1) + \ldots + \left(10^{2016} - 1\right)$$
$$= \left(10 + 10^2 + \ldots + 10^{2016}\right) - 2016 = \frac{10 \cdot \left(10^{2016} - 1\right)}{10 - 1} - 2016$$
$$= \frac{10^{2017} - 10}{9} - 2016.$$

Finally, dividing by 3, we obtain $S = \frac{10^{2017} - 10}{27} - \frac{2016}{3} = \frac{10^{2017} - 10}{27} - 672.$

Answer. $S = \frac{10\left(10^{2016} - 1\right)}{27} - 672.$

Problem 23 (Kolmogorov) Is it possible that 100, 101, and 102 are the terms (not necessarily consecutive) of a geometric progression?

Solution. Assume that it is possible and that the following is valid:

$$b_i = 100 = b_1 \cdot r^{i-1}, \ b_j = 101 = b_1 \cdot r^{j-1}, \ b_k = 102 = b_1 \cdot r^{k-1}.$$

From the above, we can easily find

$$\frac{101}{100} = r^{j-i},$$
$$\frac{102}{101} = r^{k-j}.$$

If raise both sides of the first equation to the power of $(k - j)$ and both sides of the second equation to the power of $(j - i)$, then the right sides of the both equations will become identical. Thus, we have $\left(\frac{101}{100}\right)^{k-j} = r^{(j-i)(k-j)} = \left(\frac{102}{101}\right)^{j-i}$. This can be rewritten as $101^{k-i} = 102^{j-i} \cdot 100^{k-j}$. This equation cannot have integer solutions because the left side is odd and the right hand side is even.

Answer. One-hundred, 101, and 102 cannot be terms of a geometric progression.

1.4 Finding the nth Term of a Sequence or Series

If we write sequence of numbers such as 1, 4, 7, 10, 13, ... or $\frac{1}{2}, \frac{1}{4}, \frac{1}{8}, \frac{1}{16}, \frac{1}{32}, \ldots,$ or 1, 4, 9, 16, 25, 36, ... we may ask questions like, "What is the 25th terms of the sequence?" or "What number of terms sums to 625 for the third sequence?" Fortunately for us it will be easy to recognize an arithmetic progression in the first sequence with $a_1 = 1$ and $d = 3$. You can also call that a sequence of all natural numbers that divided by 3, give a remainder of 1 and write it with the formula $a_n = 3n + 1$, $n = 0, 1, 2, 3, \ldots$. The second sequence is a geometric progression with $B_1 = \frac{1}{2}$ and $r = \frac{1}{2}$ or you may see that each term of this sequence is some power of ½ so that each term can be written as $a_n = \left(\frac{1}{2}\right)^n$, $n = 1, 2, 3 \ldots$. The third sequence is a sequence of consecutive squares with the nth term of $a_n = n^2$, $n = 1, 2, 3, \ldots$.

All sequences are different. For example, the sequence 2, 3, 5, 7, 11, 13, 17, 19, 23, ... that represents a sequence of prime numbers, the formula for a_n does not exist (in any fixed length closed form). We will discuss here only such sequences that have a fixed length formula for the nth term, even though finding this formula can sometimes be a challenging task.

1.4.1 Finding the nth Term of a Fibonacci Type Sequence

Let us look at the problem that was offered at the Volgograd District Math Olympiad in 1977.

Problem 24 Find the nth term of a sequence 1, 1, 2, 3, 5, 8, 13, 21, 34, ...

Solution. I remember that I looked at the numbers and noticed that each of them is a sum of the two preceding numbers. Then I reformulated the problem as

A certain sequence $a_1, a_2, \ldots, a_n$ satisfies the conditions $a_0 = a_1 = 1$, $a_{n+2} = a_{n+1} + a_n$. Find a_n.

When I had this problem in 1977, I was in the 9th grade and did not know anything about the Fibonacci sequence or about methods of solving recurrent sequences. However, I knew geometric progression and I started thinking: "Is there any geometric progression, the terms of which would satisfy the condition of the problem?" So I started. Assume that it does exist, then

$a_n = a_1 r^{n-1}$, $a_{n+1} = a_1 r^n$, $a_{n+2} = a_1 r^{n+1}$. Substituting this into the recursion formula $a_{n+2} = a_{n+1} + a_n$, I obtained the equation, $a_1 r^{n+1} = a_1 r^n + a_1 r^{n-1}$. After canceling common terms, I ended up solving a quadratic equation, $r^2 - r - 1 = 0$, with two irrational roots: $r_1 = \frac{1-\sqrt{5}}{2}$, $r_2 = \frac{1+\sqrt{5}}{2}$.

Because the answer did not depend on the value of the first term, I knew that to the given recursion would satisfy infinitely many geometric progressions with common ratios above, as well as their linear combinations. In fact, if $\{a_n\}\{b_n\}$ are geometric progressions with common ratios r_1, r_2 respectively, then

$$a_1 r_1^{n+1} = a_1 r_1^n + a_1 r_1^{n-1}$$
$$b_1 r_2^{n+1} = b_1 r_2^n + b_1 r_2^{n-1}$$

and a sequence $\{c_n\}$ such that $c_n = A \cdot a_n + B \cdot b_n$ will also satisfy the given recursion because

$$c_{n+1} = A \cdot a_{n+1} + B \cdot b_{n+1}$$
$$c_{n+2} = A \cdot a_{n+2} + B \cdot b_{n+2}$$

Let us show that $c_{n+2} = c_n + c_{n+1}$:

$$c_{n+2} = c_n + c_{n+1}$$
$$= (A \cdot a_n + B \cdot b_n) + (A \cdot a_{n+1} + B \cdot b_{n+1})$$
$$= A(a_n + a_{n+1}) + B(b_n + b_{n+1})$$
$$= A \cdot a_{n+2} + B \cdot b_{n+2}$$
$$= c_{n+2}$$

Next, we can try to get the general formula for the nth term as a combination of the roots,

$$a_n = A\left(\frac{1-\sqrt{5}}{2}\right)^{n-1} + B\left(\frac{1+\sqrt{5}}{2}\right)^{n-1}. \qquad (1.18)$$

We can use the values for the first and second terms to create a system and solve for A and B,

$$\begin{cases} A + B = 1, \\ -A\sqrt{5} + B\sqrt{5} = 1 \end{cases} \Leftrightarrow \begin{cases} A = -\dfrac{(1-\sqrt{5})}{2\sqrt{5}} B = \dfrac{1+\sqrt{5}}{2\sqrt{5}}. \end{cases}$$

Substituting the values for A and B into Eq. 1.18 and simplifying, I obtained the formula for the nth term of the Fibonacci sequence

$$a_n = \frac{1}{\sqrt{5}} \left(\left(\frac{1 + \sqrt{5}}{2} \right)^{n+1} - \left(\frac{1 - \sqrt{5}}{2} \right)^{n+1} \right). \tag{1.19}$$

You can verify this formula for the nth term, by checking some of the terms listed above

$$a_1 = \frac{1}{\sqrt{5}} \left(\left(\frac{1 + \sqrt{5}}{2} \right)^{2} - \left(\frac{1 - \sqrt{5}}{2} \right)^{2} \right) = 1$$

$$a_3 = \frac{1}{\sqrt{5}} \left(\left(\frac{1 + \sqrt{5}}{2} \right)^{4} - \left(\frac{1 - \sqrt{5}}{2} \right)^{4} \right)$$

$$= \frac{1}{\sqrt{5}} \left(\left(\frac{1 + \sqrt{5}}{2} \right)^{2} - \left(\frac{1 - \sqrt{5}}{2} \right)^{2} \right) \cdot \left(\left(\frac{1 + \sqrt{5}}{2} \right)^{2} + \left(\frac{1 - \sqrt{5}}{2} \right)^{2} \right) = 3.$$

Fibonacci's sequence was discovered by Leonardo Pisano (born in Italy, circa 1170) and shows up in a lot of real-life occurrences such as the growth of flowers, plants, and pine cones. The series starts with zero and one and then the next number is the former number and its former number added together. The sequence is 0, 1, 1, 2, 3, 5, 8, 13, 21, 34 ... where $a_1 = 1$, $a_2 = 1$, and $a_{n+2} = a_n + a_{n+1}$.

Pisano came up with this sequence when he was asked a problem about the breeding of rabbits and how many would be around if you started with a pair of rabbits and each pair had one female and one male baby rabbit and then the breeding continued with the new pair of rabbits. See Appendix 1 for a MAPLE program written for Fibonacci's rabbit reproduction. This problem, however, was unrealistic because it assumed that each pair of rabbits had one female and one male rabbit as well as siblings breeding together to produce the next offspring. The problem also assumed that the birth would happen every month and no rabbit would die. Although Fibonacci's sequence was not realistic with the mating of rabbits, the sequence can be seen in other parts of nature.

Fibonacci found an interesting comparison between a term and its neighbor in the sequence. He took the ratios of a term and its following neighbor and produced a sequence that converged to numbers called Phi. Using the formula from above to determine the ratios of the R_n term, the ratio is $\frac{a_{n+1}}{a_n}$. Looking at the formula from above as well, dividing both sides by a_{n+1}, it becomes $\frac{a_{n+2}}{a_{n+1}} = \frac{a_n}{a_{n+1}} + \frac{a_{n+1}}{a_{n+1}}$ Substituting R_n in, the new formula is $R_{n+1} = \frac{1}{R_n} + 1$. Now letting n approach infinity, it reaches some limit, L, allowing for the following equation $L = \frac{1}{L} + 1$. Multiplying everything by L,

$$L^2 = 1 + L$$
$$L^2 - L - 1 = 0.$$

Using the quadratic equation, the two roots are $\Phi = \frac{1+\sqrt{5}}{2}$ and $\varphi = \frac{1-\sqrt{5}}{2}$. Taking the first 11 terms we have $\frac{1}{1}, \frac{2}{1}, \frac{3}{2}, \frac{5}{3}, \frac{8}{5}, \frac{13}{8}, \frac{21}{13}, \frac{34}{21}, \frac{55}{34}, \frac{89}{55}, \frac{144}{89}, \ldots$ or 1, 2, 1.5, 1.6667, 1.6, 1.625, 1.6153, 1.619, 1.6176, 1.618, 1.618, ... These ratios end up converging to a number known as Phi, which is equal to $\frac{\sqrt{5}+1}{2}$. This number is known as the "golden ratio."

To get phi, the same process is done, but going backwards with the sequence. Earlier, the numbers were added together to get the next term, here they are subtracted to get the next term going backwards. So the terms are 1, -1, 2, -3, 5, -8, 13, -21, 34, -55, ... As ratios they are $\frac{1}{-1}, -\frac{1}{2}, \frac{2}{-3}, -\frac{3}{5}, \frac{5}{-8}, -\frac{8}{13}, \frac{13}{-21}, -\frac{21}{34}, \frac{34}{-55}, \ldots$ or -1, -0.5, -0.6667, -0.6, -0.625, -0.6154, -0.619, -0.6176, -0.6182, ... The ratios end up converging to a number known as phi, which is equal to $\frac{1-\sqrt{5}}{2}$. Going back to the quadratic equations roots, the numbers known as Phi and phi can be produced either by ratios or by the quadratic equation that we obtained above.

Binet was a French mathematician who is credited with finding the nth term of the Fibonacci sequence in 1843,

$$F(n) = \frac{\Phi^n - \varphi^n}{\sqrt{5}}$$

where Phi and phi come from the golden ratio explained in the above paragraph. Binet's formula is the only working formula other than Fibonacci and Lucas' that can represent the sequence correctly. A proof for Binet's formula is as follows:

$$L^2 = L + 1$$
$$L^3 = L \cdot L^2 = L(L+1) = L^2 + L = (L+1) + L = 2L + 1$$
$$L^4 = L \cdot L^3 = L(2L+1) = 2L^2 + L = 2(L+1) + L = 3L + 2$$
$$L^5 = L^4 + L^3 = 5L + 3$$

$$\cdots$$

$$L^n = F(n)L + F(n-1).$$

Let $\Phi = \frac{(1+\sqrt{5})}{2}$, $\varphi = \frac{(1-\sqrt{5})}{2}$, then

$$\begin{cases} \Phi^n = F(n)\Phi + F(n-1) \\ \varphi^n = F(n)\varphi + F(n-1) \end{cases} \Rightarrow \Phi^n - \varphi^n = F(n)(\Phi - \varphi)$$

$$F(n) = \frac{\Phi^n - \varphi^n}{\Phi - \varphi} = \frac{\Phi^n - \varphi^n}{\sqrt{5}}.$$

However, the result was known to Daniel Bernoulli, Leonard Euler, and Abraham de Moivre more than a century before Binet.

A French mathematician, Lucas, spent some of his time studying the Fibonacci sequence and related his sequence of numbers to Fibonacci's sequence. The Fibonacci sequence in the following formula defines the Lucas sequence, $L_n = F_{n-1} + F_{n+1}$. The first few Lucas numbers are 1, 3, 4, 7, 11, 18, 29, 47, 76, etc. Lucas numbers also have that property for $n > 2$, $L_n = L_{n-1} + L_{n-1}$. Something to notice, a Lucas number is always greater than its corresponding Fibonacci numbers except for L_1. The Lucas numbers also converge to the golden ratio. In the Homework you are asked to find the formula for the nth Lucas number, $L_n = \left(\frac{1+\sqrt{5}}{2}\right)^n + \left(\frac{1-\sqrt{5}}{2}\right)^n$.

Many new Fibonacci type sequences can be created and continue to fascinate mathematicians. The following problem appeared in "Quant" magazine that published many challenging problems. Russian teachers recommended their high school students to try solving the problems. The magazine would come out monthly but solutions to selected problems would be given only in the following issue.

Problem 25 Consider the sequence, the terms of which are given by the following recursive formula $u_n = 7u_{n-1} - 6u_{n-2}$, $u_1 = 1$, $u_2 = 2$. Find the formula for the nth term of the sequence.

Solution. First we should evaluate several terms of the sequence

$$u_n = 7u_{n-1} - 6u_{n-2}, \quad u_1 = 1, \quad u_2 = 2 \tag{1.20}$$

as 1, 2, 8, 44, 260, ... A pattern cannot be seen right away. It would be nice if the formula for the nth term looked like an exponential function with some base, similar to a geometric progression, so let us try $u_n = r^n, u_{n-1} = r^{n-1}, u_{n-2} = r^{n-2}$ and then substitute it into Eq. 1.20 to produce $r^n = 7r^{n-1} - 6r^{n-2}$. Dividing both sides by r^{n-2}, we obtain the following quadratic equation and its zeros,

$$r^2 - 7r + 6 = 0$$
$$r_1 = 1, \quad r_2 = 6.$$

Next, let us try a linear combination of both r values for the nth term,

$$u_n = Ar_1^n + Br_2^n. \tag{1.21}$$

Substituting this into Eq. 1.20 we obtain $Ar_1^n + Br_2^n = 7\left(Ar_1^{n-1} + Br_2^{n-1}\right)$ $-6\left(Ar_1^{n-2} + Br_2^{n-2}\right)$ which could be written as $A(r_1{}^n - 7r_1{}^{n-1} + 6r_1{}^{n-2}) = -B(r_2{}^n - 7r_2{}^{n-1} + 6r_2{}^{n-2})$. Since the expression inside each parentheses is zero then Eq. 1.21 is a true representation for the nth term of the recursive sequence of Eq. 1.20. We have only to find coefficients A and B by using the values for the first and second terms,

$$u_n = A \cdot 1^n + B \cdot 6^n = A + B \cdot 6^n$$
$$u_1 = 1, \qquad\qquad A + 6B = 1$$
$$u_2 = 2, \qquad\qquad A + 36B = 2.$$

Solving the system we obtain that $A = 24/30 = 4/5$ and $B = 1/30$. Therefore,

$$u_n = \frac{4}{5} + \frac{1}{30} \cdot 6^n = \frac{4 + 6^{n-1}}{5}. \tag{1.22}$$

Trying several values for n in Eq. 1.22, any member of the sequence can be obtained:

$$u_1 = \frac{4 + 6^{1-1}}{5} = \frac{4 + 1}{5} = 1$$

$$u_2 = \frac{4 + 6^{2-1}}{5} = \frac{4 + 6}{5} = 2$$

$$u_3 = \frac{4 + 6^{3-1}}{5} = \frac{4 + 36}{5} = \frac{40}{5} = 8, \text{ etc.}$$

Answer. $u_n = \frac{4+6^{n-1}}{5}; n = 2, 3, 4 \dots$.

A second order recursion depends on the roots of a quadratic (characteristic) equation. What if the equation has two equal roots, like $r^2 - 4r + 4 = (r - 2)^2 = 0$, $r_{1,2} = 2$? This would represent a recursion as the one of Problem 26.

Problem 26 Find several terms of the sequence and the formula for its nth term. Given $x_{n+1} = 4x_n - 4x_{n-1}$, $x_0 = 1$, $x_1 = 6$.

Solution. We can see that $x_n = 2^n$ alone would not work. You might try several other things and then come up with the idea that $x_n = 2^n(A + nB) = A \cdot 2^n + nB \cdot 2^n$. Substituting the values for the first two terms of the sequence into this formula, we can evaluate A and B,

$$A = 1$$
$$2A + 2B = 6, B = 2$$

so $x_n = 2^n(2n + 1)$, $n = 0, 1, 2, 3, \dots$. Using this, we can find any term of the sequence.

Answer. 1, 6, 20, 56, 144, 832, ...

Remark. How do we know that the answer is correct? From the condition of the problem, all terms, starting from the third must be divisible by 4. It works (20, 56, 144, 832, ...). Moreover, each term, starting from the third is the difference between two preceding terms times 4. It also works because $20 = 4 \cdot (6 - 1)$, $56 = 4 \cdot (20 - 6)$, $144 = 4 \cdot (56 - 30)$, etc.

1.4.2 Finding Recurrent Formula for a Known Sequence

If a sequence is given and its nth term is known, we can attempt to describe such a sequence by recursion, i.e., to describe the nth term of the sequence as a function of the previous terms, for example as $a_n = a_{n-1} + a_{n-2}$. Thus, Fibonacci type sequence is given if we know its first and second term, etc.

> **Definition.** In general, a **recursion** is a relationship $a_n = F(n, a_{n-1}, a_{n-1}, \dots, a_{n-k})$ that allows evaluating all terms of a sequence by knowing its first k terms.

Often the nth term of a recurrent sequence is given by

$$a_n = \psi(a_{n-1}, a_{n-1}, \dots, a_{n-k}) + f(n), \quad n \geq k. \qquad (1.23)$$

This is called a recursion of the kth order. If $f(n) = 0$, then such a recursion is called homogeneous of the kth order. Any sequence $\{y_n\}$ that makes Eq. 1.23 true is called a solution of a recurrent relationship.

The first k terms of a recurrent sequence can be arbitrary. Assume that we have initial conditions:

$$a_0 = \alpha_0, \ a_1 = \alpha_1, \ \dots, a_{k-1} = \alpha_{k-1} \qquad (1.24)$$

Next, we can solve Eq. 1.23 subject to initial conditions of Eq. 1.24. This problem is similar to solving a Cauchy initial value problem for a differential equation of the kth order. If a function ψ is linear then Eq. 1.23 can be rewritten as a linear recursion, $a_n + p_1(n)a_{n-1} + p_2(n)a_{n-2} + \dots + p_k(n)a_{n-k} = f(n)$, where $p_i(n) \ i = 1,2,3,\dots, k$ are the coefficients. Here we discuss linear homogeneous recursions with constant coefficients.

> **Definition.** A recursion, $a_n + p_1(n)a_{n-1} + p_2(n)a_{n-2} + \dots + p_k(n)a_{n-k} = 0$ is **linear** and **homogeneous** and the corresponding sequence $a_0, a_1, a_2, a_3, \dots$ is a recurrent sequence of the kth order.

If the coefficients $p_i(n) = p_i$ are constants, then

$$a_n + p_1 a_{n-1} + p_2 a_{n-2} + \ldots + p_k a_{n-k} = 0 \qquad (1.25)$$

is a linear homogeneous recursion with constant coefficients.

Lemma 1.1 Let y_1^n, y_2^n be solutions of recursion Eq. 1.25, then any of their linear combinations is also a solution of Eq. 1.25.

Proof. Denote $z_n = c_1 y_n^1 + c_2 y_n^2$ and substitute it into Eq. 1.25,

$$c_1 y_n^1 + c_2 y_n^2 + p_1 \left(c_1 y_{n-1}^1 + c_2 y_{n-1}^2 \right) + \ldots + p_k \left(c_1 y_{n-k}^1 + c_2 y_{n-k}^2 \right)$$
$$= c_1 \left(y_n^1 + p_1 y_{n-1}^1 + p_2 y_{n-2}^1 + \ldots + p_k y_{n-k}^1 \right)$$
$$+ c_2 \left(y_n^2 + p_1 y_{n-1}^2 + p_2 y_{n-2}^2 + \ldots + p_k y_{n-k}^2 \right)$$
$$= 0.$$

The statement is proven.

As we do in the previous section, we can look for a solution of Eq. 1.25 in the form $y_n = r^n$.

Definition. A polynomial equation,

$$r^k + p_1 r^{k-1} + p_2 r^{k-2} + \ldots + p_{k-1} r + p_k = 0 \qquad (1.26)$$

is a **characteristic equation** for a recurrent sequence of order k of the type of Eq. 1.25. A sequence $r^{(n)}$ is called its **solution**.

For example, the Fibonacci sequence, 1, 1, 2, 3, 5, 8, 13, 21, ... given by the second order recursion as $a_{n+2} = 1 \cdot a_{n+1} + 1 \cdot a_n$ is a linear recurrent sequence of the second order. We can rewrite it as $a_{n+2} - a_{n+1} - a_n = 0$. The characteristic equation for a Fibonacci sequence is $r^2 - r - 1 = 0$ and it is a quadratic. As you remember, we found a formula for its nth term in the previous section.

Theorem 1.1 If $r_1, r_2, \ldots, r_m$ are the roots of characteristic equation Eq. 1.26 of multiplicities $s_1, s_2, \ldots, s_m$, respectively, such that $s_1 + s_2 + \ldots + s_m = k$, then the solutions $r_1^n, n r_1^n, \ldots, n^{s_1-1} r_1^n, \ldots, r_m^n, n r_m^n, \ldots, n^{s_m-1} r_m^n$ are fundamental solutions of the recursion of Eq. 1.25.

Many of the sequences explored above are recurrent type sequences of a different order. It may surprise you that well-known arithmetic and geometric sequences are also examples of recurrent sequences!

Let us show that a geometric progression is a linear recurrent sequence of the first order and that an arithmetic sequence is a linear recurrent sequence of the second order. For a geometric progression each its terms is a product of the previous term and its common ratio, r, so that $a_{n+1} = r \cdot a_n$. The characteristic equation is linear; therefore, a geometric progression that is a recurrent sequence of the first order.

For an arithmetic progression we know that each term, starting from the second is an arithmetic mean of its neighbors, so $a_{n+1} = \frac{a_n + a_{n+2}}{2}$, that can be solved for a_{n+2} as $a_{n+2} = 2a_{n+1} - a_n$. Now we know that any arithmetic progression is a recurrent sequence of second order with characteristic equation $r^2 - 2r + 1 = 0$, which means that the fundamental solutions to the recursion are $y_1 = 1, y_2 = n \Rightarrow a_n = c_1 \cdot 1 + c_2 \cdot n$.

Let us demonstrate that this indeed describes the arithmetic sequence, for example, for numbers 2, 5, 8, 11, ... that divided by 3 give a remainder of 2. Substituting $n = 1$, $a_1 = 2$, and $n = 2$, $a_2 = 5$ into the formula for the nth term above, we have a system,

$$\begin{cases} c_1 + c_2 = 2 \\ c_1 + 2c_2 = 5 \end{cases} \Rightarrow c_1 = -1, c_2 = 3 \Rightarrow a_n = 3n - 1.$$

On the other hand, we can obtain the same expression by using the formula for the nth term of an arithmetic sequence, $a_1 = 2$, $d = 3 \Rightarrow a_n = 2 + (n-1) \cdot 3 = 3n - 1$. Next, let us consider a sequence of the squares of natural numbers. For this sequence we know that $a_1 = 1$, $a_2 = 2^2 = 4$, $a_3 = 3^2 = 9$, ... $a_n = n^2$. We will evaluate three consecutive squares, next to n^{th} term,

$$a_{n+1} = (n+1)^2 = n^2 + 2n + 1 = a_n + 2n + 1 \Rightarrow \boxed{a_{n+1} - a_n = 2n + 1}$$

$$a_{n+2} = (n+2)^2 = n^2 + 4n + 4 = (n+1)^2 + 2n + 3 \Rightarrow \boxed{a_{n+2} - a_{n+1} = 2n + 3}$$

$$a_{n+2} = a_{n+1} + (2n+1) + 2 = 2a_{n+1} - a_n + 2 \Rightarrow \boxed{a_{n+2} - 2a_{n+1} + a_n = 2}$$

$$a_{n+3} = (n+3)^2 = (n+2)^2 + 2n + 5$$
$$= a_{n+2} + (2n+3) + 2$$
$$= a_{n+2} + (a_{n+2} - a_{n+1}) + (a_{n+2} - 2a_{n+1} + a_n)$$
$$\boxed{a_{n+3} = 3a_{n+2} - 3a_{n+1} + a_n}.$$

We extracted the difference between two consecutive terms of the sequence and substituted it into the next step. We did it until the difference between two consecutive terms did not contain anything besides a liner combination of the previous terms. Indeed, a sequence of squares of natural numbers is a recurrent sequence of the third order with a characteristic equation: $r^3 - 3r^2 + 3r - 1 = 0$.

Let us solve Problem 27.

Problem 27 Describe a sequence of consecutive cubes by a recurrent formula.

Solution. Using the technique described and the formula for the nth term of the sequence, let us write down several terms starting from n:

$$a_n = n^3$$
$$a_{n+1} = (n+1)^3 = n^3 + 3n^2 + 3n + 1 = a_n + 3n^2 + 3n + 1$$
$$\boxed{a_{n+1} - a_n = 3n^2 + 3n + 1}$$
$$a_{n+2} = (n+2)^3 = a_{n+1} + 3(n+1)^2 + 3(n+1) + 1$$
$$= a_{n+1} + (3n^2 + 3n + 1) + 6n + 6$$
$$= 2a_{n+1} - a_n + 6(n+1) \ \Rightarrow$$
$$\boxed{a_{n+2} - 2a_{n+1} + a_n = 6(n+1)}$$
$$a_{n+3} = 2a_{n+2} - a_{n+1} + 6(n+2)$$
$$= 2a_{n+2} - a_{n+1} + 6(n+1) + 6$$
$$a_{n+3} = 2a_{n+2} - a_{n+1} + a_{n+2} - 2a_{n+1} + a_n + 6 \ \Rightarrow$$
$$\boxed{a_{n+3} - 3a_{n+2} + 3a_{n+1} - a_n = 6}.$$

Because six is just a number, we know that finding the next term will be sufficient to find a recursion,

$$a_{n+4} = 3a_{n+3} - 3a_{n+2} + a_{n+1} + 6$$
$$= 3a_{n+3} - 3a_{n+2} + a_{n+1} + a_{n+3} - 3a_{n+2} + 3a_{n+1} - a_n$$
$$\boxed{a_{n+4} = 4a_{n+3} - 6a_{n+2} + 4a_{n+1} - a_n}.$$

Therefore, a sequence of natural cubes is a recurrent sequence of the 4th order with characteristic equation, $r^4 - 4r^3 + 6r^2 - 4r + 1 = 0$.

Answer. $a_{n+4} = 4a_{n+3} - 6a_{n+2} + 4a_{n+1} - a_n$.

Remark. Recurrent formulas are very useful in computer programming. In order to evaluate some quantity given by a recursion, we can evaluate its value on the $(n+1)$ iteration by knowing the value of a quantity on the nth and $(n-1)$ step.

When solving problems on evaluating finite or infinite sums of series, or if we are trying to evaluate a limit of a sequence of partial sums, it is important to have a recurrent formula of the nth term of a series. Often it is not easy to establish a relationship between consecutive terms of a sequence or series. For example, consider $\frac{10}{3}, \frac{10^2}{9}, \frac{10^3}{27}, \ldots, \frac{10^n}{3^n}, \ldots$. Obviously, the nth term is $a_n = \left(\frac{10}{3}\right)^n$ and two consecutive terms are connected by the formula, $a_{n+1} = \frac{10}{3} \cdot a_n$. It is a geometric progression and linear recursion of the first order.

Problem 28 Find a recursion for the sequence $\frac{x^2}{1\cdot2\cdot3}, \frac{x^3}{2\cdot3\cdot4}, \frac{x^4}{3\cdot4\cdot5}, \ldots$

Solution. Let us find the formula for the nth and $(n+1)$ terms of $a_n = \frac{x^{n+2}}{(n-1)n(n+1)}$, $a_{n+1} = \frac{x^{n+3}}{n(n+1)(n+2)}$. Next, we will find the ratio of two terms,

$$\frac{a_{n+1}}{a_n} = \frac{x\cdot(n-1)}{n+2} \Rightarrow \boxed{a_{n+1} = a_n \cdot \frac{x(n-1)}{n+2}}.$$

If you remember at the beginning of the book, we introduced several sequences on a calculator. The Fibonacci sequence was introduced by a recursion. You can try to find the terms of this sequence by using the same ideas and by first setting $x = \frac{1}{2}$ and then for example $x = 2$.

Problem 29 (Moscow Math Olympiad 1993) Evaluate the 100th term of a sequence with $x_1 = 4$, $x_2 = 6$ such that x_n is a minimal natural composite number greater than $2x_{n-1} - x_{n-2}$, $\forall n \geq 3$. Derive the formula for the nth term of the sequence.

Solution. Consider some of the terms of the sequence,

$$2x_2 - x_1 = 8 \Rightarrow x_3 = 9 > 8$$
$$2x_3 - x_2 = 12 \Rightarrow x_4 = 14 > 12.$$

The 4th term cannot be 13 because 13 is prime. Let us continue by analogy

$$2x_4 - x_3 = 28 - 9 = 19 \Rightarrow x_5 = 20 > 19$$
$$2x_5 - x_4 = 40 - 14 = 26 \Rightarrow x_6 = 27 > 26$$
$$2x_6 - x_5 = 54 - 20 = 34 \Rightarrow x_7 = 35 > 34.$$

Notice there is some pattern in

$$x_3 = 9$$
$$x_4 = 14 = x_3 + 5$$
$$x_5 = 20 = x_4 + 6$$
$$x_6 = 27 = x_5 + 7$$
$$x_7 = 35 = x_6 + 8.$$

It looks like we can predict the formula $x_n = x_{n-1} + n + 1, \ n > 3$. Using this formula for different n starting from $n = 3$, we obtain

$$x_5 = x_4 + 6 = x_3 + 5 + 6$$
$$x_6 = x_5 + 7 = x_3 + 5 + 6 + 7$$
$$x_7 = x_6 + 8 = x_3 + 5 + 6 + 7 + 8$$
$$x_8 = x_7 + 9 = x_3 + 5 + 6 + 7 + 8 + 9$$
$$\cdots$$
$$x_n = x_{n-1} + n + 1 = x_3 + (5 + 6 + 7 + 8 + 9 + \ldots + n + n + 1).$$

The sum inside parentheses is easy to recognize if we replace the third term (9) by $2 + 3 + 4 = 9$, then the nth term can be calculated as

$$x_n = 2 + 3 + 4 + 5 + \ldots + n + n + 1 = \frac{(n+1)(n+2)}{2} - 1$$
$$= \frac{n(n+3)}{2}.$$

Since $x_n = \frac{n(n+3)}{2} \Rightarrow x_{100} = \frac{100 \cdot 103}{2} = 5150$.

Now, by mathematical induction let us prove that $x_n = \frac{n(n+3)}{2}, \ n \geq 4$ satisfies the problem.

1. Assume that the formula is correct. The statement is true for $n = 4$ because $x_4 = 14 = \frac{4 \cdot 7}{2}$.

2. Assume that the formula is true for $n = k$, i.e., $x_k = \frac{k(k+3)}{2}$.

3. Let us show that it is true for $n = k + 1$, i.e., $x_{k+1} = \frac{(k+1)(k+4)}{2}$. Using recursion, we obtain $2x_k - x_{k-1} = 2 \cdot \frac{k(k+3)}{2} - \frac{(k-1)(k+2)}{2} = \frac{(k+1)(k+4)}{2} - 1$. By the condition of the problem x_{k+1} must be the first composite number greater than $\frac{(k+1)(k+4)}{2} - 1$. Clearly, the first part of this number $\frac{(k+1)(k+4)}{2}$ is a composite number itself and it is precisely one more than $\frac{(k+1)(k+4)}{2} - 1$.

Case 1. If k is odd number, then $(k + 1)$ is an even number and so $x_{k+1} = \frac{(k+1)}{2} \cdot (k + 4)$.

Case 2. If k is an even $(k + 4)$ is also even number and $x_{k+1} = (k + 1) \cdot \frac{(k+4)}{2}$.

The proof is complete.

Answer. 5150.

1.4.3 Other Sequences

If a sequence or series are neither arithmetic nor geometric, then finding its nth term can be a challenging task. For example, let us consider the following five terms of a sequence: $4, -\frac{1}{3}, \frac{2}{7}, \frac{5}{11}, \frac{8}{15}, \ldots$. Because most of the terms are fractions, it helps to find the formula separately for the numerator and for the denominator, so $u_n = \frac{a_n}{b_n}$. In this particular case, we see that the numerators form an arithmetic progression with common difference of 3, first term 4, and the denominator numbers form arithmetic progression with a common difference of 5 and first term 1. So it can be written as $u_n = \frac{3n-7}{4n-5}$.

Problem 30 Find the formula the nth term of the series, $\frac{3}{2} + \frac{9}{64} + \frac{1}{24} + \frac{81}{4096} + \ldots$.

Solution. At first, it looks like the numerators are formed by powers of three and that the denominator by the powers of two, but the third term (1/24) seems to be not following this "rule." Our first idea, $\frac{3^n}{2^n}$, however, might help us in finding correct formula for the nth term of the series. Can we assume that this first guessed formula may be multiplied by an unknown factor, such as $u_n = \frac{1}{v_n}\frac{3^n}{2^n}$? Just write down several known terms,

$$n = 1 \quad u_1 = \frac{3}{2} = \frac{1}{v_1}\cdot\frac{3}{2} \Rightarrow v_1 = 1 = 1^4$$

$$n = 2 \quad u_2 = \frac{9}{64} = \frac{1}{v_2}\cdot\frac{9}{4} \Rightarrow v_2 = 16 = 2^4$$

$$n = 3 \quad u_3 = \frac{1}{24} = \frac{1}{v_3}\cdot\frac{27}{8} \Rightarrow v_3 = 81 = 3^4$$

$$n = 4 \quad u_4 = \frac{81}{4096} = \frac{1}{v_4}\cdot\frac{81}{16} \Rightarrow v_4 = 256 = 4^4$$

Based on the four terms, we can see that the assumption was correct and that $v_n = n^4$ so $u_n = \frac{1}{n^4}\cdot\left(\frac{3}{2}\right)^n$.

Answer.

$$u_n = \frac{1}{n^4}\cdot\left(\frac{3}{2}\right)^n.$$

Problem 31 Given an infinite series, $\frac{1}{2} + 2\cdot\frac{1}{4} + 3\cdot\frac{1}{8} + 4\cdot\frac{1}{16} + 5\cdot\frac{1}{32} + \ldots +$, find its nth term and the sum of its first n terms.

Solution. The nth term is $u_n = n\cdot\frac{1}{2^n}$ and the nth sum is $S_n = 1\cdot\frac{1}{2} + 2\cdot\frac{1}{4} + 3\cdot\frac{1}{8} + 4\cdot\frac{1}{16} + 5\cdot\frac{1}{32} + \ldots + n\cdot\frac{1}{2^n}$. Since all terms are positive, we will benefit from regrouping terms of this sum,

$$S_n = \left(\frac{1}{2} + \frac{1}{4} + \frac{1}{8} + \ldots + \frac{1}{2^n}\right) + \left(\frac{1}{4} + \frac{1}{8} + \frac{1}{16} + \ldots + \frac{1}{2^n}\right)$$
$$+ \left(\frac{1}{8} + \frac{1}{16} + \ldots + \frac{1}{2^n}\right) + \ldots + \left(\frac{1}{2^{n-1}} + \frac{1}{2^n}\right) + \frac{1}{2^n}.$$

Now the partial sum can be found as the sum of n geometric series with common difference of ½ and the first term of ½, ¼, etc. The first sum is

$$\frac{1}{2} \cdot \left(\frac{1 - \left(\frac{1}{2}\right)^n}{1 - \frac{1}{2}}\right) = 1 - \left(\frac{1}{2}\right)^n.$$

The second sum is

$$\frac{1}{4} + \frac{1}{8} + \frac{1}{16} + \ldots \frac{1}{2^n} = \frac{1}{2} \cdot \left(\frac{1}{2} + \frac{1}{4} + \frac{1}{8} + \ldots \frac{1}{2^{n-1}}\right)$$
$$= \frac{1}{2} \cdot \frac{1}{2}\left(\frac{1 - \left(\frac{1}{2}\right)^{n-1}}{1 - \frac{1}{2}}\right) = \frac{1}{2}\left(1 - \left(\frac{1}{2}\right)^{n-1}\right).$$

The third sum is

$$\frac{1}{8} + \frac{1}{16} + \ldots + \frac{1}{2^n} = \frac{1}{4}\left(\frac{1}{2} + \frac{1}{4} + \frac{1}{8} + \ldots \frac{1}{2^{n-2}}\right) = \frac{1}{4} \cdot \frac{1}{2}\left(\frac{1 - \left(\frac{1}{2}\right)^{n-2}}{1 - \frac{1}{2}}\right) = \frac{1}{4}\left(1 - \left(\frac{1}{2}\right)^{n-2}\right), \text{ etc.}$$

Adding terms inside all parentheses, we obtain

$$S_n = 1 - \left(\frac{1}{2}\right)^n + \frac{1}{2}\left(1 - \left(\frac{1}{2}\right)^{n-1}\right) + \frac{1}{4}\left(1 - \left(\frac{1}{2}\right)^{n-2}\right)$$
$$+ \frac{1}{8}\left(1 - \left(\frac{1}{2}\right)^{n-3}\right) + \ldots + \frac{1}{2^{n-1}}\left(1 - \frac{1}{2}\right).$$

The expression on the right can be rewritten as

$$1 + \frac{1}{2} + \frac{1}{4} + \frac{1}{8} + \ldots + \frac{1}{2^{n-1}} - \left(\frac{1}{2^n} + \frac{1}{2} \cdot \frac{1}{2^{n-1}} + \frac{1}{4} \cdot \frac{1}{2^{n-2}} + \ldots + \frac{1}{2^{n-1}} \cdot \frac{1}{2}\right).$$

Notice that each term inside parentheses is the same, $\frac{1}{2^n}$.

Finally, we get $S_n = 1 \cdot \dfrac{\left(1 - \frac{1}{2^n}\right)}{\frac{1}{2}} - n \cdot \dfrac{1}{2^n} = 2 - \dfrac{1}{2^{n-1}} - \dfrac{n}{2^n}.$ The limit of the partial sums obviously converges to $S_\infty = 2$.

Answer. $u_n = \frac{n}{2^n}$, $S_n = 2 - \frac{1}{2^{n-1}} - \frac{n}{2^n}$, $S_\infty = 2$.

What if each term of a sequence is a partial sum of an arithmetic progression? Then

$$a_1 = S_1, \quad S_1 = b_1$$
$$a_2 = S_2, \quad S_2 = b_1 + b_2, \quad b_2 = b_1 + 1 \cdot d$$
$$a_3 = S_3, \quad S_3 = b_1 + b_2 + b_3, \quad b_3 = b_1 + 2 \cdot d$$
$$\ldots$$
$$a_n = S_n, \quad S_n = b_1 + b_2 + b_3 + \ldots + b_n, \quad b_n = b_1 + (n-1) \cdot d.$$

Let us examine this by solving Problem 32.

Problem 32 Find the nth term of a sequence 3, 13, 30, 54, 85, 123, ... Evaluate its 57th term.

Solution. Clearly this is neither a geometric nor an arithmetic sequence. However, if we subtract pairs of consecutive terms, we obtain,

$$13 - 3 = 10$$
$$30 - 13 = 17$$
$$54 - 30 = 24$$
$$85 - 54 = 31$$
$$123 - 85 = 38.$$

The given sequence is not an arithmetic sequence, however, the differences of two consecutive terms, 10, 17, 24, 31, 38, ... are in an arithmetic progression with common difference $d = 7$. This means that the given sequence of numbers 3, 13, 30, 54, ... is the sequence of partial sums of this arithmetic progression and that its nth term can be evaluated as

$$a_n = S_n = \frac{2 \cdot 3 + (n-1) \cdot 7}{2} \cdot n$$
$$= \frac{(7n - 1) \cdot n}{2}.$$

In order to find the 57th term, we will substitute $n = 57$, $a_{57} = \frac{(7 \cdot 57 - 1) \cdot 57}{2} = 199 \cdot 57 = 200 \cdot 57 - 57 = 11,343$.

Answer. $a_n = \frac{(7n-1)n}{2}$, $a_{57} = 11,343$.

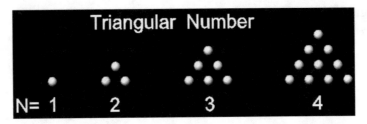

Figure 1.4 Triangular numbers

Problem 33 Find the formula for n^{th} term of a sequence 1, 3, 6, 10, 15, 21, 28, 36, 45, ...

Solution. Notice that $1 = 1, \ 3 = 1 + 2, \ 6 = 1 + 2 + 3, \ 10 = 1 + 2 + 3 + 4,$ $15 = 1 + 2 + 3 + 4 + 5$, etc. By induction, the nth term is a sum of the first n natural numbers. Therefore, its formula can be found as $a_n = S_n = 1 + 2 + 3 + \ldots + n = \frac{2 \cdot 1 + \boxed{1} \cdot (n-1)}{2} \cdot n = \frac{n(n+1)}{2}$. It is easy to check that $a_7 = S_7 = 1 + 2 + \ldots + 7 = \frac{7 \cdot 8}{2} = 28$. We can evaluate the series using the formula for the sum of the arithmetic progression, emphasizing the value of the common difference.

Answer. $a_n = \frac{n(n+1)}{2}$.

I need to mention that the numbers 1, 3, 6, 10, 15, 21, 25, . . . have a special name—they are triangular numbers. Ancient Greeks knew about these numbers and even gave them this name. Greeks tried to solve problems geometrically. Imagine a triangle where each side is formed by n dots or n billiard balls. If we arrange four such triangular numbers as in Figure 1.4, we can see how the number of the "balls" in each case denoted by $T(n)$ can be calculated. For example, we can add the balls by the rows.

Lemma 1.2 A triangular number can be evaluated as

$$T(n) = \frac{n(n+1)}{2} \tag{1.27}$$

Proof. $T(n) = 1 + 2 + 3 + \ldots + n = \frac{n(n+1)}{2}$.

What was unusual about the sequence of triangular numbers is that each term is the partial sum of an arithmetic progression with first term one and common difference one. Are there other sequences like this? Yes, there are infinitely many sequences with similar properties.

Lemma 1.3 Given a sequence $\{a_i\}$ of natural numbers and consider another sequence $\{b_i\}$, such that

$$b_1 = a_1$$
$$b_2 = a_2 - a_1$$
$$b_3 = a_3 - a_2$$
$$\dots$$
$$b_n = a_n - a_{n-1}$$

If $\{b_i\}$ is an arithmetic sequence with common difference, d, i.e., $d = b_i - b_{i-1}$, $b_i = b_1 + (n-1) \cdot d$, $i > 1$, $i \in \mathbb{N}$ then $\{a_i\}$ is the sequence of partial sums for this arithmetic progression. Its nth term can be calculated as $a_n = S_n = \frac{2a_1 + (n-1)d}{2} \cdot n$.

Let us start from some arithmetic sequences with the same first term 1 and common difference $d = 1$, $d = 2$, $d = 3$, and $d = 4$, respectively and put next to them the corresponding sequence of its partial sums:

$1, 2, 3, 4, 5, \dots,$ $b_1 = 1, d = 1 \Rightarrow S_1 = 1, S_2 = 3, S_3 = 6, S_4 = 10, S_5 = 15, \dots$

$1, 3, 5, 7, 9, \dots,$ $b_1 = 1, d = 2 \Rightarrow S_1 = 1, S_2 = 4, S_3 = 9, S_4 = 16, S_5 = 25, \dots$

$1, 4, 7, 10, 13, 16, \dots,$ $b_1 = 1, d = 3 \Rightarrow S_1 = 1, S_2 = 5, S_3 = 12, S_4 = 22, S_5 = 35, \dots$

$1, 5, 9, 13, 17, 21, \dots,$ $b_1 = 1, d = 4 \Rightarrow S_1 = 1, S_2 = 6, S_3 = 15, S_4 = 28, S_5 = 45, \dots$

Since we have already worked with the first sequence, consider the second sequence: $1, 4, 9, 16, 25, \dots$. If we did not have a discussion above, we could easily predict any term of this sequence, because obsiously each term is a perfect square of a number of the term n, i.e., $a_n = n^2$. On the other hand, a sequence of the squares is a sequence of partial sums for the arithmetic sequence: $1, 3, 5, 7, 9, \dots$ and the following is true:

$$a_1 = 1^2 = 1$$
$$a_2 = 2^2 = 1 + 3$$
$$a_3 = 3^2 = 1 + 3 + 5$$
$$a_4 = 4^2 = 1 + 3 + 5 + 7$$
$$a_5 = 5^2 = 1 + 3 + 5 + 7 + 9$$
$$\dots$$
$$a_n = n^2 = 1 + 3 + 5 + \dots + (2n - 1)$$

Hence, additionally, we can state Lemma 1.4.

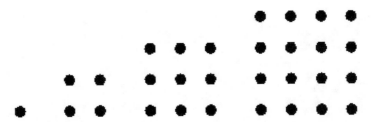

Figure 1.5 A sequence of square numbers

Lemma 1.4 Each square can be represented as a sum of n consecutive odd numbers, starting from 1.

Ancient Greeks called this a sequence of square numbers and tried to prove everything geometrically. For example, they knew that if one takes one stone and three stones, then one can make a square with side two. If one takes additionally five stones and places them around the square of side two, then one will make a square of side three, etc. (Figure 1.5). Read more about these numbers in Section 1.5.

Problem 34 Given a sequence 1, 5, 12, 22, 35, 51, 70, 92, 117, ... What is the 50th term of the sequence and find the formula for the nth term.

Solution. We can see that each term can be described by $a_n = \frac{n(3n-1)}{2}$, $n \in \mathbb{N}$. So $a_{50} = \frac{50 \cdot (3 \cdot 50 - 1)}{2} = 3725$. Did you recognize this sequence as a sequence of partial sums of an arithmetic progression with one as the first term and common difference of 3? Thus, the following is true:

$$a_1 = S_1 = \boxed{b_1 = 1} \qquad a_1 = 1$$
$$a_2 = S_2 = b_1 + b_2 = 5 \Rightarrow \quad b_2 = 4, \ d = b_2 - b_1 = 4 - 1 = 3, \quad a_2 = 1 + 4$$
$$a_3 = S_3 = b_1 + b_2 + b_3 = 12 \Rightarrow \quad b_3 = 7, \ d = b_3 - b_2 = 7 - 4 = 3, \quad a_3 = 1 + 4 + 7$$
$$a_4 = S_4 = S_3 + b_4 = 22 \Rightarrow \quad b_4 = 10, \ \boxed{d = 3}, \qquad a_3 = 1 + 4 + 7 + 10$$

$$\ldots$$

$$\boxed{a_n} = S_n = \frac{2 \cdot 1 + (n-1) \cdot \boxed{3}}{2} \cdot n = \boxed{\frac{(3n-1)n}{2}}$$

Answer. The 50th term is 3725.

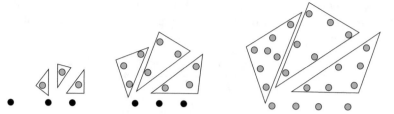

Figure 1.6 Pentagonal numbers

Remark. These numbers also have a geometric interpretation and are called pentagonal numbers. The nth number represents the number of balls that can fit a side of a regular pentagon with side n (Figure 1.6).

Ancient Greeks constructed pentagonal numbers using stones, which in this book we call "balls." First, we have one ball. Next, we put two balls on each side of a second pentagon for a total of five. Then we put three balls on each side on the third pentagon for a total of 12, etc. It is interesting that each pentagon can be developed from triangular numbers. For example, for the third one, we can put four balls on the bottom and then add to it three triangular numbers, each containing three balls, and so on. This allows us to derive the formula for the nth pentagonal number as $P(n) = n + 3 \cdot T(n-1)$.

The formula for the pentagonal numbers can also be derived algebraically. By evaluating the nth term as the nth partial sum of the arithmetic progression with the first term 1 and common difference of 3, we obtained a formula that can be rewritten as $a_n = \frac{2+3(n-1)}{2} \cdot n = n + 3 \cdot \boxed{\frac{(n-1)n}{2}}$, where the expression inside the box is the formula for the $(n-1)$ term of the sequence of triangular numbers. In general, any k^{th} figurate number can be constructed out of triangular numbers by $n + (k-2) \cdot T(n-1)$. The nth term of a k-angular number (or k^{th} figurate number) is the partial sum of an arithmetic progression with the first term of 1 and the common difference of $d = k - 2$. You could see that for a triangular number $d = 1 = 3 - 2$, for a pentagonal number $d = 3 = 5 - 2$, and hence an n^{th}hexagonal number would correspond to the partial sum of an arithmetic progression with first term 1 and the common difference of $d = 6 - 2 = 4$, etc.

We can create a general formula for the nth term of a k-angular number,

$$\boxed{a_n^k} = \frac{2 \cdot 1 + (n-1)(k-2)}{2} \cdot n$$

$$= \boxed{n + (k-2) \cdot \frac{(n-1)n}{2}} \tag{1.28}$$

$$= n + (k-2) \cdot T(n-1).$$

Remark. K-polygonal numbers (or k-angular) are nonnegative integers constructed geometrically from the regular polygons. The kth polygonal number can be calculated by

$$P_k(n) = \frac{(k-2)n(n-1)}{2} + n$$

$$= \frac{(k-2)n(n-1) + 2n}{2}.$$

Thus, the nth terms for first 8 polygonal numbers can be evaluated by

$$P_3(n) = T(n) = \frac{n(n+1)}{2}, \ P_4(n) = n^2, \ P_5(n) = \frac{n(3n-1)}{2},$$

$$P_6(n) = n(2n-1), \ P_7(n) = \frac{n(5n-3)}{2}, \ P_8(n) = n(3n-2).$$

Generalized octagonal numbers can be found from $P_8(x) = x(3x-2)$, $x \in Z$. While regular octagonal numbers 1, 8, 21, 40, 65,... are obtained for natural x, the sequence of the generalized octagonal numbers includes additional numbers obtained for a negative value of x in such a way that each octagonal number is surrounded by generalized numbers: 0, $\boxed{1}$, 5, $\boxed{8}$, 16, $\boxed{21}$, 33, $\boxed{40}$, 56, $\boxed{65}$, 85, ...

> **Problem 35** Find the 25th term and a formula for the nth term for the sequence, 1, 7, 19, 37, 61, 91,...

Solution. Inspecting each term, we notice that each one is the difference of two consecutive cubes. Thus, $7 = 8 - 1$, $19 = 27 - 8$, $37 = 64 - 27$, etc. Therefore, $a_n = n^3 - (n-1)^3 = 3n^2 - 3n + 1$. The 25th term is 1801.

These numbers also have a geometric interpretation and are called centered hexagonal numbers. Imagine a hexagonal shape and let us fit it by balls without empty spaces including the center. Then the total number of balls that fit a hexagon of side n will be given by $a_n = 3n^2 - 3n + 1$ and for example in a hexagon with side 2 we can place 7 balls, with side 3 there are 19 balls, and with side 4 there are 37 balls, etc. (Figure 1.7). From this figure, you can see that the 5th centered hexagonal number (61) is one more than six times the 4th triangular number (10). Hence, the nth centered hexagonal number is precisely one more than six times the $(n-1)$ triangular number.

We can prove this statement algebraically, if we rewrite the formula for the nth term as $a_n = 3n(n-1) + 1 = 6 \cdot \frac{n(n-1)}{2} + 1$.

Answer. $a_n = 3n^2 - 3n + 1$, $a_{25} = 1801$.

Figure 1.7 Centered
hexagonal number

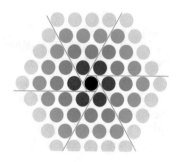

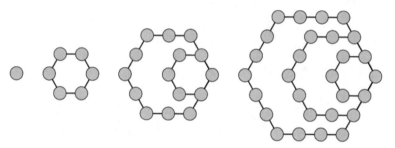

Figure 1.8 Hexagonal numbers

Problem 36 Find the nth term of the sequence 1, 6, 15, 28, 45, 66, 91, ...

Solution. By manipulating the numbers, we notice that each is a product of n and
corresponding odd number $(2n - 1)$,

$$a_1 = 1 = 1 \cdot 1$$
$$a_2 = 6 = 2 \cdot 3$$
$$a_3 = 15 = 3 \cdot 5$$
$$a_4 = 28 = 4 \cdot 7$$
$$\cdots$$
$$a_n = n(2n - 1)$$

Hence, any term of the given sequence can be written as $a_n = 2n^2 - n$.

This sequence also has a geometric interpretation and is called the sequence of
hexagonal numbers. The nth term of the sequence represents the number of balls
that can fit a hexagon of side n. Four hexagonal numbers: 1, 6, 15 and 28 are shown
in Figure 1.8. Moreover, these numbers are represented by the partial sums of an
arithmetic sequence with the first term 1 and the common difference of
4. Therefore each term of this sequence can be calculuated using Eq. 1.28 by the
substitution of $k = 6$.

Problem 37 Find the formula for a general term of the sequence 1, 4, 10, 20, 35, 56, 84, 120,...and find its 20th term.

Solution. Some numbers in the sequence are odd and some are even. Let us multiply each number by 6, so we will obtain another sequence of the numbers that is related to the given one: 6, 24, 60, 120, 210, 336, 504,..... Notice that each term of this sequence is the product of three consecutive integers,

$$6 = 1 \cdot 2 \cdot 3$$
$$24 = 2 \cdot 3 \cdot 4$$
$$60 = 3 \cdot 4 \cdot 5$$
$$120 = 4 \cdot 5 \cdot 6$$
$$\cdots$$
$$b_n = n(n+1)(n+2), \quad a_n = \frac{b_n}{6}.$$

Finally, a general formula and its 20th term of the given sequence are $a_n = \frac{n(n+1)(n+2)}{6}$, $a_{20} = 1540$.

The sequence of numbers of Problem 37 also have a geometric interpretation and are called tetrahedral numbers. Each number equals the total number of balls that can fit a tetrahedron of side n. In Figure 1.9, three consecutive tetrahedral numbers are constructed. If $n = 4$, we can put one additional ball inside the triangle of the base. Denote an nth tetrahedral number by $Tr(n)$. Using Figure 1.9 we can find a geometric way to calculate $Tr(n)$,

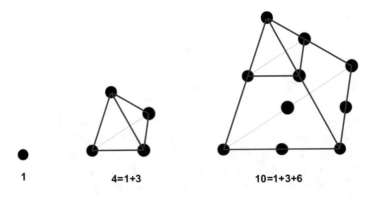

1   4=1+3   10=1+3+6

Figure 1.9 Tetrahedral numbers

$$Tr(1) = 1$$
$$Tr(2) = 1 + 3 = 4$$
$$Tr(3) = 1 + 3 + 6 = 10$$

$$\cdots$$

$$Tr(n) = T(1) + T(2) + \ldots + T(n),$$

where $T(n)$ is the corresponding triangular number. Substituting Eq. 1.27 for a triangular number in the formula above, we obtain $Tr(n) = \sum_{n=1}^{n} T(n) =$ $\boxed{\sum_{n=1}^{n} \frac{n(n+1)}{2}} = \frac{n(n+1)(n+2)}{6}$. We will learn how to evaluate the sum of the box by solving Problem 40.

Pythagoras (Greek mathematician, 570–490 BC) created the so-called triangular pyramidal numbers that differ from the tetrahedral numbers starting from the 4th term. Imagine the 4th tetrahedral term constructed without a ball inside the base triangle. This would generate a sequence of pyramidal numbers: 1, 4, 10, 19, 31, 46, 64, ... (Figure 1.10). The nth number is the sum of balls that can fit triangles with sides of 1, 2, 3, 4, 5, ..., n balls. Thus, the base triangle will contain $3 \cdot (n - 1)$ balls. To get the nth pyramidal number, we need to add the balls in all n triangles,

Figure 1.10 The 4th pyramidal number

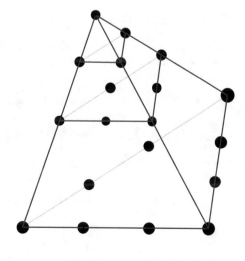

19 = 1 + 3 + 6 + 9

$$p(n) = 1 + 3 + 6 + 9 + 12 + 15 + \ldots + 3(n-1)$$
$$= 1 + 3 \cdot (1 + 2 + 3 + 4 + 5 + \ldots + (n-1))$$
$$= 1 + \frac{3(n-1)n}{2}.$$

1.5   Summation Formulas Known to Ancient Babylonians and Greeks

Remember that when we deal with sequences, we often have to find a formula for the nth term of a sequence. It was quite easy for geometric and arithmetic sequences. Any geometric sequence $\{u_n\}$ can be written as $u_{n+1} = r \cdot u_n$, $n \in \mathbb{N}$, where r is the common ratio. For example if $u_1 = 1$, $r = 2$, then any term of the sequence will be some power of 2, e.g., $u_k = 2^{k-1}$. If you determine an explicit formula for the nth term of a sequence, then each series can be written in a compact form using sigma notation, $\sum$ (summation) $a_1 + a_2 + a_3 + \ldots + a_n = \sum_{k=1}^{n} a_k$. Summation means that for the sequence, the kth term described by a_k, we add all terms between $k = 1$ and $k = n$.

Sigma notation benefits us a great deal. For example, when we want to find the sum of all numbers between 8 and 503 that divided by 5 leave a remainder of 3, then the nth term of the series is $a_n = 5n + 3$. Substituting the values of the last term, 503, and the first term, 8, we will evaluate the number of the upper and lower index of summation as 100 and 1, respectively, and write the series as $\sum_{n=1}^{100} (5n + 3) = 8 + 13 + 18 + 23 + \ldots + 503$. Moreover, using properties of summation and some known summation formulas, we can evaluate many previously unknown sums easily.

The Basic Properties of Sigma Notation

1. $\sum_{k=1}^{n} a \cdot b_k = a \cdot \sum_{k=1}^{n} b_k$ (a constant can be put before the summation)

2. $\sum_{m}^{n} a_k \pm \sum_{m}^{n} b_k = \sum_{m}^{n} (a_k \pm b_k)$

3. $\sum_{1}^{n} (b_k \pm m) = \sum_{1}^{n} b_k \pm n \cdot m$ (because the number m appears n times)

4. $\sum_{k=1}^{n} a_k = \sum_{k=2}^{n+1} a_{k-1} = \sum_{k=m+1}^{n+m} a_{k-m}$

The last property shifts the summation index and is often helpful in finding sums using other known sums that can be extracted and for simplification of terms. In the following example, $a_k = \frac{1}{k}$.

Example. Evaluate $\sum\limits_{n=1}^{100} \frac{1}{n} - \sum\limits_{n=2}^{98} \frac{1}{n-1}$.

The second sum can be rewritten as $\sum\limits_{n=1}^{97} \frac{1}{n}$, and then using the second and fourth properties of summation, we can simplify the given expression as follows:

$$\sum_{n=1}^{100} \frac{1}{n} - \sum_{n=2}^{98} \frac{1}{n-1}$$

$$= \sum_{n=1}^{100} \frac{1}{n} - \sum_{n=1}^{97} \frac{1}{n}$$

$$= \sum_{n=1}^{97} \frac{1}{n} + \frac{1}{98} + \frac{1}{99} + \frac{1}{100} - \sum_{n=1}^{97} \frac{1}{n}$$

$$= \frac{1}{98} + \frac{1}{99} + \frac{1}{100}$$

Note that property 4 can be very useful in finding infinite sums as well. However, we do not have to worry about upper index of the summation; it will remain as ∞.

Example. $\sum\limits_{n=1}^{\infty} \left(\frac{1}{n} - \frac{1}{n+2} \right) = \sum\limits_{n=3}^{\infty} \frac{1}{n} + 1 + \frac{1}{2} - \sum\limits_{n=3}^{\infty} \frac{1}{n} = \frac{3}{2}$.

Here are some useful summation formulas:

$$\sum_{k=1}^{n} k = \frac{n(n+1)}{2} \tag{1.29}$$

$$\sum_{k=1}^{n} k^2 = \frac{n(n+1)(2n+1)}{6} \tag{1.30}$$

$$\sum_{k=1}^{n} k^3 = \left(\sum_{k=1}^{n} k \right)^2 = \frac{n^2(n+1)^2}{4} \tag{1.31}$$

The sum of all numbers between 8 and 503 that gives remainder 3 when divided by 5 is evaluated as $\sum\limits_{n=1}^{100} (5n+3) = 5 \sum\limits_{n=1}^{100} n + 3 \cdot 100 = 5 \cdot \frac{100 \cdot 101}{2} + 300 = 25,550$.
Alternatively, the same sum can be obtained using the formula for the sum of the first 100 terms of an arithmetic progression with the first term 8 and common difference $d = 5$, $S_{100} = \frac{(8+503)}{2} \cdot 100 = 25,550$.

Example. $\displaystyle\sum_1^{100} (n^2 + 5) = \sum_1^{100} n^2 + 100 \cdot 5 = 500 + \sum_1^{100} n^2 = 500 + \frac{100 \cdot 101 \cdot (2 \cdot 100 + 1)}{6}$

$= 338,850.$

Problem 38 Evaluate the sum, $\displaystyle\sum_{k=1}^{n} \left(4k^3 - 6k^2 + 3\right)$.

Solution. Applying the properties of the summation and Eqs. 1.29–1.31, we have

$$\sum_{k=1}^{n} \left(4k^3 - 6k^2 + 3\right) = 4\sum_{k=1}^{n} k^3 - 6\sum_{k=1}^{n} k^2 + 3n$$
$$= n^2(n+1)^2 - n(n+1)(2n+1) + 3n$$
$$= n^4 - 2n^2 + 2n.$$

Moreover, for a sequence like 2, 9, 28, 65, 126, ... we can recognize the pattern and find the formula for the nth term as $(n^3 + 1)$, then use sigma notation and the well-known summation formulas to evaluate the exact partial sum of the first k terms, $S_k = \displaystyle\sum_1^{k} (n^3 + 1) = \left(\frac{k(k+1)}{2}\right)^2 + k = \frac{k\left(k^3 + 2k^2 + k + 4\right)}{4}$.

Let us see how knowledge of sigma notation and the formulas can help us with the Problem 39.

Problem 39 Evaluate the sum, $S = 2 + 6 + 12 + 20 + 30 + 42 + \ldots + 2550$.

Solution. First, we have a finite sum. Notice that each term is a product of consecutive natural numbers such as $n(n + 1)$. For example, $2 = 1 \cdot (1 + 1)$, $6 = 2 \cdot (2 + 1)$, $12 = 3 \cdot (3 + 1)$, $30 = 5 \cdot (5 + 1)$, $\ldots, 2550 = 50 \cdot (50 + 1)$. $S = \displaystyle\sum_{n=1}^{50}$

$n(n + 1) = \displaystyle\sum_{n=1}^{50} (n^2 + n) = \sum_{n=1}^{50} n^2 + \sum_{n=1}^{50} n = \frac{50 \cdot 51 \cdot (2 \cdot 50 + 1)}{6} + \frac{50 \cdot 51}{2} = 44,200.$ You could also recognize in this sum a double sum of triangular numbers: 1, 3, 6, 10, 15, 21, 28, etc.

Answer. $S = 44,200$.

Remark. We could evaluate the sum of the first n terms of the series above as follows:

$$S_n = \sum_{k=1}^{n} \left(k^2 + k\right)$$
$$= \sum_{k=1}^{n} k^2 + \sum_{k=1}^{n} k$$
$$= \frac{n \cdot (n+1) \cdot (2n+1)}{6} + \frac{n \cdot (n+1)}{2}$$
$$= \frac{n(n+1)}{6} (2n+1+3)$$
$$= \frac{n(n+1)(n+2)}{3}$$

We can see that this partial sum can be found exactly for any n. For $n = 50$, we could verify our answer above, $S = 44,200$. However, this series is divergent and the sum will increase without bound as n increases. It is unlikely that you would ever see any contest problem like Problem 38 because its solution is straightforward. Usually you will need at least to recognize a pattern in series such as the one in Problem 39 and then decide what approach to use and how to evaluate the sum. Sigma notation can help when you are faced with similar problems.

A standard method of proving Eqs. 1.29–1.31 is by mathematical induction. It is important to mention that ancient Babylonians (2000 BC) and ancient Greeks (1000 BC) knew these formulas but derived them from a geometric point of view. Equation 1.29 is the sum of the first n natural numbers and can be reformulated as in Lemma 1.5.

Lemma 1.5 The sum of all natural numbers from 1 to N equals $1+2+3+4+5+6+\ldots+N = \frac{N(N+1)}{2}$.

This statement can be proven using Gauss's approach, using the formula for the sum of the nth term of an arithmetic progression, or geometrically as it was done by ancient Greeks. Let us briefly describe all four proofs.

Proof 1. (Carl Friedrich Gauss's approach—see also Section 1.2)

$$1 + 2 \quad + \ldots + (N-1) + N = S$$
$$+$$
$$N + (N-1) + \ldots + \quad 2 \quad + 1 = S$$
$$\ldots$$
$$(N+1) \cdot N = 2S$$
$$S = 1 + 2 + \ldots + N = \frac{N(N+1)}{2}$$

Proof 2. Consider the left side of the formula again. $1 + 2 + 3 + \ldots + N$ is the sum of the first N terms of an arithmetic progression where both the first term and the common difference equals 1. Using Eq. 1.8 for the sum of an arithmetic series, we obtain $S_N = \frac{2 \cdot 1 + (N-1)}{2} \cdot N = \frac{(N+1)N}{2}$.

Proof 3. (Using mathematical induction) The statement is obviously true for $N = 1$. Assume that it is true for $N = k$, i.e., $1 + 2 + 3 + \ldots + k = \frac{(k+1)k}{2}$. Let us demonstrate that this statement is also true for $N = k + 1$ and that $1 + 2 + 3 + \ldots + k + k + 1 = \frac{(k+2)(k+1)}{2}$. Consider the left side of the formula. Extracting the sum of the first k terms and working only with the left side,

$$(1 + 2 + 3 + \ldots + k) + k + 1 = \frac{(k+1)k}{2} + k + 1$$
$$= \frac{(k+1)k + 2(k+1)}{2} = \frac{(k+1)(k+2)}{2}.$$

The proof is complete.

Proof 4. (Approach known to ancient Greeks) Consider Figure 1.11. Such a construction can be reproduced using billiard balls. Imagine a right triangle with the legs of length 6 made by the white balls. Make a similar right triangle out of red balls and assuming that such a creation keeps its shape, we can stick two triangles together as shown in Figure 1.15. It is clear that two triangles together forms a rectangle with one (vertical) side of 6 and the other (horizontal) side of 7. The entire rectangle of the billiard balls now has $6 \cdot 7 = 4 \, 2$ balls. If we look closely at this construction, we can see that starting from the very left corner (1 white ball) and by

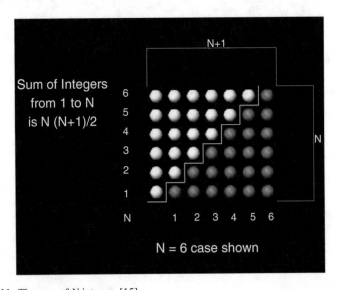

Figure 1.11 The sum of N integers [15]

moving up to 2 balls, 3 balls,..., 6 balls, we can this way add all the balls $1+2+3+4+5+6$ inside the white triangle. On the other hand, the same answer can be obtained by dividing 42 by 2.

If instead of 6 we have N rows, then the answer for the sum of all natural numbers between 1 and N is $\frac{N(N+1)}{2}$. Additionally, it is known that ancient Greeks also geometrically proved a modification of Eq. 1.29 as stated in Lemma 1.6.

Lemma 1.6 The sum of odd consecutive numbers from 1 to $(2N-1)$ is $N \cdot N$.

Using sigma notation this statement can be written as

$$1+3+5+7+\ldots+2N-1 = \sum_{n=1}^{N}(2n-1) = N^2. \qquad (1.32)$$

Using Figure 1.12, you may quickly see a geometric proof of Lemma 1.8, but first consider an auxiliary statement in Lemma 1.7. This result was known to Pythagoras.

Lemma 1.7 A square, n^2 and the corresponding odd number $(2n+1)$ make the next higher square, $(n+1)^2$, i.e., $N^2 + (2N+1) = N^2 + 2N + 1 = (N+1)^2$.

Definition. A **square number** is a perfect square or a product of two the same natural numbers.

Figure 1.12 Square numbers [15]

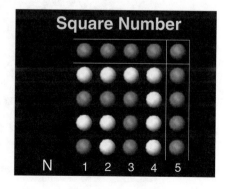

Lemma 1.8 Each square number can be written as a sum of two consecutive triangular numbers.

For example, $4 = 1 + 3, 9 = 3 + 6, 16 = 6 + 10$, etc.. It is easy to see that, if we have a number of balls filling up a square (say 16 balls, as in the Figure 1.12), the next higher square, the square of 5 balls, can be formed by adding rows of balls around two sides of the original square as shown. Starting from the very left corner and by going up, this process of forming successive squares can lead us to the following correct statements:

$$1^2 + (1 + 1 + 1) = 1^2 + 2 \cdot 1 + 1 = (1 + 1)^2 = 4 = 2^2$$
$$2^2 + (2 + 2 + 1) = 2^2 + 2 \cdot 2 + 1 = (2 + 1)^2 = 9 = 3^2$$
$$3^2 + (3 + 3 + 1) = 3^2 + 2 \cdot 3 + 1 = (3 + 1)^2 = 16 = 4^2$$
$$4^2 + (4 + 4 + 1) = 4^2 + 2 \cdot 4 + 1 = (4 + 1)^2 = 25 = 5^2$$

In general, each row can be written as $N^2 + N + N + 1 = N^2 + (2N + 1) = N^2 + 2N + 1 = (N + 1)^2$. However, if we add the very left and very right sides of the equations, we get $1 + 3 + 5 + 7 + \ldots + 2N - 1 + 2N + 1 = (N + 1)^2$. The successive numbers added to 1 are 3, 5, 7, ..., (Figure 1.13) that is to say, the successive odd numbers. The method of construction shows that the sum of any number of consecutive terms of the series of the odd numbers 1, 3, 5, 7 ....(starting from 1) is a square, and in fact $1 + 3 + 5 + \ldots + (2n - 1) = n^2$, while the addition of the next odd number $(2n + 1)$ makes the next higher square, $(n + 1)^2$, e.g.

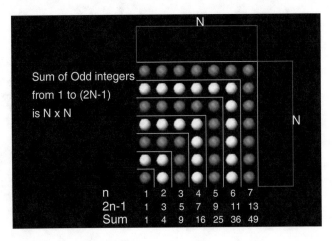

Figure 1.13 Sum of odd numbers [15]

$$1 + 3 = 4 = 2^2$$
$$1 + 3 + 5 = 9 = 3^2$$
$$1 + 3 + 5 + 7 = 16 = 4^2$$
$$1 + 3 + 5 + 7 + 9 = 25 = 5^2$$

$$\cdots$$

$$1 + 3 + 5 + 7 + 9 + \ldots + (2n - 1) = n^2.$$

An algebraic proof of this formula can be obtained in several ways, similar to the proofs of Eq. 1.29. For example, using sigma notation,

$$1 + 3 + 5 + \ldots + 2n - 1 = \sum_{k=1}^{n}(2k - 1) = 2 \cdot \frac{n(n + 1)}{2} - n = n^2.$$

Let us solve Problem 40.

Problem 40 Evaluate the sum $1 + 3 + 6 + 10 + 15 + \ldots + 5050$.

Solution. It would be great to find the formula for the nth term of the series. If you did every exercise in this book, then you might notice that each term of this series is precisely ½ times the corresponding term of the series solved in Problem 39 $(2 + 6 + 12 + 20 + 30 + \ldots)$. Hence, you can assume that $a_n = \frac{n(n + 1)}{2}$ and then apply the summation formulas. However, we can solve this problem independently from the known solution of Problem 39. Let me ask you the following questions: "Did you notice that the sum of any two consecutive terms is a perfect square?" Thus

$$a_1 = 1 = 1^2$$
$$a_1 + a_2 = 1 + 3 = 4 = 2^2$$
$$a_2 + a_3 = 3 + 6 = 9 = 3^2$$
$$a_3 + a_4 = 6 + 10 = 16 = 4^2$$
$$a_4 + a_5 = 10 + 15 = 25 = 5^2$$

$$\ldots$$

We can state the following hypotheses:
1. The sum of two consecutive terms of the given series is a perfect square and $a_{n-1} + a_n = n^2$.
2. Each term with index n can be composed of the sum of the first natural numbers from 1 to n, i.e., $a_n = 1 + 2 + 3 + \ldots + n$ i.e.

$$a_1 = 1$$
$$a_2 = 3 = 1 + 2 = a_1 + 2$$
$$a_3 = 6 = 1 + 2 + 3 = a_2 + 3$$
$$a_4 = 10 = 1 + 2 + 3 + 4 = a_3 + 4$$
$$a_5 = 15 = 1 + 2 + 3 + 4 + 5 = a_4 + 5.$$

It is time to state another hypothesis:

3. The nth term can be written as the sum of previous term and index n, $a_n = a_{n-1} + n$. We can check if both our hypotheses are correct by solving the system,

$$\begin{cases} a_{n-1} + a_n = n^2 \\ a_n = a_{n-1} + n \end{cases} \Rightarrow 2a_n = n^2 + n \Rightarrow \boxed{a_n = \frac{n(n+1)}{2}} \tag{1.33}$$

It is easy to check that Eq. 1.33 is indeed the formula for the nth term of the given series. (Please check it yourself). We can also find how many terms there are by using the last term of the series.

I do not know if you noticed that 5050 is the number obtained by young Gauss when he added 100 natural numbers. This observation ($5050 = 100 \cdot 101/2$) could help us to find the formula for the nth term as well, $a_n = \dfrac{n(n+1)}{2} = 5050$ $= \dfrac{100 \cdot 101}{2} \Rightarrow n = 100$. The sum of n terms given by Eq. 1.33 can be found using the formulas of summation:

$$1 + 3 + 6 + 10 + \ldots + \frac{n(n+1)}{2} = \sum_{i=1}^{n} \frac{i^2 + i}{2} = \frac{1}{2}\left(\sum_{i=1}^{n} i^2 + \sum_{i=1}^{n} i\right) =$$
$$= \frac{1}{2}\left(\frac{n(n+1)(2n+1)}{6} + \frac{n(n+1)}{2}\right) = \frac{1}{2} \cdot \frac{n(n+1)}{2}\left(\frac{2n+1+3}{3}\right) =$$
$$\boxed{1 + 3 + 6 + 10 + \ldots + \frac{n(n+1)}{2} = \frac{n(n+1)(n+2)}{6}} \tag{1.34}$$

Obviously, a sum of natural numbers is a natural number. Although Eq 1.34 looks like a fraction, it is an integer, because the product of three natural numbers within the numerator is always divisible by 6. Moreover, Eq. 1.33 is the formula for the nth triangular number and as we mentioned in the previous section, the sum of n consecutive triangular numbers is the nth tetrahedral number given by Eq. 1.34. Finally, substituting 100 for n, we can find the sum of 100 such numbers, $1 + 3 + 6 + 10 + \ldots + 5050 = \frac{100 \cdot 101 \cdot 102}{6} = 171,700$.

Figure 1.14 Sum of two consecutive triangular numbers [15]

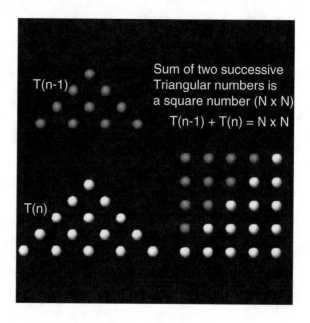

Answer. 171,700.

Ancient Greeks also geometrically proved Lemma 1.9 by forming a square number out of two consecutive triangular numbers (Figure 1.14).

Lemma 1.9 The sum of two consecutive triangular numbers is a square number, i.e., $T(n-1) + T(n) = n^2$.

The ancient Greeks proved everything visually, but this formula can be easily derived analytically as well. By adding numbers in each triangle by the rows, it is obvious that $T(1) = 1$, $T(2) = 1 + 2 = 3$, $T(3) = 1 + 2 + 3 = 6$, $T(4) = 1 + 2 + 3 + 4 = 10, \ldots$ Since each sum can be evaluated analytically, we have $T(n) = 1 + 2 + 3 + \ldots + (n-1) + n = \frac{n(n+1)}{2}$ and $T(n-1) = 1 + 2 + 3 + \ldots + (n-1) = \frac{(n-1)n}{2}$. Therefore, $T(n-1) + T(n) = \frac{(n-1)n}{2} + \frac{n(n+1)}{2} = \frac{(n+1+n-1)n}{2} = n^2$.

Lemma 1.10 An infinite series of the numbers that are reciprocals of triangular numbers is convergent and its sum is 2.

Proof. Consider $S_\infty = \frac{1}{1} + \frac{1}{3} + \frac{1}{6} + \frac{1}{10} + \frac{1}{15} + \ldots + \frac{1}{T(n)} + \ldots$ and rewrite it using sigma notation substituting formula for $T(n)$ i.e.,

$$S_n = \sum_{n=1}^{n} \frac{1}{T(n)} = \sum_{n=1}^{n} \frac{2}{n(n+1)} = 2 \cdot \sum_{n=1}^{n} \frac{1}{n(n+1)}$$

$$= 2 \cdot \sum_{n=1}^{n} \left(\frac{1}{n} - \frac{1}{n+1} \right) = 2 \cdot \left(1 - \frac{1}{2} + \frac{1}{2} - \frac{1}{3} + \frac{1}{3} - \frac{1}{4} + \ldots - \frac{1}{n+1} \right)$$

$$= 2 - \frac{2}{n+1}$$

$$S_\infty = \lim_{n \to \infty} S_n = 2.$$

Remark. Other properties of reciprocals of the triangular or tetrahedral numbers will be exposed later in the second chapter.

Problem 41 Evaluate the sum of the four consecutive triangular numbers, starting from the nth term.

Solution.

$$T_n + T_{n+1} + T_{n+2} + T_{n+3} = (T_n + T_{n+1}) + (T_{n+2} + T_{n+3})$$
$$= (n+1)^2 + (n+3)^3 = 2n^2 + 8n + 10.$$

Answer. $T_n + T_{n+1} + T_{n+2} + T_{n+3} = 2n^2 + 8n + 10.$

Using the previous results, obtained in Problem 40 for the sum of triangular numbers, we can prove Eq. 1.30 here restated as Lemma 1.11.

Lemma 1.11 $1^2 + 2^2 + 3^2 + 4^2 + \ldots + n^2 = \frac{n(n+1)(2n+1)}{6}.$

Proof. It is known that this formula was derived by Archimedes (287–212 BC) while he tried to solve some geometric and mechanic problems. Each square number is a sum of two consecutive triangular numbers e.g.,

$$1^2 = 1$$
$$2^2 = 1 + 3$$
$$3^2 = 3 + 6$$
$$4^2 = 6 + 10$$
$$5^2 = 10 + 15$$

$$\ldots\ldots$$

$$n^2 = T(n-1) + T(n) = \frac{(n-1)n}{2} + \frac{n(n+1)}{2}.$$

Adding the left and right sides, we will obtain the sum of first n squares on the left and double the sum of the first $(n-1)$ triangular numbers plus the nth triangular number that is not paired with any other number, $\sum_{k=1}^{n} k^2 = 2 \cdot \sum_{k=1}^{n-1} T(k) + T(n)$.

Replacing the sum of $(n-1)$ triangular numbers by Eq. 1.34 derived in Problem 40, we finally obtain the required sum,

$$\sum_{k=1}^{n} k^2 = \frac{2 \cdot (n-1)n(n+1)}{6} + \frac{n(n+1)}{2} = \frac{n(n+1)(2(n-1)+3)}{6}$$

$$= \frac{n(n+1)(2n+1)}{6}.$$

The proof is complete.

Let us now prove that the sum of the first n cubes of naturals numbers can be evaluated as $1^3 + 2^3 + 3^3 + \ldots + n^3 = (1 + 2 + 3 + \ldots + n)^2$. Hopefully, you recognized in it the third summation, Eq. 1.31, which was known to the ancient Babylonians.

Consider, 36 cubes of side 1 and arrange them in six layers as shown in Figure 1.15. In order to see the idea clearly, we will draw cubes in different colors: 27 pink cubes ($3^3 = 27$), 8 yellows cubes ($2^3 = 8$), and one blue cube (top).

Let us prove the sum

$$1^3 + 2^3 + 3^3 = (1 + 2 + 3)^2 = 36. \tag{1.35}$$

First, rearrange our construction so that all cubes are in one layer with a square top. Assume that Figure 1.16 represents the top of the box and that we, starting from the

Figure 1.15 The sum of cubes

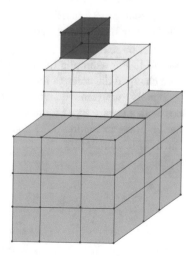

Figure 1.16 Top of the box

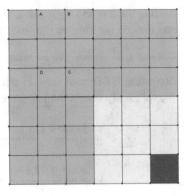

Figure 1.17 The 4th layer
of "pink" cubes

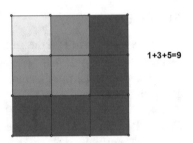

1+3+5=9

right lower corner, first, place the blue cube, then as Babylonians did, we will attach to it a corner piece, made out of three yellow cubes to form a square of side 2 and then attach to it another corner piece of five yellow cubes, finishing a square of side three. Next we will start using pink cubes, putting in corners of 7, 9 and 11 cubes.

On the other hand, if we take Figure 1.15 apart layer by a layer, starting from the top, we can count all the cubes. First, we remove the top blue cube. Next, remove the second layer (4 cubes $= 1 + 3$), then the third layer which will also have four cubes, totaling 8 cubes, $2^3 = 8 = 2 \cdot (1 + 3) = 3 + 5$. Finally, we remove the fourth layer of nine unit cubes. There are three layers of 9 cubes each so that the sum can be seen as $1 + 3 + 5 + 3^3$ unit cubes where $3^3 = 27 = 3(1 + 3 + 5) = 7 + 9 + 11$. For example, the fourth layer of pink cubes will look like Figure 1.17. (We changed the color of the cubes in order to better see the pattern $1 + 3 + 5 = 9$).

Finally, we removed all unit cubes and rearranged them into a rectangular box with a square top (Figure 1.16) with area $(1 + 2 + 3)(1 + 2 + 3) = 36$. Indeed, the volumes in Figures 1.15 and 1.16 are the same so the geometric proof of the Eq. 1.35 is complete.

We can also prove Eq. 1.35 algebraically,

$$1^3 = 1$$
$$2^3 = 2(1 + 1 + 2) = 2 \cdot 1 + 2(1 + 2)$$
$$3^3 = 3(1 + 1 + 2 + 2 + 3) = 3 \cdot 1 + 3(1 + 2) + 3(2 + 3)$$

$$\dots$$

$$1^3 + 2^3 + 3^3 = 1 \cdot (1 + 2 + 3) + 2 \cdot (1 + 2 + 3) + 3 \cdot (1 + 2 + 3)$$
$$= (1 + 2 + 3) \cdot (1 + 2 + 3) = (1 + 2 + 3)^2.$$

If we expand our idea for the sum of n cubes, we obtain

$$1 = 1^3$$
$$2(1 + 3) = 2^3$$
$$3(1 + 3 + 5) = 3^3$$
$$4(1 + 3 + 5 + 7) = 4^3$$
$$5(1 + 3 + 5 + 7 + 9) = 5^3$$

$$\dots$$

$$n(1 + 3 + 5 + \dots + 2n - 1) = n^3.$$

Therefore, the sum of all n cubes can be written as

$$\sum_{i=1}^{n} i^3 = 1^3 + 2^3 + 3^3 + \dots + n^3$$

$$= \sum_{i=1}^{n} i + 2 \cdot \sum_{i=1}^{n} i + 3 \cdot \sum_{i=1}^{n} i + \dots + n \cdot \sum_{i=1}^{n} i \qquad (1.36)$$

$$= \left(\sum_{i=1}^{n} i \right)^2.$$

Remember, that while doing the geometric proof we noticed an interesting pattern,

$$1^3 = 1$$
$$2^3 = 8 = 3 + 5$$
$$3^3 = 27 = 7 + 9 + 11 \qquad (1.37)$$
$$4^3 = 64 = 13 + 15 + 17 + 19$$
$$5^3 = 125 = 21 + 23 + 25 + 27 + 29.$$

This observation is written as Lemma 1.12.

Lemma 1.12 A cube of n can be written as the sum of precisely n consecutive odd numbers.

Let us prove it by solving the following problem.

Problem 42 Prove that a cube of a natural number n can be written as a sum of precisely n odd consecutive numbers.

Proof.

1. Consider our finding given by Eq. 1.37 above. If a cube of any natural number k is represented by a group of k consecutive odd numbers, then it is one number in the first group, two odd numbers in the second group, three odd numbers in the 3rd group, etc. How many odd numbers are before the kth group? $1 + 2 + 3 + 4 + \ldots + k - 1 = \frac{k(k-1)}{2}$ odd numbers. If we add the kth consecutive odd number, we obtain $\frac{k(k-1)}{2} + k = \frac{k(k+1)}{2}$, which represents the total number of the consecutive odd numbers before and including group k.

2. In order to evaluate the sum of the k consecutive odd numbers in group k, say $S(k)$, we can subtract the sum of $\frac{k(k+1)}{2}$ odd numbers and the sum of $\frac{k(k-1)}{2}$ odd numbers:

$$S(k) = \sum_{n=1}^{\frac{k(k+1)}{2}} (2n - 1) - \sum_{n=1}^{\frac{k(k-1)}{2}} (2n - 1) = \frac{k^2(k+1)^2}{4} - \frac{k^2(k-1)^2}{4}$$

$$= \frac{k^2}{4} \cdot \left((k+1)^2 - (k-1)^2 \right) = k^3.$$

Obviously, an nth odd number can be written as $a_n = 2n - 1$. Therefore, a cube of a natural number k can be written as the sum of k odd consecutive numbers. The proof is complete.

Moreover, by adding all consecutive cubes from 1 to n, we can state Lemma 1.13.

Lemma 1.13 A sum of all consecutive cubes between 1 and n equals the sum of all consecutive odd numbers between 1 and $\frac{n(n+1)}{2}$.

This allows us to prove Eq. 1.31 again, $\sum_{k=1}^{n} k^3 = \sum_{k=1}^{\frac{n(n+1)}{2}} (2k - 1) = \left(\frac{n(n+1)}{2} \right)^2 = \frac{n^2(n+1)^2}{4}.$

Problem 43 Find the formula for the first and last odd number in the kth group of consecutive odd numbers representing k^3. Give an example for representation of 7^3 as a sum of seven consecutive odd numbers.

Solution. There are precisely $\frac{k(k-1)}{2}$ consecutive odd numbers before the first number in the kth group. Then the first odd number in the kth group is $2n - 1$, $n = \frac{k(k+1)}{2} + 1$ and $a_1(k) = 2 \cdot \left(\frac{k(k-1)}{2} + 1\right) - 1 = k(k-1) + 1 = \boxed{k^2 - k + 1}$. The last, k^{th}, odd number in the kth group corresponds to the index $n = \frac{k(k+1)}{2}$ and it can be evaluated as $a_k(k) = 2 \cdot \left(\frac{k(k+1)}{2}\right) - 1 == \boxed{k^2 + k - 1}$.

The number $7^3 = 343$ can be written by a sum of seven odd consecutive numbers, starting with $a_1(7) = 7 \cdot (7 - 1) + 1 = 43$ and ending in $a_7(7) = 7 \cdot 6 + 2 \cdot 7 - 1 = 55$. Indeed, $7^3 = 43 + 45 + 47 + 49 + 51 + 53 + 55 = 343$.

Answer. $k^3 = k^2 - k - 1 + \ldots + k^2 + k - 1$; $7^3 = 343 = 43 + 45 + 47 + 49 + 51 + 53 + 55$.

Problem 44 Evaluate $S = 1^2 + 3^2 + 5^2 + \ldots + (2n - 1)^2$.

Solution. This sum can be seen as the difference of the sum of the squares of all natural numbers between 1 and $(2n)$ and the sum of the squares of all even numbers from 1 to $(2n)$. Then use the fact that a square of any even number is divisible by 4,

$$S = \{1^2 + 2^2 + 3^2 + 4^2 + \cdots + (2n)^2\} - \{2^2 + 4^2 + 6^2 + 8^2 + \cdots + (2n)^2\}$$
$$= (1^2 + 2^2 + 3^2 + 4^2 + \cdots + (2n)^2) - 4 \cdot (1^2 + 2^2 + 3^2 + 4^2 + \cdots + n^2)$$

Each sum now can be evaluated using the same Eq. 1.30 for the sum of first natural squares with a different upper summation index. Finally, we have

$$S = \sum_{k=1}^{2n}(k^2) - 4 \cdot \sum_{k=1}^{n}(k^2) = \frac{2n(2n+1)(4n+1)}{6} - \frac{4 \cdot n(n+1)(2n+1)}{6}$$
$$= \frac{n(2n-1)(2n+1)}{3} = \frac{4n^3 - n}{3}.$$

Answer. $S = \dfrac{n(2n-1)(2n+1)}{3} = \dfrac{4n^3 - n}{3}.$

Remark. While an infinite series of the squares of natural numbers is divergent, the series consisting of reciprocals of the squares is convergent. The proof of this fact, called the Basel Problem, was first given by Euler in 1735 and it is not trivial. We will discuss it later, in Chapter 3. The solution is $\sum_{n=1}^{\infty}\frac{1}{n^2}=\frac{1}{1^2}+$
$\frac{1}{2^2}+\frac{1}{3^2}+\ldots+\frac{1}{n^2}+\ldots+=\frac{\pi^2}{6}.$

Problem 45 Prove that the sum of the first n triangular numbers is the nth tetrahedral number.

Proof. Consider the sum of the first triangular numbers,

$$T(1)+T(2)+\ldots+T(n)=\sum_{1}^{n}\frac{n(n+1)}{2}=\frac{1}{2}\cdot\left(\sum_{1}^{n}n^2+\sum_{1}^{n}n\right)$$
$$=\frac{n(n+1)(2n+1)}{12}+\frac{n(n+1)}{4}=\frac{n(n+1)}{12}\cdot(2n+1+3)$$
$$=\frac{n(n+1)(n+2)}{6}=Tr(n).$$

Problem   46 Evaluate   the nth partial   sum   for   the   series, $3+13+30+54+85+123+\ldots.$

Solution. I hope you recognized these numbers, they are the partial sums of an arithmetic progression with common difference $d=7$. Then each term of the given series can be written as $a_n=\frac{(7n-1)\cdot n}{2}$. Hence, the requested sum can be evaluated as

$$S_n=\sum_{k=1}^{n}\frac{(7k-1)k}{2}=\frac{1}{2}\cdot\left(7\sum_{k=1}^{n}k^2-\sum_{k=1}^{n}k\right)$$
$$=\frac{1}{2}\cdot\left(\frac{7\cdot n(n+1)(2n+1)}{6}-\frac{n(n+1)}{2}\right)$$
$$=\frac{n(n+1)(7n+2)}{6}$$

We can check this formula by adding the first four terms of the series,

$$S_4=3+13+30+54=100$$
$$S_4=\frac{4\cdot5(7\cdot4+2)}{6}=100.$$

Answer. $S_n = \frac{n(n+1)(7n+2)}{6}$.

Although the ancient Greeks and Babylonians knew how to add integers and proved their formulas using geometry, there were limitations to their techniques. For example, they did not know how to represent the fourth power of a number because we all live in a three dimensional space. Additionally, there are infinitely many examples of sequences and series that involve fractional expressions or combination of exponential and trigonometric functions that of course was unknown in ancient times. In Chapter 2 you will learn different methods of finding exact sums and will prove many formulas using other creative ideas.

Chapter 2
Further Study of Sequences and Series

As you would see earlier in Chapter 1, some problems would ask you to add the first ten terms or even evaluate the sum of the first k terms of a sequence or maybe investigate whether the limit of such sum exists. Expressions such as

$$1 + 4 + 7 + 10 + 13 + \ldots \tag{2.1}$$

$$\tfrac{1}{2} + \tfrac{1}{4} + \tfrac{1}{8} + \tfrac{1}{16} + \tfrac{1}{32} + \ldots \tag{2.2}$$

$$1 + 4 + 9 + 16 + 25 + 36 + \ldots \tag{2.3}$$

are called series and in all three cases can be evaluated exactly for the sum of any finite number of terms. Since Eq. 2.1 represents an arithmetic series with first term 1 and common difference 3, we can use the formula for the sum of the first n terms that is derived in the earlier section. We can write the sum as

$$S_n = 1 + 4 + 7 + 10 + \ldots = \frac{2a_1 + (n-1)d}{2} \cdot n = \frac{2 \cdot 1 + (n-1)3}{2} \cdot n$$
$$= \frac{(3n-1)n}{2}. \tag{2.4}$$

Since Eq. 2.2 represents a geometric series with the first term 1/2 and common ratio 1/2, then the formula for the sum of the first n terms is known. We have

$$S_n = \frac{1}{2} + \frac{1}{4} + \frac{1}{8} + \frac{1}{16} + \ldots = \frac{b_1(1 - r^n)}{1 - r} = \frac{\frac{1}{2}\left(1 - \left(\frac{1}{2}\right)^n\right)}{1 - \frac{1}{2}} = 1 - \left(\frac{1}{2}\right)^n. \tag{2.5}$$

The sum of the last series of Eq. 2.3 can be evaluated exactly as well. We prove this formula in Chapter 1 and prove it in a different way in the following subsection,

© Springer International Publishing Switzerland 2016
E. Grigorieva, *Methods of Solving Sequence and Series Problems*,
DOI 10.1007/978-3-319-45686-7_2

$$S_n = 1^2 + 2^2 + 3^2 + \ldots + n^2 = \frac{n(n+1)(2n+1)}{6}. \qquad (2.6)$$

What do these series of Eqs. 2.1–2.3 have in common? Their partial sums can be evaluated exactly for any number of terms n. So we could add the first 25, the first 100 or even the first 2011 terms and get an exact answer for the sum using Eqs. 2.1–2.3 by replacing n by 25, 100, or 2011, respectively. However, if the number of terms, n, were to become infinitely large, then we would see some differences. For example, if we increase n then the partial sums of Eqs. 2.4 and 2.6 would increase without limit. The result is different for the sum of Eq. 2.5; it will approach its limit of one since the second term will approach zero. This behavior is typical for any infinite geometric series with common ratio less than one as we established earlier.

We say that the series of Eqs. 2.1 and 2.3 diverge and the series of Eq. 2.2 converges. Serious study of convergence and divergence is a subject of mathematical analysis. For now we simply determine whether or not the series are divergent or convergent and why. Many challenging math contest problems are dedicated to finding an exact sum of the first n terms of a series. The determination of the partial and infinite sums is the topic of the first section of this chapter.

2.1   Methods of Finding Partial and Infinite Sums

Let us derive again Eq. 2.6 for the sum of squares of the first n natural numbers and Eq. 1.31 for the sum of the cubes of n natural numbers.

> **Problem 47** Prove that $\sum_{k=1}^{n} k^2 = \frac{n(n+1)(2n+1)}{6}$

Proof. We need to prove that the following relationship is true:

$$N = 1^2 + 2^2 + 3^2 + 4^2 + \ldots + (n-2)^2 + (n-1)^2 + n^2 = \frac{n(n+1)(2n+1)}{6}.$$

Arranging sums in ascending and descending order does not help. We need to find a different approach. If you have read Chapter 1 of the book then you probably have an idea of how to start. Let us consider the difference of two consecutive cubes,

$$\boxed{n^3 - (n-1)^3 = 3n^2 - 3n + 1}.$$

$$1^3 - 0^3 = 3 \cdot 1^2 - 3 \cdot 1 + 1 = 1$$
$$2^3 - 1^3 = 3 \cdot 2^2 - 3 \cdot 2 + 1$$
$$3^3 - 2^3 = 3 \cdot 3^2 - 3 \cdot 3 + 1$$

$$\cdots$$

$$(n-2)^3 - (n-3)^3 = 3 \cdot (n-2)^2 - 3 \cdot (n-2) + 1$$
$$(n-1)^3 - (n-2)^3 = 3 \cdot (n-1)^2 - 3(n-1) + 1$$
$$n^3 - (n-1)^3 = 3n^2 - 3n + 1$$

Adding the left and the right sides, we obtain $n^3 = 3(1^2 + 2^2 + 3^2 + \ldots + n^2) - 3(1 + 2 + 3 + \ldots + n) + 1 \cdot n$. This can be written using sigma notation as $n^3 = 3\sum_{k=1}^{n} k^2 - 3\sum_{k=1}^{n} k + n$. Solving this for $\sum_{k=1}^{n} k^2$ and assuming that we know the formula for the sum of the first n natural numbers we obtain

$$\sum_{k=1}^{n} k^2 = \frac{2n^3 - 2n + 3n(n+1)}{6} = \frac{n(2n^2 + 3n + 1)}{6}$$
$$\sum_{k=1}^{n} k^2 = \frac{n(2n+1)(n+1)}{6}.$$

The statement is proven.

Problem 48 Prove that $\sum_{k=1}^{n} k^3 = \left(\frac{n(n+1)}{2}\right)^2$

Solution. Try to use a similar approach so consider the difference of the fourth powers of two consecutive integers $n^4 - (n-1)^4 = 4n^3 - 6n^2 + 4n - 1$. Write this out for the first few terms and then for the values as we reach n,

$$1^4 - 0^4 = 4 \cdot 1^3 - 6 \cdot 1^2 + 4 \cdot 1 - 1$$
$$2^4 - 1^4 = 4 \cdot 2^3 - 6 \cdot 2^2 + 4 \cdot 2 - 1$$
$$3^4 - 2^4 = 4 \cdot 3^3 - 6 \cdot 3^2 + 4 \cdot 3 - 1$$

$$\cdots$$

$$(n-2)^4 - (n-3)^4 = 4(n-2)^3 - 6(n-2)^2 + 4 \cdot (n-2) - 1$$
$$(n-1)^4 - (n-2)^4 = 4(n-1)^3 - 6(n-1)^2 + 4 \cdot (n-1) - 1$$
$$n^4 - (n-1)^4 = 4n^3 - 6n^2 + 4 \cdot n - 1.$$

Next, we add the left and the right sides together as we did in the previous problem using sigma notation and solve the equation for the unknown sum,

$$n^4 = 4\sum_{k=1}^{n} k^3 - 6\sum_{k=1}^{n} k^2 + 4 \cdot \sum_{k=1}^{n} k - n$$

$$\sum_{k=1}^{n} k^3 = \frac{n^4 + n(n+1)(2n+1) + n - 2n(n+1)}{4}$$

$$\sum_{k=1}^{n} k^3 = \frac{n(n^3+1) + n(n+1)(2n-1)}{4} = \frac{n(n+1)(n^2 - n + 1 + 2n - 1)}{4}$$

$$\sum_{k=1}^{n} k^3 = \frac{n(n+1)(n^2+n)}{4} = \left(\frac{n(n+1)}{2}\right)^2 = \left(\sum_{k=1}^{n} k\right)^2.$$

This is a very interesting relationship because we established again that the sum of the first n cubes equals the square of the sum of the first n natural numbers. For example, $1^3 + 2^3 + 3^3 + 4^3 = (1 + 2 + 3 + 4)^2 = 100$.

Remark. Earlier we proved the same formula using the geometric approaches of ancient Babylonians and Greeks to demonstrate that the sum of the first n cubes equals the sum of the first $m = \frac{n(n+1)}{2}$ odd consecutive numbers.

Problem 49 Find the sum, $1 + 11 + 111 + 1111 + \ldots + 11\ldots.111$, where the last number consists of n repetitions of the digit 1. Evaluate the sum for $n = 9$.

Solution. We solve this problem in three different ways so you can compare the different methods.

Method 1. At first glance, we notice that 1, 11, 111, 1111,…. is neither an arithmetic nor a geometric sequence. Hence, we have to rewrite the sum in another form. For example,

$$1 = 1$$
$$11 = 1 + 10$$
$$111 = 1 + 10 + 100$$
$$1111 = 1 + 10 + 100 + 1000$$
$$111...11 = 1 + 10 + 10^2 + 10^3 + 10^4 + \ldots + 10^{n-2} + 10^{n-1}$$

Each number on the left containing digit 1 repeated n times can be written as a sum of the first n terms of a geometric sequence with the first term equals 1 and a common ratio 10. Thus,

$$1 = S_1 = 1$$
$$11 = S_2 = 1 + 10$$
$$111 = S_3 = 1 + 10 + 100$$
$$\cdots$$
$$111...11 = S_n = \frac{1 \cdot (10^n - 1)}{10 - 1} = \frac{10^n - 1}{9}$$

Adding over the left and right sides, $1 + 11 + 111 + \ldots + 111\ldots11 = S_1 + S_2 + \ldots + S_n$ and using the formula for the sum of n terms of a geometric sequence and properties of $\sum$ - notation we have

$$S = \sum_{k=1}^{n} \frac{10^k - 1}{9} = \sum_{k=1}^{n} \frac{10^k}{9} - \frac{1}{9}\sum_{k=1}^{n} 1 = \frac{1}{9}\left(\sum_{k=1}^{n} 10^k - n\right) \qquad (2.7)$$

Let us consider the first term of difference of Eq. 2.7, $\sum_{k=1}^{n} 10^k = 10 + 10^2 + 10^3 + \ldots + 10^n$. The expression on the right is again a geometric sequence with $b_1 = 10$ and $r = 10$ and

$$\sum_{k=1}^{n} 10^k = \frac{10 \cdot (10^n - 1)}{9} = \frac{10^{n+1} - 10}{9} \qquad (2.8)$$

Substituting Eq. 2.8 into Eq. 2.7 we obtain a formula for S, $S = \frac{10^{n+1}-10-9n}{81}$.

This formula can be used in order to find a sum like $1 + 11 + 111 + \ldots + 111\ldots11$ for any specific number n. Thus, when $n = 9$, $S = 1 + 11 + 111 + \ldots + 111111111 = \frac{10^{10}-10-9\cdot9}{81} = 123,456,789$.

Method 2. Denote the total sum by S as $S = 1 + 11 + 111 + 1111 + 11111 + \ldots + 11\ldots1$. Multiplying S by 10, we obtain $10S = 10 + 110 + 1110 + 11110 + 111110 + \ldots$. If we subtract the first sum from the second, we obtain (It may help to rewrite S as $S = 1 + (10+1) + (110+1) + (1110+1) + \ldots$).

Then $9S = \overbrace{111...1}^{n \text{ times}}0 - n \cdot 1$ which leads us to the answer, $S = \dfrac{\overbrace{111...1}^{n \text{ times}}0 - n \cdot 1}{9}$.

Method 3. We can notice that $9 = 10 - 1$, $99 = 100 - 1$, $999 = 1000 - 1$, etc. If we multiply and divide the given sum by 9 we can easily evaluate it using a formula for geometric series.

$$S = \frac{1}{9}(10 - 1 + 100 - 1 + 1000 - 1 + 10000 - 1 + \ldots + 100\ldots0 - 1)$$

$$= \frac{(10 + 10^2 + \ldots + 10^n - n)}{9}$$

$$S = \frac{1}{9}\left[\frac{10(10^n - 1)}{9} - n\right]$$

Our series is divergent because S increases without bound as n increases.

As we mentioned above, evaluating an exact sum for a finite series or a partial sum for an infinite series can be a challenging task, and this is why many such problems appear in different contests. Each problem is unique but we are going to share with you some ideas of finding such sums; you may find them helpful and applicable to other or similar problems.

Problem 50 Find the sum: $\frac{1}{1\cdot2} + \frac{1}{2\cdot3} + \frac{1}{3\cdot4} + \ldots + \frac{1}{1998\cdot1999} + \frac{1}{1999\cdot2000}$

Solution. Sometimes it is a good idea to rewrite a sum in a different but equivalent form by noticing something that the terms have in common, some pattern. One thing you might notice is that the denominator of each fraction is a product of two consecutive natural numbers. How can we obtain a product of two such numbers within a denominator? What operation can give us a product? Answer: When we put together (add or subtract) two fractions with different denominators, that have no common factors, the least common denominator is going to be a product of these numbers. In general,

$$\frac{1}{c} + \frac{1}{d} = \frac{d + c}{c \cdot d}$$

$$\frac{1}{c} - \frac{1}{d} = \frac{d - c}{c \cdot d}$$

Looking at the second formula above, we can find the way of solving the problem. If c and d differ by 1, i.e., $d - c = 1$, then

$$\frac{1}{c} - \frac{1}{d} = \frac{1}{c \cdot d}$$

$$\frac{1}{1} - \frac{1}{2} = \frac{1}{1 \cdot 2}$$

$$\frac{1}{2} - \frac{1}{3} = \frac{1}{2 \cdot 3}$$

$$\frac{1}{3} - \frac{1}{4} = \frac{1}{3 \cdot 4}$$

$$\cdots$$

$$\frac{1}{1999} - \frac{1}{2000} = \frac{1}{1999 \cdot 2000}.$$

Using these, we replace each fraction on the right by the difference on the left obtaining

$$\frac{1}{1 \cdot 2} + \frac{1}{2 \cdot 3} + \frac{1}{3 \cdot 4} + \ldots + \frac{1}{1998 \cdot 1999} + \frac{1}{1999 \cdot 2000} =$$
$$1 - \frac{1}{2} + \frac{1}{2} - \frac{1}{3} + \frac{1}{3} - \frac{1}{4} + \ldots + \frac{1}{1998} - \frac{1}{1999} + \frac{1}{1999} - \frac{1}{2000}$$

In this sum all middle terms cancel each other except the first term, 1, and the last term, $-\frac{1}{2000}$. This gives us $S_{1999} = 1 - \frac{1}{2000} = \frac{1999}{2000}$. Evaluating this sum when $n = 1999$ (a big number), we see that $S_{1999} = \frac{1999}{2000}$ is almost 1. On the other hand, $S_4 = \frac{1}{1 \cdot 2} + \frac{1}{2 \cdot 3} + \frac{1}{3 \cdot 4} + \frac{1}{4 \cdot 5} = 1 - \frac{1}{5} = \frac{4}{5} = 0.8$. Four is not a "big" number, hence 0.8 is not as close to 1. Using the same technique, we can find the sum to infinity of the series:

$$S = \frac{1}{1 \cdot 2} + \frac{1}{2 \cdot 3} + \ldots + \frac{1}{n(n+1)} + \ldots \quad \text{so} \quad S_n = 1 - \frac{1}{n+1} = \frac{n}{n+1} \text{ and also have that}$$
$$\lim_{n \to \infty} S_n = \lim_{n \to \infty} \frac{n}{n+1} = 1.$$

Remark. In order to be considered for possible convergence, the series must first pass the necessary condition for the limit of its nth term, that is, does $\lim_{n \to \infty} u_n = 0$. If we try to look at the nth term of this sum, $\frac{1}{n(n+1)}$, we can see that $\lim_{n \to \infty} \frac{1}{n(n+1)} = 0$. We also find that the limit of the partial sums exists, $\lim_{n \to \infty} S_n = S$ where S is a finite number 1. However, in general, satisfying the necessary condition is not sufficient.

Convergence or divergence of series is established with the use of sufficient convergence theorems. We list some of these rules in Chapter 3.

Why didn't we use a calculator approach? A calculator can be used to find a sum like $\frac{1}{1 \cdot 2} + \frac{1}{2 \cdot 3} + \frac{1}{3 \cdot 4}$, i.e., $sum(seq(1/(x(x+1)), x, 1, 3) = 0.75$ This is an exact answer. A calculator can evaluate this as $\frac{1}{1 \cdot 2} + \frac{1}{2 \cdot 3} + \ldots + \frac{1}{100 \cdot 101} = 0.990094$, i.e., $sum(seq(1/x/(x+1), x, 1, 100) = 0.990094$. Even this: $\frac{1}{1 \cdot 2} + \ldots + \frac{1}{500 \cdot 501} = 0.99800$. But if we have more than 100 terms in summation, for example, $x = 1999$, such as our original problem, TI83/84 graphing calculators cannot be used. We might have some idea that this number gets closer and closer to 1. But how close? What if we need to find the exact answer or figure out the value of S_n, the sum of the first n terms for any n? Remember that since $S_n = 1 - \frac{1}{n+1} = \frac{n}{n+1}$, we evaluated its limit analytically as $\lim_{n \to \infty} S_n = 1$.

In the preceding problem numbers within each denominator differed by 1. But the idea of replacing each fraction by a difference is so elegant, we wonder, "What happens if two numbers in each fraction differ by the same number but not by 1? Can we use the same technique here?"

Problem 51 Evaluate $\frac{1}{1 \cdot 5} + \frac{1}{5 \cdot 9} + \frac{1}{9 \cdot 13} + \ldots + \frac{1}{197 \cdot 201}$.

Solution. Look at the sequence of the first numbers of each denominator: $1, 5, 9, \ldots, 197$. They are terms of an arithmetic sequence with $a_1 = 1$ and $d = 4$. Let us find the number of the term that is 197.

$$a_n = a_1 + (n-1)d$$
$$197 = 1 + (n-1)4$$
$$n = 50$$

This means that we have to add 50 fractions together. Look at the differences:

$$1 - \frac{1}{5} = \frac{5-1}{1 \cdot 5} = \frac{4}{1 \cdot 5} = 4 \cdot \frac{1}{1 \cdot 5}$$
$$\frac{1}{5} - \frac{1}{9} = \frac{9-5}{5 \cdot 9} = \frac{4}{5 \cdot 9} = 4 \cdot \frac{1}{5 \cdot 9}$$
$$\cdots$$
$$\frac{1}{197} - \frac{1}{201} = \frac{4}{197 \cdot 201} = 4 \cdot \frac{1}{197 \cdot 201}$$

Now the given sum can be written in the form:

$$S_{50} = \frac{1}{4}\left(1 - \frac{1}{5} + \frac{1}{5} - \frac{1}{9} + \frac{1}{9} - \frac{1}{13} + \ldots + \frac{1}{193} - \frac{1}{197} + \frac{1}{197} - \frac{1}{201}\right)$$
$$S_{50} = \frac{1}{4}\left(1 - \frac{1}{201}\right) = \frac{50}{201}.$$

Notice that the nth term of the series can be written as $\frac{1}{(4n-3)(4n+1)}$. We can evaluate the partial sum (the sum of the first n terms) as $S_n = \frac{1}{4}\left(1 - \frac{1}{4n+1}\right) = \frac{n}{4n+1}$. If $n \to \infty$, $S_n \to \frac{1}{4}$. Therefore, the series is convergent.

Answer. $\frac{50}{201}$

Now we can make a trivial but very useful conclusion. For any real c and d such that $c \neq d$

$$\boxed{\frac{1}{c \cdot d} = \frac{1}{d-c} \cdot \left[\frac{1}{c} - \frac{1}{d}\right]} \tag{2.9}$$

Problem 52 Numbers $a_1, a_2, \ldots, a_n, a_{n+1}$ are terms of an arithmetic sequence. Prove that $\frac{1}{a_1 \cdot a_2} + \frac{1}{a_2 \cdot a_3} + \ldots + \frac{1}{a_n \cdot a_{n+1}} = \frac{n}{a_1 \cdot a_{n+1}}$

Proof. $a_1, a_2, \ldots, a_n, a_{n+1}$ are terms of an arithmetic sequence, then $a_2 - a_1 = a_3 - a_2 = \ldots = a_{n+1} - a_n = d$, where d is a common difference of the sequence. Using (Eq. 2.9) we can state the following:

$$\frac{1}{a_1 a_2} = \left(\frac{1}{a_1} - \frac{1}{a_2}\right) \cdot \frac{1}{d}$$

$$\frac{1}{a_2 a_3} = \left(\frac{1}{a_2} - \frac{1}{a_3}\right) \cdot \frac{1}{d}$$

$$\cdots$$

$$\frac{1}{a_n a_{n+1}} = \left(\frac{1}{a_n} - \frac{1}{a_{n+1}}\right) \cdot \frac{1}{d}$$

Replacing each term on the left of the given expression by formulas above and factoring out $^1/_d$ we obtain

$$S = \frac{1}{d}\left\{\frac{1}{a_1} - \frac{1}{a_2} + \frac{1}{a_2} - \frac{1}{a_3} + \ldots + \frac{1}{a_n} - \frac{1}{a_{n+1}}\right\}$$
$$= \frac{1}{d} \cdot \frac{(a_{n+1} - a_1)}{a_1 \cdot a_{n+1}} \tag{2.10}$$

But $a_{n+1} = a_1 + nd$, then

$$a_{n+1} - a_1 = nd \tag{2.11}$$

Replacing Eq. 2.11 into Eq. 2.10 we have the required expression for S, $S = \frac{nd}{d(a_1 \cdot a_{n+1})} = \frac{n}{a_1 \cdot a_{n+1}}$.

The proof is complete.

Problem 53 Prove that $\frac{1}{1^2} + \frac{1}{2^2} + \frac{1}{3^2} + \ldots + \frac{1}{n^2} < 2$.

Proof. Denote the given sum by $S = \frac{1}{1^2} + \frac{1}{2^2} + \frac{1}{3^2} + \ldots + \frac{1}{n^2}$. In addition, consider another series, made of one that we have already seen and evaluated:

$$\Sigma = 1 + \left(\frac{1}{1 \cdot 2} + \frac{1}{2 \cdot 3} + \frac{1}{3 \cdot 4} + \ldots + \frac{1}{(n-1)n} \right)$$

Each term of this auxiliary series, starting from the second term, is greater than the corresponding term of the given series, such that

$$\frac{1}{n^2} < \frac{1}{(n-1)n} = \frac{1}{n-1} - \frac{1}{n}, \quad n \geq 2, \ n \in \mathbb{N}$$

Hence, the sum of all terms of the given series is less than the sum of the auxiliary series:

$$S < \Sigma = 1 + 1 - \frac{1}{n} = 2 - \frac{1}{n}, \quad n \in \mathbb{N}.$$

Therefore, we can state that $S < 2 - \frac{1}{n} < 2, \ n \in \mathbb{N}$. The statement is proven.

An interesting approach of rewriting a fraction as a difference of two other fractions can be applied to many other math problems. For example, we can use this approach in calculus when evaluating integrals like this: $\int \frac{du}{u^2 - 1}$ or any integral of the form: $\int \frac{du}{u^2 - m^2}$, where m is any integer. Let us do the following problem.

Problem 54 Evaluate the integral, $\int \frac{du}{u^2 - 1}$.

Solution. Consider the rational expression under a symbol of an integral. Because the quantities, $(u-1)$ and $(u+1)$ differ by 2, we can use the same technique (Eq. 2.9) of rewriting this as a difference of two fractions multiplied by (1/2):

$$\frac{1}{u^2 - 1} = \frac{1}{(u-1)(u+1)} = \left(\frac{1}{u-1} - \frac{1}{u+1} \right) \cdot \frac{1}{2}$$

and

$$\int \frac{du}{u^2 - 1} = \frac{1}{2} \cdot \left(\int \frac{du}{u-1} - \int \frac{du}{u+1} \right)$$

$$= \frac{1}{2} (\ln|u-1| - \ln|u+1|) + C = \frac{1}{2} \ln \left| \frac{u-1}{u+1} \right| + C$$

Answer. $\frac{1}{2}\ln\left|\frac{u-1}{u+1}\right| + C$.

Problem 55 Prove that $\frac{1^2}{1\cdot3} + \frac{2^2}{3\cdot5} + \ldots + \frac{n^2}{(2n-1)(2n+1)} = \frac{n(n+1)}{2(2n+1)}$.

Proof. Would it be nice to have the sum of the first n squares or the sum of n fractions with those denominators but unit in each numerator? Yes. We would evaluate such sums without any troubles. These little observations can help us to prove the statement. Denote the unknown sum by $S = \frac{1^2}{1\cdot3} + \frac{2^2}{3\cdot5} + \ldots + \frac{n^2}{(2n-1)(2n+1)}$ and then rewrite it using sigma notation and by applying the difference of squares formula to the nth term, $\sum\limits_{n=1}^{n} \frac{n^2}{4n^2-1} = S$. Let us multiply both sides by 4 and put 4 inside the summation:

$$4\cdot\sum_{n=1}^{n}\frac{n^2}{4n^2-1} = 4S$$

$$\sum_{n=1}^{n}\frac{4n^2}{4n^2-1} = 4S$$

Would it be nice to add just n units instead? We do not have it but the following operation will make it possible

$$\sum_{n=1}^{n}\frac{4n^2}{4n^2-1} - \sum_{n=1}^{n}\frac{1}{4n^2-1} = 4S - \sum_{n=1}^{n}\frac{1}{4n^2-1}$$

$$\sum_{n=1}^{n}\frac{4n^2-1}{4n^2-1} = 4S - \sum_{n=1}^{n}\frac{1}{(2n-1)(2n+1)}$$

$$n = 4S - \sum_{n=1}^{n}\frac{1}{(2n-1)(2n+1)}$$

The sum on the right hand side looks familiar to you because denominator of each term consists of a product of two consecutive odd numbers that differ by 2.

$$\sum_{n=1}^{n}\frac{1}{(2n-1)(2n+1)} = \frac{1}{1\cdot3} + \frac{1}{3\cdot5} + \ldots + \frac{1}{(2n-1)(2n+1)}$$

$$= \frac{1}{2}\left(\frac{1}{1} - \frac{1}{3} + \frac{1}{3} - \ldots - \frac{1}{2n+1}\right) = \frac{2n+1-1}{2(2n+1)}$$

$$= \frac{n}{2n+1}$$

Finally, we have $n = 4S - \frac{n}{2n+1}$. Solving this equation for S, we obtain the requested quantity:

$$4S = n + \frac{n}{2n+1}$$

$$S = \frac{2n^2 + 2n}{4(2n+1)} = \frac{n(n+1)}{2(2n+1)}$$

The proof is complete.

Problem 56 demonstrates another approach for finding sums.

Problem 56 Find the sum $S = \frac{1}{1\cdot3\cdot5} + \frac{1}{3\cdot5\cdot7} + \frac{1}{5\cdot7\cdot9} + \dots$

Solution. Notice that the nth term of the series can be represented as $u_n = \frac{1}{(2n-1)(2n+1)(2n+3)}$.

Let us rewrite it as follows:

$$u_n = \frac{A}{2n-1} + \frac{B}{2n+1} + \frac{C}{2n+3} = \frac{1}{(2n-1)(2n+1)(2n+3)} \tag{2.12}$$

where A, B, and C are some constants to be determined.

If we put expressions on the left side of Eq. 2.12 over the common denominator, and equate both sides, we can find these constants:

$$A(4n^2 + 8n + 3) + B(4n^2 + 4n - 3) + C(4n^2 - 1) = 1$$
$$4n^2(A + B + C) + 4n(2A + B) + (3A - 3B - C) = 1$$

Since $n \neq 0$ we have to solve the system:

$$\begin{cases} A + B + C = 0 \\ 2A + B = 0 \\ 3A - 3B - C = 1 \end{cases} \Leftrightarrow A = C = 1/8, \ B = -\frac{1}{4} \tag{2.13}$$

Using Eq. 2.13 the given sum can be written as

$$\frac{1}{8}(1 + 1/3 + 1/5 + 1/7 + \dots) - \frac{1}{4}(1/3 + 1/5 + 1/7 + 1/9 + \dots)$$

$$+ \frac{1}{8}(1/5 + 1/7 + 1/9 + \dots)$$

$$= \frac{1}{8}(1 + 1/3) - \frac{1}{4}\cdot\frac{1}{3} - \frac{1}{4}(1/5 + 1/7 + \dots) + \frac{1}{4}(1/5 + 1/7 + \dots)$$

$$= \frac{1}{12} \approx 0.0833$$

$$\tag{2.14}$$

Answer. 1/12.

Remark. The required sum can be evaluated using properties of sigma notation as

$$8S = \sum_{n=1}^{\infty}\frac{1}{2n-1} - 2\sum_{n=1}^{\infty}\frac{1}{2n+1} + \sum_{n=1}^{\infty}\frac{1}{2n+3}$$

$$= \left(\sum_{n=3}^{\infty}\frac{1}{2n-1} - 2\sum_{n=3}^{\infty}\frac{1}{2n-1} + \sum_{n=3}^{\infty}\frac{1}{2n-1} \right) + 1 + \frac{1}{3} - 2\cdot\frac{1}{3} = \frac{2}{3}$$

$$S = \frac{1}{12}.$$

Additionally, notice that 1/12 in Eq. 2.14 is the sum of the infinite series. If the number of terms, k, is some counting number we can evaluate the sum exactly as $\quad S_k = \frac{1}{1\cdot3\cdot5} + \frac{1}{3\cdot5\cdot7} + \frac{1}{5\cdot7\cdot9} + \cdots \frac{1}{(2k-1)(2k+1)(2k+3)} = \frac{1}{12} + \frac{1}{8}\left(\frac{1}{2k+3} - \frac{1}{2k+1} \right) = \frac{1}{12} + \frac{1}{2(2k+1)(2k+3)} \to \frac{1}{12}$ as $k \to \infty$. We say that the series is convergent to 1/12. $k \to \infty$. However, for small k and sums up to, for example $\frac{1}{11\cdot13\cdot15}$ ($k = 6$ and we have to add only six terms), we should use the exact formula for the partial sum above, that yields $\frac{1}{12} + \frac{1}{2\cdot13\cdot15} = \frac{201}{2340} \approx \frac{1}{12} + 0.002564 \approx .0859$

Problem 57 Find the sum $S_n = \frac{1}{1\cdot2\cdot3} + \frac{1}{2\cdot3\cdot4} + \cdots + \frac{1}{n(n+1)(n+2)}.$

Solution. Let us rewrite the kth term as

$$\frac{1}{k(k+1)(k+2)} = \frac{1}{k+1}\cdot\frac{1}{k}\cdot\frac{1}{k+2} = \frac{1}{k+1}\cdot\frac{1}{2}\left[\frac{1}{k} - \frac{1}{k+2} \right]$$
$$= \frac{1}{2}\left[\frac{1}{k(k+1)} - \frac{1}{(k+1)(k+2)} \right]$$

Therefore, the partial sum is

$$S_n = \frac{1}{2}\left[\frac{1}{1\cdot2} - \frac{1}{2\cdot3} + \frac{1}{2\cdot3} - \frac{1}{3\cdot4} + \frac{1}{3\cdot4} - \frac{1}{4\cdot5} + \cdots + \frac{1}{n(n+1)} - \frac{1}{(n+1)(n+2)} \right]$$
$$= \frac{1}{2}\left[\frac{1}{2} - \frac{1}{(n+1)(n+2)} \right] = \frac{n(n+3)}{4(n+1)(n+2)}$$

Notice that $\lim_{n\to\infty} S_n = \frac{1}{4}$. The series is convergent.

Answer. $S_n = \frac{n(n+3)}{4(n+1)(n+2)}$

Here is another example of how these ideas can be applied in Calculus when taking integrals:

Problem 58 Evaluate $\int \dfrac{dx}{x(x+1)(x+2)}$ for all positive x.

Solution. Noticing that $\dfrac{2}{(x)(x+1)(x+2)} = \dfrac{1}{x} - \dfrac{2}{x+1} + \dfrac{1}{x+2}$ we can evaluate the integral as

$$\int \dfrac{dx}{x(x+1)(x+2)} = \dfrac{1}{2}\left\{\ln[x(x+2)] - \ln(x+1)^2\right\} + C = \dfrac{1}{2}\ln\dfrac{x(x+2)}{(x+1)^2} + C.$$

Answer. $\ln\dfrac{\sqrt{x(x+2)}}{x+1} + C$

Problem 59 Evaluate $\dfrac{1}{2} + \dfrac{3}{2^2} + \dfrac{5}{2^3} + \ldots + \dfrac{2n-1}{2^n}$.

Solution. Denote $S_n = \dfrac{1}{2} + \dfrac{3}{2^2} + \dfrac{5}{2^3} + \ldots + \dfrac{2n-1}{2^n}$. Multiplying this by two and regrouping terms, we obtain $2 \cdot S_n = 1 + \dfrac{3}{2^1} + \dfrac{5}{2^2} + \ldots + \dfrac{2n-1}{2^{n-1}}$. Within this sum, we recognize a geometric series and the original sum minus the last term, The first term is 1 and the common ratio is ½.

$$1 + \left(\dfrac{2}{2} + \dfrac{1}{2}\right) + \left(\dfrac{2}{2^2} + \dfrac{3}{2^2}\right) + \left(\dfrac{2}{2^3} + \dfrac{5}{2^3}\right) + \ldots + \left(\dfrac{2}{2^{n-1}} + \dfrac{2n-3}{2^{n-1}}\right)$$

$$2S_n = 1 + \dfrac{1 - \dfrac{1}{2^{n-1}}}{1 - \dfrac{1}{2}} + S_n - \dfrac{2n-1}{2^n}$$

Solving this for S_n, $S_n = 3 - \dfrac{2n+3}{2^n}$. This series is convergent because if n increases the second term will approach zero and the limit of partial sums will approach 3, i.e., $\lim\limits_{n\to\infty} S_n = 3$.

Answer. $S_n = 3 - \dfrac{2n+3}{2^n}$.

Problem 60 Evaluate the sum: $S = \dfrac{1}{1+\sqrt{2}} + \dfrac{1}{\sqrt{2}+\sqrt{3}} + \dfrac{1}{\sqrt{3}+\sqrt{4}} + \ldots \dfrac{1}{\sqrt{1977}+\sqrt{1978}}$ $+ \ldots + \dfrac{1}{\sqrt{2016}+\sqrt{2017}}$.

Solution. This problem was given at the Volgograd District Math Olympiad, with the only difference being that the last term ended in $\dfrac{1}{\sqrt{1977}+\sqrt{1978}}$, because the current year was 1978. Despite the different last term, the method of solving this problem is the same: we rationalize the denominator of each fraction:

$$\frac{1}{\sqrt{n-1}+\sqrt{n}} = \frac{\sqrt{n-1}-\sqrt{n}}{\left(\sqrt{n-1}+\sqrt{n}\right)\cdot\left(\sqrt{n-1}-\sqrt{n}\right)} = \sqrt{n}-\sqrt{n-1}$$

S can be written as

$S = \sqrt{2}-1+\sqrt{3}-\sqrt{2}+\sqrt{4}-\sqrt{3}+\ldots+\sqrt{2017}-\sqrt{2016} = \sqrt{2017}-1$. Next,
we can easily add the first n terms of the series and find S_n:

$$S_n = \frac{1}{1+\sqrt{2}} + \frac{1}{\sqrt{2}+\sqrt{3}} + \frac{1}{\sqrt{3}+\sqrt{4}} + \ldots + \frac{1}{\sqrt{n}+\sqrt{n+1}} = \sqrt{n+1}-1$$

This partial sum can be evaluated precisely for any natural n. The series is divergent because this sum will increase without bound.

Answer. $\sqrt{2017}-1$.

Next, using similar idea, let us solve the following problem.

Problem 61 Positive numbers $a_1, a_2, \ldots, a_n$ form an arithmetic progression. Prove the following: $S_n = \frac{1}{\sqrt{a_1}+\sqrt{a_2}} + \frac{1}{\sqrt{a_2}+\sqrt{a_3}} + \ldots + \frac{1}{\sqrt{a_{n-1}}+\sqrt{a_n}} = \frac{n-1}{\sqrt{a_1}+\sqrt{a_n}}$.

Proof. Since this looks similar to the sum we just evaluated in the previous problem, let us try the same idea: we rationalize each denominator,

$$S_n = \frac{\sqrt{a_2}-\sqrt{a_1}}{a_2-a_1} + \frac{\sqrt{a_3}-\sqrt{a_2}}{a_3-a_2} + \ldots + \frac{\sqrt{a_n}-\sqrt{a_{n-1}}}{a_n-a_{n-1}}$$

For any arithmetic progression the differences in these denominators are the differences between consecutive terms of the arithmetic sequence and must be the same. We denote it by d. Next, after substitution and eliminating opposite terms, this expression will be written as

$$S_n = \frac{\sqrt{a_2}-\sqrt{a_1}}{d} + \frac{\sqrt{a_3}-\sqrt{a_2}}{d} + \ldots + \frac{\sqrt{a_n}-\sqrt{a_{n-1}}}{d} = \frac{\sqrt{a_n}-\sqrt{a_1}}{d}$$

Since the original problem does not have any information about common difference of the progression, then we can find d from the formula that connects the first and the nth term of any arithmetic progression:

$$a_n = a_1 + (n-1)d$$
$$d = \frac{a_n - a_1}{n-1}$$

Therefore, $S_n = \frac{\sqrt{a_n}-\sqrt{a_1}}{d} = \frac{(n-1)\left(\sqrt{a_n}-\sqrt{a_1}\right)}{a_n-a_1} = \frac{n-1}{\sqrt{a_n}+\sqrt{a_1}}$

The proof is complete.

Problem 62 Find the sum $S_n = 1 \cdot 1! + 2 \cdot 2! + 3 \cdot 3! + \ldots + n \cdot n!$

Solution. The following is true for the nth term of the series

$$n \cdot n! = (n+1)! - n! = n!(n+1) - n! = n!n$$

The given sum can be written as

$$S_n = 2! - 1! + 3! - 2! + 4! - 3! + \ldots + n! - (n-1)! + (n+1)! - n!$$
$$= (n+1)! - 1.$$

The series is divergent since the limit of the partial sums does not exist.

Answer. $(n+1)! - 1$.

Problem 63 Evaluate the sum: $1 + 2 \cdot 2 + 3 \cdot 2^2 + 4 \cdot 2^3 + \ldots + 100 \cdot 2^{99}$.
Find a general formula for the sum of the first N terms of series
$S_N = 1 + 2 \cdot 2 + 3 \cdot 4 + 4 \cdot 8 + 5 \cdot 16 + \ldots + N \cdot 2^{N-1}$.

Solution. Method 1.
Denote the required sum as S and multiply it by 2,
$2S = 2 + 2 \cdot 2^2 + 3 \cdot 2^3 + 4 \cdot 2^4 + \ldots + 99 \cdot 2^{99} + 100 \cdot 2^{100}$. Next, we subtract
S from $2S$,

$$S = 100 \cdot 2^{100} - \left(1 + 2 + 2^2 + 2^3 + \ldots + 2^{99}\right)$$
$$= 100 \cdot 2^{100} - \left(2^{100} - 1\right)$$
$$= 99 \cdot 2^{100} + 1$$

Clearly, a general formula is $S_N = (N-1) \cdot 2^N + 1$.

Method 2.
We can rewrite this series as follows:

$$1 + 2 \cdot 2 + 3 \cdot 2^2 + \ldots + N \cdot 2^{N-1} = \left(1 + 2 + 2^2 + \ldots + 2^{N-1}\right)$$
$$+ \left(2 + 2^2 + \ldots + 2^{N-1}\right) + \left(2^2 + 2^3 + \ldots + 2^{N-1}\right)$$
$$+ \left(2^3 + 2^4 + \ldots + 2^{N-1}\right) + \ldots + 2^{N-1}$$
$$= \left(2^N - 1\right) + 2\left(2^{N-1} - 1\right) + 2^2\left(2^{N-2} - 1\right)$$
$$+ 2^3\left(2^{N-3} - 1\right) + \ldots + 2^{N-1}(2 - 1)$$
$$= N \cdot 2^N - \left(1 + 2 + 2^2 + \ldots + 2^{N-1}\right)$$
$$= N \cdot 2^N - \left(2^N - 1\right)$$
$$S = (N-1)2^N + 1.$$

Therefore, we obtain $\displaystyle\sum_{N=1}^{N} N \cdot 2^{N-1} = 1 + 2^N(N-1)$.

Method 3. (Using a derivative).

Consider a polynomial $P(x) = x + x^2 + x^3 + \ldots + x^N$ and its first derivative $P'(x) = 1 + 2x + 3x^2 + \ldots + N \cdot x^{N-1}$. We can evaluate the sum of all terms of the polynomial as the sum of the N terms of geometric series, $P(x) = \frac{x(x^N-1)}{x-1} = \frac{x^{N+1}-x}{x-1}$. The derivative of this sum will be

$$P'(x) = \frac{\left(x^{N+1} - x\right)' \cdot (x-1) - \left(x^{N+1} - x\right)(x-1)'}{(x-1)^2} =$$
$$= \frac{N \cdot x^{N+1} - (N+1)x^N + 1}{(x-1)^2}$$

If we replace $x = 2$, we obtain that the given sum is

$$P'(x = 2) = N \cdot 2^{N+1} - (N+1)2^N + 1$$
$$S = 1 + 2^N \cdot (N-1).$$

Answer. $S_N = 1 + 2^N(N-1)$.

Problem 64 Evaluate $S = 1 \cdot 2^2 + 2 \cdot 3^2 + 3 \cdot 4^2 + \ldots + n(n+1)^2$.

Solution. Notice that $2 \cdot 3^2 + 3^2 = 3^3$, $3 \cdot 4^2 + 4^2 = 4^3$, $\ldots, n(n+1)^2 + (n+1)^2 = (n+1)^3$. Hence, $a_n = (n+1)^3 - (n+1)^2$. We can evaluate the series as follows:

$$1 \cdot 2^2 + 2 \cdot 3^2 + 3 \cdot 4^2 + \ldots + n(n+1)^2$$
$$+$$
$$2^2 + \quad 3^2 + \quad 4^2 + \ldots + \quad (n+1)^2$$
$$= 2^3 + 3^3 + 4^3 + \ldots + (n+1)^3$$

The sum above can be rewritten as

$$S + \sum_{n=2}^{n+1} n^2 = \sum_{n=2}^{n+1} n^3$$

$$S + \frac{(n+1)(n+2)(2(n+1)+1)}{6} - 1 = \frac{(n+1)^2(n+2)^2}{4} - 1$$

$$S = \frac{(n+1)(n+2)}{2} \left(\frac{2n+3}{3} - \frac{(n+1)(n+2)}{2} \right)$$

$$= \frac{n(n+1)(n+2)(3n+5)}{12}$$

For example, we can check this formula as $S_4 = 1 \cdot 2^2 + 2 \cdot 3^2 + 3 \cdot 4^2 + 4 \cdot 5^2 = 170 = \frac{4 \cdot 5 \cdot 6 \cdot (4 \cdot 3 + 5)}{12} = 170.$

Answer. $S = \frac{n(n+1)(n+2)(3n+5)}{12}$.

Problem 65 Evaluate the sum: $S = \frac{1}{2!} + \frac{2}{3!} + \frac{3}{4!} + \ldots + \frac{2015}{2016!}$.

Solution. Let us find the formula for the nth term. We can see that $a_n = \frac{n}{(n+1)!}$. Notice that $\frac{n}{(n+1)!} + \frac{1}{(n+1)!} = \frac{n+1}{(n+1)!} = \frac{1}{n!}$. Hence $a_n + \frac{1}{(n+1)!} = \frac{1}{n!}$. Since $a_{n-1} = \frac{n-1}{n!}$, then $a_{n-1} + \frac{1}{n!} = \frac{n-1}{n!} + \frac{1}{n!} = \frac{n}{n!}$ and $a_{n-1} + \frac{1}{n!} = \frac{1}{(n-1)!}$, which can be continued until we have the last term $\frac{1}{2!} + \frac{1}{2!} = \frac{2}{2!} = 1$. Therefore, if we add $\frac{1}{2016!}$ to the given sum and start adding the terms by pairing them from right to left, we obtain

$$S + \frac{1}{2016!} = 1$$

$$\boxed{S = 1 - \frac{1}{2016!}}$$

In general, we can evaluate the partial sum for any number of terms n, $S_n = 1 - \frac{1}{(n+1)!}$.

It is clear that the series is convergent because the limit of the partial sum equals 1.

Answer. $S = 1 - \frac{1}{2016!}$.

Problem 66 Prove that

$$S = 1 \cdot 2 \cdot 3 + 2 \cdot 3 \cdot 4 + 3 \cdot 4 \cdot 5 + \ldots + n(n+1)(n+2) = \frac{n(n+1)(n+2)(n+3)}{4}.$$

Proof.

Method 1. Consider the nth term of the series and rewrite it as $a_n = n(n+1)(n+2) = n^3 + 3n^2 + 2n$. Hence using sigma notation we can rewrite this sum as

$$\sum_{n=1}^{n} n(n+1)(n+2) = \sum_{n=1}^{n} n^3 + 3 \cdot \sum_{n=1}^{n} n^2 + 2 \cdot \sum_{n=1}^{n} n.$$ If we substitute Eqs. 1.29–1.31, the right hand side is rewritten as

$$S = \frac{(n+1)^2 n^2}{4} + \frac{3n(n+1)(2n+1)}{6} + \frac{2n(n+1)}{2} = \frac{n(n+1)(n^2 + 5n + 6)}{4}$$
$$= \frac{n(n+1)(n+2)(n+3)}{4}.$$

Method 2. On the other hand, the nth term and the corresponding partial sum can be evaluated as

$$a_n = (n+1)[(n+2)n] = (n+1) \cdot (n^2 + 2n) = (n+1)\left((n+1)^2 - 1\right)$$
$$= (n+1)^3 - (n+1).$$
$$S = \left(\frac{(n+1)(n+2)}{2}\right)^2 - \frac{(n+1)(n+2)}{2} = \frac{(n+1)(n+2)((n+1)(n+2) - 2)}{4}$$
$$= \frac{n(n+1)(n+2)(n+3)}{4}.$$

Which allows us to evaluate the requested sum as a difference between the sum of cubes and the sum of all natural numbers from 1 to $(n+1)$. The proof is complete.

Problem 67 Prove that for any natural $n \geq 2$, $n \in \mathbb{N}$, the sum $\frac{1}{n+1} + \frac{1}{n+2} + \frac{1}{n+3} + \ldots + \frac{1}{2n}$ is greater than ½ but less than 1.

Proof. Consider the chain of true inequalities,

$$\frac{1}{2n} < \frac{1}{n+1} < \frac{1}{n}$$

$$\frac{1}{2n} < \frac{1}{n+2} < \frac{1}{n}$$

$$\frac{1}{2n} < \frac{1}{n+3} < \frac{1}{n}$$

$$\cdots$$

$$\frac{1}{2n} \leq \frac{1}{2n} < \frac{1}{n}$$

Adding all these inequalities, we obtain $\frac{n}{2n} = \frac{1}{2} < \frac{1}{n+1} + \frac{1}{n+2} + \ldots + \frac{1}{2n} < \frac{n}{n} = 1$.
The proof is complete.

Problem 68 Prove the following statements:

a) $\dfrac{1}{1} = \dfrac{1}{2} + \dfrac{1}{6} + \dfrac{1}{12} + \dfrac{1}{20} + \dfrac{1}{30} + \ldots$

b) $\dfrac{1}{2} = \dfrac{1}{3} + \dfrac{1}{12} + \dfrac{1}{30} + \dfrac{1}{60} + \dfrac{1}{105} + \ldots$

c) $\dfrac{1}{3} = \dfrac{1}{4} + \dfrac{1}{20} + \dfrac{1}{60} + \dfrac{1}{140} + \dfrac{1}{280} + \ldots$

Proof.
a) The partial and infinite sums for the first infinite series can be rewritten and
evaluated as:

$$S_n = \frac{1}{1\cdot 2} + \frac{1}{2\cdot 3} + \frac{1}{3\cdot 4} + \frac{1}{4\cdot 5} + \frac{1}{5\cdot 6} + \ldots + \frac{1}{n(n+1)} = 1 - \frac{1}{n+1}$$

$$S = \sum_{n=1}^{\infty} \frac{1}{n(n+1)} = \lim_{n\to\infty} S_n = 1.$$

b) Consider the second sum: $\frac{1}{3} + \frac{1}{12} + \frac{1}{30} + \frac{1}{60} + \frac{1}{105} + \ldots$

Method 1. Would it be nice to recognize a similar pattern here? Can we rewrite
each term as a difference of two other terms? Let us rewrite this sum by factoring
out two from each fraction:

$$2\left(\frac{1}{6}+\frac{1}{24}+\frac{1}{60}+\frac{1}{120}+\frac{1}{210}+\cdots\right)$$

$$=2\cdot\left(\frac{1}{1\cdot2\cdot3}+\frac{1}{2\cdot3\cdot4}+\frac{1}{3\cdot4\cdot5}+\frac{1}{4\cdot5\cdot6}+\frac{1}{5\cdot6\cdot7}+\cdots+\right)$$

$$=2\sum_{n=1}^{\infty}\frac{1}{n(n+1)(n+2)}$$

This formula must look familiar to you (Prob. 57). The sum above can be found as

$$S_n=\sum_{k=1}^{n}\left(\frac{1}{k(k+1)}-\frac{1}{(k+1)(k+2)}\right)$$

$$=\frac{1}{1\cdot2}-\frac{1}{2\cdot3}+\frac{1}{2\cdot3}-\frac{1}{3\cdot4}+\frac{1}{3\cdot4}-\frac{1}{4\cdot5}+\cdots-\frac{1}{(n+1)(n+2)}$$

$$=\frac{1}{2}-\frac{1}{(n+1)(n+2)}.$$

Therefore, the sum of infinite series is ½.

Method 2. One could also notice the following:

$$\frac{1}{3}=\frac{1}{2}-\frac{1}{6}$$

$$\frac{1}{12}=\frac{1}{6}-\frac{1}{12}$$

$$\frac{1}{30}=\frac{1}{12}-\frac{1}{20}$$

$$\frac{1}{60}=\frac{1}{20}-\frac{1}{30}$$

$$\frac{1}{105}=\frac{1}{30}-\frac{1}{42}$$

$$\cdots$$

It looks like if we add the left and right sides of the relationships, we can evaluate the corresponding sums of the first two, first three, first four and first five terms of the series as follows:

$$S_2 = \frac{1}{3} + \frac{1}{12} = \frac{1}{2} - \frac{1}{12} = \frac{1}{2} - \frac{1}{3 \cdot 4}$$

$$S_3 = \frac{1}{3} + \frac{1}{12} + \frac{1}{30} = \frac{1}{2} - \frac{1}{20} = \frac{1}{2} - \frac{1}{4 \cdot 5}$$

$$S_4 = \frac{1}{3} + \frac{1}{12} + \frac{1}{30} + \frac{1}{60} = \frac{1}{2} - \frac{1}{30} = \frac{1}{2} - \frac{1}{5 \cdot 6}$$

$$S_5 = \frac{1}{3} + \frac{1}{12} + \frac{1}{30} + \frac{1}{60} + \frac{1}{105} = \frac{1}{2} - \frac{1}{42} = \frac{1}{2} - \frac{1}{6 \cdot 7}$$

By induction, the formula for the sum of the first n term is

$$S_n = \frac{1}{2} - \frac{1}{(n+1) \cdot (n+2)} \tag{2.15}$$

Using Eq. 2.15 and subtracting the sum of the first n terms and the sum of the first $(n-1)$ terms we obtain the formula for the nth term:

$$a_n = S_n - S_{n-1} = \frac{1}{n+1} \left(\frac{1}{n} - \frac{1}{n+2} \right) = \frac{2}{n(n+1)(n+2)} \tag{2.16}$$

By replacing n by 1, 2, 3, 4, and 5, we obtain correct values of the terms. For example,

$$a_3 = \frac{2}{3 \cdot 4 \cdot 5} = \frac{1}{30}$$

$$a_5 = \frac{2}{5 \cdot 6 \cdot 7} = \frac{1}{105}$$

Now we can predict any term of the series, $a_6 = \frac{1}{168}$, $a_7 = \frac{1}{252}$, $a_8 = \frac{1}{360}, \ldots$. Therefore, Eq. 2.15 is correct and then the infinite series sum is ½.

The proof is complete.

The second method of proof can help us to introduce the so-called **Leibniz triangle**.

The Leibniz harmonic triangle is a triangular arrangement of fractions in which each row starts with the reciprocal of the row number and every entry of the triangle is equal to the sum of the two fractions below it. For example, $\frac{1}{42} = \frac{1}{56} + \frac{1}{168}$ or $\frac{1}{4} = \frac{1}{5} + \frac{1}{20}$, etc.. In order to see a connection between Leibniz and Pascal's triangles, we place them together as in Figure 2.2. Instead of showing the fractions as in Figure 2.1, we record only the denominators of the fractions in the Leibniz triangle. Note that the first row for both triangles corresponds to $i = 0$.

Whereas each entry in Pascal's triangle is the sum of the two entries in the above row, each entry in the Leibniz triangle is the sum of the two entries in the row below it. Denote by $P(i,j)$, $L(i,j)$, $z(i,j)$ the entries of Pascal, Leibniz, and modified Leibniz triangles, respectively. For example, in the 5th row of Pascal triangle, the

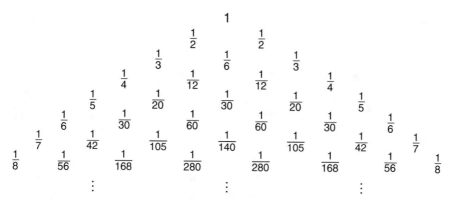

Figure 2.1 Leibniz triangle

entry $P(5,2) = 10$ is the sum of 4 and 6 in the 4th row. On the other hand, in the 5th row of the Leibniz triangle the corresponding entry $L(5,2) = 1/60$ is the sum of 1/105 and 1/140 in the 6th row. Just as Pascal's triangle can be computed by using binomial coefficients, so can Leibniz's triangle. The connection between the entries of three triangles is summarized by Eq. 2.17.

$$L(i,j) = \frac{1}{z(i,j)}$$
$$z(i,j) = \left(\frac{1}{z(i+1,j)} + \frac{1}{z(i+1,j+1)}\right)^{-1}$$
$$P(i,j) = P(i-1,j-1) + P(i-1,j) \tag{2.17}$$
$$P(i,j) = C_i^j = \frac{i!}{j!(i-j)!}$$
$$z(i,j) = (i+1) \cdot P(i,j) \quad i = 0,\,1,\,2,\,\ldots$$

Because any Leibniz triangle entry $L(n-1,k-1)$ is the sum of two entires, $L(n,k-1)$ and $L(n,k)$, the following is true:

$$\frac{1}{n \cdot C_{n-1}^{k-1}} = \frac{1}{(n+1)C_n^{k-1}} + \frac{1}{(n+1)C_n^k}$$
$$L(n-1,k-1) = L(n,k-1) + L(n,k) \tag{2.18}$$

Please prove it yourself by using Eq. 2.17 for binomial coefficients and by adding fractions.

Consider $P(6,2) = 15$ in Pascal's triangle, $P(6,2) = \frac{6!}{(6-2)!2!} = 15$. Corresponding to it Leibniz number is $L(6,2) = \frac{1}{(6+1)P(6,2)} = \frac{1}{7\cdot15} = \frac{1}{105}$ (Please use Figure 2.2 to see that $z(6,2) = 105$).

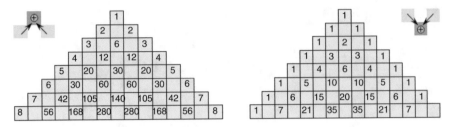

Figure 2.2 Modified Liebnitz (*left*) and Pascal (*right*) triangles

Moreover, each diagonal of Leibniz triangle does not only relate to the corresponding Pascal's triangle diagonals but also relates to a certain modification of the figurate numbers. Consider a sequence of the numbers in the second diagonal, just z numbers presented by the left diagram in Figure 2.2: 2, 6, 12, 20, 42, 56, ... Each term of this sequence is 2 times the corresponding triangular number 1, 3, 6, 10, 15, 21, 28, ... and can be written as $b_n = 2T_n = 2 \cdot \frac{n(n+1)}{2}$. Hence, an nth entry of the second Leibniz diagonal is its reciprocal, $a_n = \frac{1}{b_n} = \frac{1}{n(n+1)}$.

Consider a sequence of the numbers in the third diagonal, just z numbers presented by the left diagram in Figure 2.2. Each term of this sequence, 3, 12, 30, 60, 105, 168, ..., is 3 times the corresponding tethrahedron numbers and can be written as $b_n = 3TH_n = 3 \cdot \frac{n(n+1)(n+2)}{6} = \frac{n(n+1)(n+2)}{2}$. Hence, the corresponding n^{th} term of the third diagonal of Leibniz triangle (Figure 2.1) is its reciprocal (Eq. 2.16), $a_n = \frac{1}{b_b} = \frac{2}{n(n+1)(n+2)}$. Therefore, we can also state that the infinite series of the reciprocals of tetrahedral numbers is convergent and its sum is 3/2,

$$\sum_{n=1}^{\infty} \frac{6}{n(n+1)(n+2)} = \frac{3}{2}.$$ The proof is complete.

Further, the first Leibniz diagonal consists of reciprocals of natural numbers, $z = 1, 2, 3, 4, 5, 6,...$ The second diagonal consists of 1/(2x triangular numbers), $z = 2 \cdot 1, \ 2 \cdot 3, \ 2 \cdot 6, \ 2 \cdot 10, \ 2 \cdot 15, \ 2 \cdot 21, \ ...$ (Here 1,3,6,10,15,21, ... are triangular numbers). The third diagonal consists of 1/(3x tetrahedral numbers) and so on.

Method 3. Consider again the sum $\frac{1}{3} + \frac{1}{12} + \frac{1}{30} + \frac{1}{60} + \frac{1}{105} + \cdots$

We can see that this infinite series represent the sum of all fractions in the third diagonal of Leibniz triangle. Hence, each fraction can be replaced by the difference of two others using Eq. 2.18,

$$L(n, k-1) = L(n-1, k-1) - L(n, k) \qquad (2.19)$$

For example, $z(3, 1) = 12, \ P(3, 1) = 3, \ L(3, 1) = \frac{1}{12} = \frac{1}{(3+1)P(3,1)} = \frac{1}{3 \cdot 4}$

For some terms of the series b) we obtain the following:

$$\boxed{\frac{1}{3} = \frac{1}{2} - \frac{1}{6}}$$

$$a_1 = L(2,0) = L(1,0) - L(2,1)$$

$$\boxed{\frac{1}{12} = \frac{1}{6} - \frac{1}{12}}$$

$$a_2 = L(3,1) = L(2,1) - L(3,2)$$

$$\boxed{\frac{1}{30} = \frac{1}{12} - \frac{1}{20}}$$ (2.20)

$$a_3 = L(4,2) = L(3,2) - L(4,3)$$

$$\boxed{\frac{1}{60} = \frac{1}{20} - \frac{1}{30}}$$

$$a_4 = L(5,3) = L(4,3) - L(5,4)$$

$$\dots$$

Additionally, for this chain of equations, by induction, we can find the formula of the nth term of this series, $\boxed{a_n = L(n+1, n-1) = L(n, n-1) - L(n+1, n)}$. We can see that if we add the left and right sides of Eq. 2.20, then on the left we have the given series and on the right, all the terms except the first one and the last one are cancelled and that the partial sum is

$$S_n = \frac{1}{2} - L(n+1, n) = \frac{1}{2} - \frac{1}{(n+2)C_{n+1}^n}$$

$$= \frac{1}{2} - \frac{1}{(n+2)(n+1)}.$$

$$S_\infty = \frac{1}{2}.$$

This matches with our other formula found earlier and proves the statement.

   c) Let us now prove that $\frac{1}{3} = \frac{1}{4} + \frac{1}{20} + \frac{1}{60} + \frac{1}{140} + \frac{1}{280} + \dots.$

Method 1. Denote the sum by $S = \frac{1}{4} + \frac{1}{20} + \frac{1}{60} + \frac{1}{140} + \dots$, and multiply and divide the right side by 6,

$$S = 6 \cdot \left(\frac{1}{4 \cdot 6} + \frac{1}{20 \cdot 6} + \frac{1}{60 \cdot 6} + \frac{1}{140 \cdot 6} + \dots \right)$$

$$S = 6 \left(\frac{1}{1 \cdot 2 \cdot 3 \cdot 4} + \frac{1}{2 \cdot 3 \cdot 4 \cdot 5} + \frac{1}{3 \cdot 4 \cdot 5 \cdot 6} + \dots \right)$$

Hence the given sum is six times the sum inside the parentheses. We have seen such a series earlier in this chapter. It can be evaluated as $S = 6 \cdot \sum_{n=1}^{\infty} \dfrac{1}{n(n+1)(n+2)(n+3)}$. In order to evaluate an infinite sum of this series, we rewrite the nth term in a different form and firstly, we multiply two inner and two outer factors of the denominator, $6 \cdot \left(\frac{1}{(n^2+3n)\cdot(n^2+3n+2)} \right)$.

We obtained a familiar structure: two quantities in the denominator differ by two, and we can rewrite the fraction again and again decompose it into two new fractions as follows:

$$a_n = \frac{6}{2} \cdot \left(\frac{1}{n(n+3)} - \frac{1}{(n+1)(n+2)} \right) = \frac{3}{3} \cdot \left(\frac{1}{n} - \frac{1}{n+3} \right) - 3 \left(\frac{1}{n+1} - \frac{1}{n+2} \right)$$

$$= \frac{1}{n} - \frac{1}{n+3} - \frac{3}{n+1} + \frac{3}{n+2}.$$

This n^{th} term can be rewritten in a little different form so we can calculate the partial sum of the series easily:

$$a_n = \left(\frac{1}{n} - \frac{1}{n+1} \right) + \left(\frac{1}{n+2} - \frac{1}{n+3} \right) - 2 \cdot \left(\frac{1}{n+1} - \frac{1}{n+2} \right)$$

$$\sum_{n=1}^{\infty} a_n = \sum_{n=1}^{\infty} \left(\left(\frac{1}{n} - \frac{1}{n+1} \right) + \left(\frac{1}{n+2} - \frac{1}{n+3} \right) - 2 \cdot \left(\frac{1}{n+1} - \frac{1}{n+2} \right) \right)$$

Now, the sum of each quantity can be evaluated separately and the final answer will be the sum of these three answers:

$$\sum_{n=1}^{n} \left(\frac{1}{n} - \frac{1}{n+1} \right) = 1 - \frac{1}{n+1}$$

$$\sum_{n=1}^{n} \left(\frac{1}{n+2} - \frac{1}{n+3} \right) = \frac{1}{3} - \frac{1}{n+3}$$

$$-2 \sum_{n=1}^{n} \left(\frac{1}{n+1} - \frac{1}{n+2} \right) = -\frac{2}{2} + \frac{2}{n+2}$$

$$\boxed{S_n = \frac{1}{3} - \frac{1}{n+1} - \frac{1}{n+3} + \frac{2}{n+2}}$$

Obviously, as n goes to infinity, the partial sum will go to 1/3. Therefore, $S_\infty = \frac{1}{3} = \frac{1}{4} + \frac{1}{20} + \frac{1}{60} + \frac{1}{140} + \frac{1}{280} + \cdots$.

Method 2. Please notice that the series is the sum of all fractions in the fourth diagonal of Leibniz triangle. Looking at that diagonal of the Leibniz triangle in Figure 2.2, and using Eq. 2.19 we have the following chain of the true relationships:

$$a_1 = \frac{1}{4} = \frac{1}{3} - \frac{1}{12} \Leftrightarrow a_1 = L(3,0) = L(2,0) - L(3,1)$$

$$a_2 = \frac{1}{20} = \frac{1}{12} - \frac{1}{30} \Leftrightarrow a_2 = L(4,1) = L(3,1) - L(4,2)$$

$$a_3 = \frac{1}{60} = \frac{1}{30} - \frac{1}{60} \Leftrightarrow a_3 = L(5,2) = L(4,2) - L(5,3)$$

$$a_4 = \frac{1}{140} = \frac{1}{60} - \frac{1}{105} \Leftrightarrow a_4 = L(6,3) = L(5,3) - L(6,4)$$

$$a_5 = \frac{1}{280} = \frac{1}{105} - \frac{1}{168} \Leftrightarrow a_5 = L(7,4) = L(6,4) - L(7,5)$$

$$\dots$$

From these relationships, by induction, we can recognize the formula for n^{th} term of the series and evaluate its nth partial sum,

$$a_n = L(n+2, n-1) = L(n+1, n-1) - L(n+2, n)$$
$$S_n = \sum_{i=1}^{n} a_i = \frac{1}{3} - L(n+2, n) \tag{2.21}$$

It follows from Eq. 2.21 that the nth partial sum of the series is $1/3$ minus the Leibniz entry $L(n+2, n)$. Additionally, we can evaluate the nth term of the series by using Eq. 2.17 for $L(i, j)$,

$$L(n+1, n-1) = \frac{1}{(n+2)C_{n+1}^{n-1}} = \frac{1}{(n+2)\dfrac{(n+1)!}{(n-1)! \cdot 2!}}$$

$$= \frac{2}{(n+2)(n+1)n}$$

$$L(n+2, n) = \frac{1}{(n+3)C_{n+2}^{n}} = \frac{1}{(n+3)\dfrac{(n+2)!}{(n)! \cdot 2!}} \tag{2.22}$$

$$= \frac{2}{(n+3)(n+2)(n+1)}$$

Subtracting the left and right sides, we obtain

$$a_n = \frac{2}{(n+2)(n+1)n} - \frac{2}{(n+3)(n+2)(n+1)}$$

$$= 2 \cdot \left(\frac{1}{(n+2)(n+1)n} - \frac{1}{(n+3)(n+2)(n+1)} \right) a_n$$

$$= \frac{2}{(n+2)(n+1)} \cdot \left(\frac{1}{n} - \frac{1}{n+3} \right)$$

$$\boxed{a_n = \frac{6}{n(n+1)(n+2)(n+3)}}$$

Making substitutions of the Leibniz entry $L(n+2, n)$ from Eq. 2.22 into Eq. 2.21, we have

$$S_n = \frac{1}{3} - \frac{2}{(n+3)(n+2)(n+1)} \tag{2.23}$$

If n increases without bound, then the partial sum above will get closer and closer to 1/3. Therefore, the sum in part (c) is 1/3. The statement is proven.

Consider again Figure 2.2. Start counting the rows from the top $i = 1$. Take the numbers of the nth row and add them. For example, for the 4th row, we have $5 + 20 + 30 + 20 + 5 = 80 = 5 \cdot 2^4$. The following statement is true.

Lemma 2.1 The sum of the numbers in the nth row of a triangle made of the denominators of Leibniz triangle equals $n \cdot 2^{n-1}$.

Proof. The sum of all numbers in the nth row is the sum of the z-numbers and hence, it can be written using a definition of a z number as

$$\sum_{k=0}^{n-1} n \cdot C_{n-1}^k = n \sum_{k=0}^{n-1} C_{n-1}^k = n \cdot 2^{n-1}.$$

2.2 Trigonometric Series

The following problems are very different from anything above. They are trigonometric series. In order to evaluate trigonometric series we need to know trigonometric identities and de Moivre's Formula. Some formulas are given by,

$$\sin x \cos y = \frac{1}{2}(\sin(x+y) + \sin(x-y))$$

$$\sin x \sin y = \frac{1}{2} \cdot (\cos(x-y) - \cos(x+y)) \qquad (2.24)$$

$$\cos x - \cos y = 2\sin\frac{x+y}{2}\sin\frac{y-x}{2}$$

---

de Moivre's Formula (Abraham de Moivre, French mathematician, 1667-1754)

$$\cos nx + i\sin nx = (\cos x + i\sin x)^n. \qquad (2.25)$$

We do not give a proof for the first three formulas because students study them in high school. de Moivre's Formula is not in the regular high school curriculum so we need to discuss it a little more. Let us see how easily it can be derived under assumption that the Euler's relationship below is true.

Euler's Formula

$$e^{ix} = \cos x + i\sin x \qquad (2.26)$$

Let us raise the left and the right side of Eq. 2.26 to the second power, then the third, fourth, and so on and apply Eq. 2.25 again each time. We obtain the following chain of the correct equations:

$$(e^{ix})^2 = (\cos x + i\sin x)^2$$
$$e^{i2x} = \cos 2x + i\sin 2x$$
$$(e^{ix})^3 = (\cos x + i\sin x)^3$$
$$e^{i3x} = \cos 3x + i\sin 3x$$
$$(e^{ix})^4 = (\cos x + i\sin x)^4$$
$$e^{i4x} = \cos 4x + i\sin 4x$$
$$\cdots$$
$$e^{inx} = \cos nx + i\sin nx$$

Problem 69 Evaluate $S_n = \cos\frac{\pi}{n} + \cos\frac{2\pi}{n} + \cos\frac{3\pi}{n} + \ldots + \cos\frac{(n-1)\pi}{n}$

Solution. Let us multiply the given sum by $\sin\frac{\pi}{2n}$. Using the first formula of trigonometric identities of Eq. 2.24 and the fact that sine is an odd function $(\sin(-y) = -\sin(y))$, we obtain,

$$2S_n \sin\frac{\pi}{2n} = 2\sin\frac{\pi}{2n}\cos\frac{\pi}{n} + 2\sin\frac{\pi}{2n}\cos\frac{2\pi}{n} + \ldots + 2\sin\frac{\pi}{2n}\cos\frac{(n-2)\pi}{n} +$$

$$2\sin\frac{\pi}{2n}\cos\frac{(n-1)\pi}{n} = \sin\frac{3\pi}{2n} - \sin\frac{\pi}{2n} + \sin\frac{5\pi}{2n} - \sin\frac{3\pi}{2n} +$$

$$\ldots + \sin\frac{(2n-3)\pi}{2n} - \sin\frac{(2n-5)\pi}{2n} + \sin\frac{(2n-1)\pi}{2n} - \sin\frac{(2n-3)\pi}{2n}.$$

After simplification and canceling opposite terms we obtain $2S_n \sin\frac{\pi}{2n} = -\sin\frac{\pi}{2n} + \sin\frac{(2n-1)\pi}{2n} = -\sin\frac{\pi}{2n} + \sin\left(\pi - \frac{\pi}{2n}\right) = 0$. Considering the expression above we notice that the second factor on the left hand side is never zero for any natural n, therefore the given sum must be zero.

Answer. $S_n = 0$.

Problem 70 Prove that

$$S_n = \sin x + \sin 2x + \sin 3x + \ldots + \sin nx = \frac{\sin\frac{nx}{2} \cdot \sin\frac{(n+1)x}{2}}{\sin\frac{x}{2}.}$$

Proof. This proof will involve only knowledge at a high school curriculum level, and trigonometric identities. Multiplying the sum by $2\sin(x/2)$ we obtain:

$$2\sin\frac{x}{2}(\sin x + \sin 2x + \sin 3x + \ldots + \sin nx)$$

$$= 2\sin\frac{x}{2}\sin x + 2\sin\frac{x}{2}\sin 2x + 2\sin\frac{x}{2}\sin 3x + \ldots + 2\sin\frac{x}{2}\sin nx$$

$$= \cos\left(\frac{x}{2}-x\right) - \cos\left(\frac{x}{2}+x\right) + \cos\left(\frac{x}{2}-2x\right) - \cos\left(\frac{x}{2}+2x\right)$$

$$+ \cos\left(\frac{x}{2}-3x\right) - \cos\left(\frac{x}{2}+3x\right) + \ldots + \cos\left(\frac{x}{2}-nx\right) - \cos\left(\frac{x}{2}+nx\right)$$

Since cosine is an even function, then $\cos(-y) = \cos(y)$ and all terms in the middle of the last formula will be eliminated as $\cos\left(\frac{x}{2}\right) - \cos\left(\frac{3x}{2}\right) + \cos\left(\frac{3x}{2}\right) - \cos\left(\frac{5x}{2}\right) + \cos\left(\frac{5x}{2}\right) + \ldots - \cos\left(\frac{x(2n+1)}{2}\right)$. Now we obtain that $S_n \cdot 2\sin\frac{x}{2} = \cos\frac{x}{2} - \cos\frac{(2n+1)x}{2}$. Apply the difference of cosines formula (3rd formula of Eq. 2.24):

$$S_n \cdot 2\sin\frac{x}{2} = 2\sin\frac{\left(\frac{1}{2} + \frac{2n+1}{2}\right)x}{2} \cdot \sin\frac{\left(-\frac{1}{2} + \frac{2n+1}{2}\right)x}{2}$$

$$= 2\sin\frac{(n+1)x}{2} \cdot \sin\frac{nx}{2}$$

Dividing the last row by $2\sin(x/2)$ we prove the formula:

$$S_n = \sin x + \sin 2x + \sin 3x + \ldots + \sin nx$$

$$= \frac{\sin\frac{nx}{2} \cdot \sin\frac{(n+1)x}{2}}{\sin\frac{x}{2}}$$

You will have a chance to prove this formula a second way in a homework problem using de Moivre's Formula. You can use the next problem as an example.

Problem 71 Evaluate $A = \dfrac{\cos\frac{\pi}{4}}{2} + \dfrac{\cos\frac{2\pi}{4}}{2^2} + \ldots + \dfrac{\cos\frac{\pi n}{4}}{2^n}$.

Solution. Denote

$$B = \frac{\sin\frac{\pi}{4}}{2} + \frac{\sin\frac{2\pi}{4}}{2^2} + \ldots + \frac{\sin\frac{\pi n}{4}}{2^n} \tag{2.27}$$

Assuming that B is imaginary part of a complex number $A + iB$, we multiply Eq. 2.27 by i and add the corresponding A:

$$A + iB = \frac{1}{2}\left(\cos\frac{\pi}{4} + i\sin\frac{\pi}{4}\right) + \frac{1}{2^2}\left(\cos\frac{2\pi}{4} + i\sin\frac{2\pi}{4}\right) + \ldots + \frac{1}{2^n}\left(\cos\frac{\pi n}{4} + i\sin\frac{\pi n}{4}\right)$$

Applying de Moivre's Formula (Eq. 2.25) to the previous expression, we have

$$A + iB = \frac{1}{2}\left(\cos\frac{\pi}{4} + i\sin\frac{\pi}{4}\right) + \frac{1}{2^2}\left(\cos\frac{\pi}{4} + i\sin\frac{\pi}{4}\right)^2 + \ldots + \frac{1}{2^n}\left(\cos\frac{\pi}{4} + i\sin\frac{\pi}{4}\right)^n \tag{2.28}$$

We can notice that Eq. 2.28 is a geometric series with both, first term and the ratio equal to $\frac{1}{2} \cdot \left(\cos\frac{\pi}{4} + i\sin\frac{\pi}{4}\right)$.

Therefore, using the sum of geometric series, Eq. 2.28 can be rewritten in a compact form as follows:

$$A + iB = \frac{1}{2} \left(\cos \frac{\pi}{4} + i \sin \frac{\pi}{4} \right) \frac{\left(1 - \frac{1}{2^n} \left(\cos \frac{\pi}{4} + i \sin \frac{\pi}{4} \right)^n \right)}{\left(1 - \frac{1}{2} \left(\cos \frac{\pi}{4} + i \sin \frac{\pi}{4} \right) \right)} \qquad (2.29)$$

Applying de Moivre's Formula (Eq. 2.25) to Eq. 2.29 again and using the fact that $\cos \frac{\pi}{4} = \sin \frac{\pi}{4} = \frac{1}{\sqrt{2}}$ we have,

$$A + iB = \frac{1}{2\sqrt{2}} (1 + i) \cdot \frac{\left(1 - \frac{1}{2^n} \left(\cos \frac{\pi}{4} + i \sin \frac{\pi}{4} \right)^n \right)}{\left(1 - \frac{1}{2\sqrt{2}} - \frac{i}{2\sqrt{2}} \right)} \qquad (2.30)$$

Rationalizing the denominator and extracting the real part of $A + iB$ in Eq. 2.30, we obtain

$$A = \frac{\left(\sqrt{2} - 1 \right) \left(2^n - \cos \frac{\pi n}{4} \right) + \sqrt{2} \sin \frac{\pi n}{4}}{2^n \left(5 - 2\sqrt{2} \right)}.$$

Answer. $A = \dfrac{\left(\sqrt{2} - 1 \right) \left(2^n - \cos \frac{\pi n}{4} \right) + \sqrt{2} \sin \frac{\pi n}{4}}{2^n \left(5 - 2\sqrt{2} \right)}$

2.3   Using Mathematical Induction for Sequences and Series

The principle of mathematical induction is very helpful in proving many statements about positive integers. According to this principle, a mathematical statement involving the variable n can be shown to be true for any positive integer n by proving the following two statements:

- The statement is true for $n = 1$
- If the statement is true for any positive integer k, then it is also true for $n = k + 1$.

Let us show how mathematical induction can help us to prove and solve some problems involving sequences and series.

Problem 72 Use mathematical induction to prove that $1 + 3 + 5 + \ldots + (2n - 1) = n^2$ is true for any positive number n.

Proof. Step 1. Replacing n by 1 in the above equality gives $2 \cdot (1) - 1 = 1$ which is true, so $n = 1$ satisfies the equation.

Step 2. Assume that the equality is true at $n = k$. And let us show that it will be true at $n = k + 1$. If $1 + 3 + 5 + \ldots + (2k - 1) = k^2$ is true, then let us show that for $n = k + 1$ the left side of the equality equals $(k + 1)^2$.

$$1 + 3 + 5 + \cdots + (2k - 1) + \left(2(k + 1) - 1\right).$$

Start with the left hand side, and notice that (because of our assumption) it is equal to k^2, plus an additional term.

$$1 + 3 + 5 + \ldots + (2k - 1) + (2(k + 1) - 1) = k^2 + 2k + 1 = (k + 1)^2.$$

The final equality proves that the equation is true for $n = k + 1$, given that it is true for $n = k$. By the principle of mathematical induction, we have proven the statement.

Problem 73 Prove that: $\displaystyle\sum_{k=1}^{n} k^3 = \left(\sum_{k=1}^{n} k\right)^2$.

Proof. Step 1. Replacing n by 1 in the above equality gives
$1^3 = 1^2$ which is true, so $n = 1$ satisfies the equation.

Step 2. Assume that the equality is true at $n = k$ and let us show that it will be true at $n = k + 1$:

If $1^3 + 2^3 + 3^3 + \ldots + k^3 = (1 + 2 + 3 + \ldots + k)^2 = \frac{k^2(k+1)^2}{4}$ is true, then let us show that for $n = k + 1$ the left side of the given equality equals $(1 + 2 + 3 + \ldots + k + 1)^2 = \frac{(k+1)^2(k+2)^2}{4}$.

We can state that $1^3 + 2^3 + 3^3 + \ldots + k^3 + (k + 1)^3 = (1 + 2 + 3 + \ldots + k)^2 + (k + 1)^3$. Replacing the right hand side, putting fractions over the common denominator and factoring, we obtain the required formula:

$$\frac{k^2(k + 1)^2}{4} + (k + 1)^3 = \frac{k^2(k + 1)^2 + 4(k + 1)^3}{4}$$

$$= \frac{(k + 1)^2(k^2 + 4k + 4)}{4}$$

$$= \frac{(k + 1)^2(k + 2)^2}{4}.$$

The final equality proves that the equation is true for $n = k + 1$, assuming that it is true for $n = k$. Using the principle of mathematical induction, we have completed our proof.

In Prob. 74, we use mathematical induction for recurrent sequences.

Problem 74 Given a sequence $\{x_n\}$ such that
$x_0 = 2$, $x_1 = \frac{3}{2}$, $x_{n+1} = \frac{3}{2}x_n - \frac{1}{2}x_{n-1}$. a) Find the exact formula for x_n.

b) Evaluate $\lim\limits_{n\to\infty} x_n$, if the limit exists. c) Is series $S_\infty = \sum\limits_{k=0}^{\infty} x_k$ convergent or divergent?

Solution.
a) Using the given recursive formula we can calculate a few terms of the sequence:

$$x_0 = 2 = 1 + 2^0$$

$$x_1 = \frac{3}{2} = 1 + \frac{1}{2} = 1 + 2^{-1}$$

$$x_2 = \frac{3}{2}\cdot\frac{3}{2} - \frac{1}{2}\cdot 2 = 1 + \frac{1}{4} = 1 + 2^{-2}$$

$$x_3 = \frac{3}{2}\cdot\frac{5}{4} - \frac{1}{2}\cdot\frac{3}{2} = \frac{9}{8} = 1 + \frac{1}{8} = 1 + 2^{-3}$$

$$x_4 = \frac{3}{2}\cdot\frac{9}{8} - \frac{1}{2}\cdot\frac{5}{4} = \frac{17}{16} = 1 + 2^{-4}$$

Notice that every n^{th} term of the sequence is obtained as a sum of 1 and 2 raised to a negative power that is equal to the number of the term. We can assume that

$$x_n = 1 + 2^{-n} \tag{2.31}$$

Using mathematical induction let us prove that Eq. 2.31 is the exact formula for the nth term of the sequence. Denote by $A(m)$ our statement for $n = m$.
Step 1. $A(1)$ is true because $x_1 = \frac{3}{2} = 1 + 2^{-1}$
Step 2. Assume that $A(k)$ and $A(k-1)$ are true (i.e., Eq. 2.31 holds for $n = k$ and for $n = k - 1$), i.e., $x_k = 1 + 2^{-k}$ and $x_{k-1} = 1 + \frac{1}{2^{k-1}}$.
Step 3. Let us prove that $A(k+1)$ is also true. That is, we need to show that
$x_{k+1} = 1 + 2^{-(k+1)} = 1 + 2^{-k-1}$. Indeed, $x_{k+1} = \frac{3}{2}x_k - \frac{1}{2}x_{k-1}$ by the condition of the problem, then

$$x_{k+1} = \frac{3}{2} \cdot \left(1 + \frac{1}{2^k}\right) - \frac{1}{2} \cdot \left(1 + \frac{1}{2^{k-1}}\right)$$

$$= \frac{1}{2} \cdot \frac{(3 \cdot 2^k + 3 - 2^k - 2)}{2^k}$$

$$= \frac{1}{2} \cdot \frac{(2 \cdot 2^k + 1)}{2^k}$$

$$= \frac{1}{2} \cdot \left(2 + \frac{1}{2^k}\right)$$

$$= 1 + \frac{1}{2^{k+1}}$$

$$= 1 + 2^{-(k+1)}$$

We proved that Eq. 2.31 is the exact formula for the nth term of the sequence. The proof is complete.

b) Now, knowing the nth term of the sequence explicitly, $x_n = 1 + 2^{-n}$, let us find the limit of $\{x_n\}$, $\lim\limits_{n \to \infty} x_n = \lim\limits_{n \to \infty} \left(1 + \frac{1}{2^n}\right) = 1$.

c) The series is divergent because, as we established above, the limit of the nth term as n goes to infinity is not zero. Therefore, the series does not pass the Necessary Condition (See Chapter 3 for clarification). We can also evaluate the partial sum for the series exactly as $S_n = 2 + 1 + \frac{1}{2^1} + 1 + \frac{1}{2^2} + 1 + \frac{1}{2^3} + \ldots + 1 + \frac{1}{2^n}$. You can see that this is the sum of $(2 + n \cdot 1)$ and the first n terms of geometric series with $b_1 = \frac{1}{2}$, $r = \frac{1}{2}$. Thus, $S_n = 2 + n + \frac{\frac{1}{2}\left(1 - \left(\frac{1}{2}\right)^n\right)}{1 - \frac{1}{2}} = 2 + n + 1 - \left(\frac{1}{2}\right)^n$ so $S_n = 3 + n - \left(\frac{1}{2}\right)^n$.

We can see that this partial sum depends on n and increases without bound as n increases.

Answer. a) $X_n = 1 + 2^{-n}$; b) $S_n = 3 + n - \left(\frac{1}{2}\right)^n$. c) The series is divergent.

In the homework chapter, you will be asked to find the formula for the nth term using the knowledge of recursion.

Problem 75 Given a sequence $\{a_n\}$, $a_n = n(3n + 1)$, $n \in \mathbb{N}$. Prove that its nth partial sum can be evaluated by formula $S_n = n(n + 1)^2$.

Proof. We prove this by induction. It is easy to see that the formula is true for $n = 1$. Indeed, $S_1 = 1 \cdot (1 + 1)^2 = a_1 = 4$. If we evaluate the sums of two, three, four of even five terms of the sequence, we see that the formula works. However, it does not prove the statement. Assume that this formula is correct for $n = k$, i.e., the sum of the first k terms of the given sequence equals $S_k = k \cdot (k + 1)^2$. Let us demonstrate that it will be also true for $n = k + 1$, i.e., $S_{k+1} = (k + 1) \cdot (k + 2)^2$.

We know that the sum of $(k + 1)$ terms of the sequence equals the sum of its first k terms plus the $(k + 1)^{st}$ term, $S_{k+1} = S_k + a_{k+1}$. Substituting here the kth sum and the value of the $(k + 1)^{st}$ term of the sequence we obtain

$$\begin{aligned} S_{k+1} &= k(k + 1)^2 + (k + 1)(3 \cdot (k + 1) + 1) \\ &= k(k + 1)^2 + (k + 1)(3k + 4) \\ &= (k + 1)(k(k + 1) + 3k + 4) \\ &= (k + 1)(k + 2)^2. \end{aligned}$$

Therefore, the formula is correct for $n = k + 1$, hence it is true any natural n. The statement is proven.

Problem 76 Prove that any term of the sequence $a_n = 4 \cdot 6^n + 5n - 4$ is divisible by 25.

Proof. We can substitute $n = 1$ and obtain that $a_1 = 25$. Yes, it is divisible by 25.

Assume that the statement is true for $n = k$ and that $a_k = 4 \cdot 6^k + 5k - 4$ is divisible by 25, then it can be written as $4 \cdot 6^k + 5k - 4 = 25m \Rightarrow 4 \cdot 6^k = 25m + 4 - 5k$. Let us prove that the next term, $k + 1$, $a_{k+1} = 4 \cdot 6^{k+1} + 5(k + 1) - 4$ is also is also divisible by 25. Next, because the kth term is divisible by 25, we extract the kth term of the sequence in the expression of the $(k + 1)$ term,

$$a_{k+1} = 6 \cdot 4 \cdot 6^k + 5k + 1 = 6 \cdot \left(4 \cdot 6^k + 5k - 4\right) - 25k + 25.$$

Each term of the sum is a multiple of 25, then the total sum or $(k + 1)^{st}$ term is divisible by 25. You could do this proof a little bit differently by replacing the kth term by $25 \cdot m$:

$$\begin{aligned} a_{k+1} &= 6 \cdot 4 \cdot 6^k + 5k - 4 + 5 \\ &= 6(25m + 4 - 5k) + 5k + 1 \\ &= 150m - 25k + 25 \\ &= 25 \cdot (6m - k + 1) \end{aligned}$$

It is clear that it is divisible by 25. The statement is proven.

Problem 77 Given a Fibonacci sequence $\{a_n\}$: $a_1 = a_2 = 1$, $a_n = a_{n-1} + a_{n-2}$, $n > 2$. Prove that the terms of the sequence satisfy the equation: $a_{n+1}^2 - a_n \cdot a_{n+2} = (-1)^n$, $\forall n \in \mathbb{N}$.

Proof. We prove this by induction.
1. Notice that the equality is true for $n = 1$ because
$a_2^2 - a_1 \cdot a_3 = 1 - 1 \cdot 2 = (-1)^1$.
2. Assume that the statement is true for $n = k$, $a_{k+1}^2 - a_k \cdot a_{k+2} = (-1)^k$. From this
it follows that $a_{k+1}^2 = (-1)^k + a_k \cdot a_{k+2}$
3. Let us demonstrate that it is also true for $n = k + 1$, i.e.,
$a_{k+2}^2 - a_{k+1} \cdot a_{k+3} = (-1)^{k+1}$.

Let us substitute the expression for the $(k + 3)^{\text{th}}$ term of Fibonacci sequence,
$a_{k+2}^2 - a_{k+1} \cdot (a_{k+1} + a_{k+2}) = a_{k+2}^2 - a_{k+1} \cdot a_{k+2} - \boxed{a_{k+1}^2}$. Substituting in this
formula the value of the term in the box, we obtain

$$a_{k+2}^2 - a_{k+1} \cdot a_{k+3} = a_{k+2}^2 - a_{k+1} \cdot a_{k+2} - (-1)^k - a_k \cdot a_{k+2}$$
$$= a_{k+2}^2 - a_{k+2}(a_k + a_{k+1}) + (-1)^{k+1}$$
$$= \underbrace{a_{k+2}^2 - a_{k+2} \cdot a_{k+2}}_{0} + (-1)^{k+1}.$$

The proof is complete.

2.4   Problems on the Properties of Arithmetic and Geometric Sequences

If three numbers form an arithmetic sequence, the middle term is called the arithmetic mean of the other two. Thus,

$$a_3 - a_2 = a_2 - a_1$$
$$2a_2 = a_1 + a_3$$
$$a_2 = \frac{a_1 + a_3}{2}$$

By analogy the arithmetic mean of two numbers is half of their sum. Therefore, $\frac{(a+b)}{2}$ is the arithmetic mean of numbers a and b, or the average of two numbers.

Similarly the average or mean value of three numbers a, b, and c is $\frac{a+b+c}{3}$. In general, the mean value of n positive numbers, $a_1, a_2, a_3, \ldots a_n$ is $\frac{a_1+a_2+\ldots+a_n}{n}$, that is the average of the sum. Besides the arithmetic mean defined above there is another form of mean value that is defined by the formula: $b = \sqrt{ac}$. Value b is called a geometric mean of numbers a and c. Recalling a geometric progression with positive terms $b_1, b_2, \ldots, b_{n-1}, b_n, b_{n+1} \ldots$ and common ratio r such that

$\frac{b_n}{b_{n-1}} = \frac{b_{n+1}}{b_n}$, $b_n = \sqrt{b_{n-1}b_{n+1}}$ or $b_n^2 = b_{n-1} \cdot b_{n+1}$. Every term of a geometric progression is a geometric mean of the preceding and consequent terms.

Now consider the following problems.

Problem 78 Peter lives near a bus stop A. The bus stops A, B, C, and D are on the same street. Peter walks for exercise every weekend. He starts at A with a speed of 5 km per hour and goes to D. Reaching D he turns back and goes to B. Walking this rout (A-D-B) requires 5 h. At B Peter takes a bus and goes home. It is known that he can cover the distance between A and C in 3 h. The distances between A and B, B and C, C and D form a geometric sequence in the given order. Find the distance between B and C.

Solution. Usually it is a good idea to draw a picture of the problem. A, B, C, and D are on the same street (Figure 2.3). It means that we can draw them as points on the same line, A and D will be the end points of the segment and B and C between them in the order A-B-C-D.

Because our unknown is the distance between B and C it seems obvious to introduce 3 variables x, y, and z as distances between A and B, B and C, and C and D respectively. Using the condition of the problem and distance $=$ speed $\cdot$ time we write, $x + y + z + z + y = 5 \cdot 5 = 25$ and $x + y = 3 \cdot 5 = 15$.

Now we are going to write the last equation of the system. Because x, y, and z are consecutive terms of a geometric sequence, then $y^2 = xz$, and we can complete a system:

$$\begin{cases} x + 2y + 2z = 25 \\ x + y = 15 \\ y^2 = xz \end{cases}$$

$$\begin{cases} y + 2z = 10 \\ x = 15 - y \\ y^2 = xz \end{cases}$$

$$\begin{cases} z = \dfrac{10 - y}{2} \\ x = 15 - y \\ y^2 = \dfrac{(15 - y)(10 - y)}{2} \end{cases}$$

Figure 2.3 Problem 78

Subtracting the second equation from the first of the first system, we can eliminate variable x in the second system. Then we express z and x in terms of y and put them into the third equation of the last system. Let us solve the last equation for y. Multiplying both sides by 2, we have

$$2y^2 = 150 - 25y + y^2$$
$$y^2 + 25y - 150 = 0$$
$$y_1 = 5, \quad y_2 = -30$$

Because y is a distance, it has to be a positive, so we choose $y = 5$.

Answer. The distance between B and C is 5 km.

Problem 79 The four numbers a, b, c, and z are given. It is known that the first three numbers form an arithmetic sequence, and the last three numbers form a geometric sequence. A sum of the outer terms is 4 and the sum of the inner terms is 2. Find the numbers.

Solution. Let us write the numbers in a row: $a\ b\ c\ z$. If variables a, b and c are terms of an arithmetic sequence, then

$$b = \frac{a + c}{2} \tag{2.32}$$

On the other hand,

$$+ \begin{cases} a + z = 4 \\ b + c = 2 \end{cases} \atop \overline{a + b + c + z = 6} \tag{2.33}$$

Replacing $(a + c)$ by $2b$ from Eq. 2.32 into Eq. 2.33, we have

$$3b + z = 6 \tag{2.34}$$

Our purpose now is to eliminate some variables. It would be nice to obtain a system of two equations in just two variables. (for example, z and b). Let us use the second part of the condition. If b, c, and z form a geometric sequence, then c is a geometric mean of b and z or

$$c^2 = b \cdot z \tag{2.35}$$

Expressing c as $(2 - b)$ from system (Eq. 2.33) and substituting into (Eq. 2.35) we derive

$$(2 - b)^2 = b \cdot z \qquad (2.36)$$

Let us combine Eqs. 2.34 and 2.36 as

$$(2 - b)^2 = b(6 - 3b)$$
$$4 - 4b + b^2 = 6b - 3b^2$$
$$2b^2 - 5b + 2 = 0$$
$$b_{1,2} = \frac{5 \pm \sqrt{5^2 - 4 \cdot 2 \cdot 2}}{2 \cdot 2} = \frac{5 \pm 3}{4}$$
$$b_1 = 2 \qquad b_2 = 0.5$$

Two different values of b will give us two sets of a, b, c, and z.
1. $b = 2$
   $z = 6 - 3b = 0$
   $a = 3b - 2 = 4$
   $c = 2 - b = 0$
2. $b = 0.5$
   $z = 4.5$
   $a = -0.5$
   $c = 1.5$

Answer. $(a, b, c, z) = \{(4, 2, 0, 0),\ (-0.5,\ 0.5,\ 1.5,\ 4.5)\}$

Problem 80 The sequence $a_1, a_2, a_3, \ldots$ satisfies $a_1 = 19$, $a_9 = 99$, and for any $n \geq 3$, a_n is the arithmetic mean of the first $(n - 1)$ terms. Find a_2.

Solution. Let us write down the formula for the nth and $(n - 1)^{st}$ terms of the sequence:

$$a_n = \frac{a_1 + a_2 + \ldots + a_{n-2} + a_{n-1}}{n - 1} \qquad (2.37)$$

$$a_{n-1} = \frac{a_1 + a_2 + \ldots + a_{n-2}}{n - 2} \qquad (2.38)$$

Using Eq. 2.38 we can find that

$$a_1 + a_2 + \ldots + a_{n-2} = (n - 2)a_{n-1} \qquad (2.39)$$

Plugging Eq. 2.39 into Eq. 2.37 we obtain $a_n \cdot (n-1) = a_{n-1} \cdot (n-2)$ $+a_{n-1} = a_{n-1} \cdot (n-1)$. Therefore,

$$a_n = a_{n-1} \text{ for any } n \geq 3 \tag{2.40}$$

1. Since $a_9 = 99$ we can rewrite Eq. 2.40 as $a_3 = a_4 = \ldots = a_9 = 99$
2. Now we can evaluate $a_2, a_3 = \frac{a_1 + a_2}{2} \Rightarrow 99 = \frac{19 + a_2}{2} \Leftrightarrow a_2 = 2 \cdot 99 - 19 = 179$.

Answer. 179

Problem 81 (MGU Entrance exam 2008). Integers x, y, z are members of a geometric progression but numbers $7x - 3$, y^2, $5z - 6$ are members of an arithmetic progression. Find x, y and z.

Solution. From the condition of the problem and with the use of geometric and arithmetic means, we have the following two equations,

$$y^2 = xz$$
$$y^2 = \frac{7x - 3 + 5z - 6}{2}$$

from which

$$zx = \frac{7x + 5z - 9}{2}$$
$$2zx = 7x + 5z - 9$$
$$x(2z - 7) = 5z - 9$$
$$x = \frac{5z - 9}{2z - 7}$$

Multiplying both sides of the last equation by 2 we obtain $2x = \frac{2 \cdot 5z - 2 \cdot 9}{2z - 7}$. Extracting the largest integer from the numerator of the last fraction we obtain $2x = \frac{5(2z-7)+17}{2z-7} = 5 + \frac{17}{2z-7}$. Since 17 is prime, then in order for $2x$ to be an integer, $(2z - 7)$ can take only the following values: $\pm 1; \pm 17$.

Consider the following cases:

a) If $2z - 7 = 1$, then $z = 4$, $x = 11$, $y = \sqrt{xz} = \sqrt{44}$, not a solution
b) If $2z - 7 = -1$, then $z = 3$, $x = -6$, $xz < 0$, not a solution
c) $2z - 7 = 17$, then $z = 12$, $x = 3$, $y = 6$ or $y = -6$
d) $2z - 7 = -17$, $z = -5$, $x = 2$, $zx < 0$, not a solution

Answer. $(x,y,z) = \{(3,6,12),(3,-6,12)\}$.

Problem 82 (Lidsky). Prove that there exists an infinite convergent geometric series $1, r, r^2, \ldots, r^n, \ldots$ each member of which divided by the sum of all terms following it, equals given number k. For what value of k does the problem have a solution?

Solution. By the condition of the problem $|r| < 1$ and we have $r^n = k(r^{n+1} + r^{n+2} + \ldots) = kr^{n+1} \cdot \frac{1}{1-r}$. From this expression $1 - r = kr$ or solving for r, $r = 1/(k+1)$. Since $|r| < 1$, then

$$\left| \frac{1}{k+1} \right| < 1$$
$$k > 0 \text{ or } k < -2$$

Answer. $k \in (-\infty, -2) \cup (0, \infty)$.

Problem 83 (AIME 2000). A sequence of numbers $x_1, x_2, \ldots, x_{100}$ has the property that, for every integer k between 1 and 100, inclusive, the number x_k is k less than the sum of the other 99 numbers. Given that $x_{50} = \frac{m}{n}$, where m and n are relatively prime positive integers, find $(m + n)$.

Solution. Because by the condition of the problem, $x_k + k = x_1 + x_2 + \ldots + x_{k-1} + x_{k+1} + \ldots + x_{100}$, then

$$x_1 + 1 = x_2 + x_3 + \ldots + x_{100}$$
$$x_2 + 2 = x_1 + x_3 + \ldots + x_{100}$$
$$\ldots$$
$$x_{50} + 50 = x_1 + x_2 + \ldots + x_{49} + x_{51} + \ldots + x_{100}$$
$$x_{51} + 51 = x_1 + x_2 + \ldots + x_{50} + x_{52} + \ldots + x_{100} \qquad (2.41)$$
$$\ldots$$
$$x_{99} + 99 = x_1 + x_2 + \ldots + x_{98} + x_{100}$$
$$x_{100} + 100 = x_1 + x_2 + \ldots + x_{99}$$

Let us add x_k to both sides of each equation, where k is the number of the equation:

$$2x_k + k = \sum_{i=1}^{100} x_i, \quad k = 1, 2, \ldots, 100. \qquad (2.42)$$

Now the right side of each equation will be the same, $\sum_{i=1}^{100} x_i$. For example, for the 50th equation we have

$$2x_{50} + 50 = \sum_{i=1}^{100} x_i \tag{2.43}$$

Method 1. (Using properties of sigma notation)
Adding the left and right sides of Eq. 2.42 for $k = 1, 2, \ldots, 100$ and after simplification, we obtain that $2\sum_{i=1}^{100} x_i + \sum_{i=1}^{100} i = 100 \sum_{i=1}^{100} x_i$, which can be simplified as follows: $\frac{100 \cdot 101}{2} = 98 \sum_{i=1}^{100} x_i$

On the other hand, replacing the sum here by the left side of Eq. 2.43 for the 50th equation, we obtain $98(2x_{50} + 50) = \frac{100 \cdot 101}{2}$. After simplification and replacement the 50th term by the formula given in the condition of the problem yields

$$50 \cdot 101 = 98\left(2 \cdot \frac{m}{n} + 50\right)$$

Solving for m/n we obtain that

$$\frac{m}{n} = \frac{75}{98}$$

Therefore, $m + n = 75 + 98 = 173$.

Method 2. (Using properties of an arithmetic sequence)
Subtracting the left and the right sides of two consecutive equations of Eq. 2.41 and then dividing both sides by 2 we obtain that $x_k - x_{k-1} = -0.5$, where $k = 2, 3, \ldots 100$. This means that the sequence $\{x_k\}$ is an arithmetic progression with the common difference $d = -0.5$. Now the 50th term of the sequence can be written as

$$\begin{aligned} x_{50} &= x_1 + 49 \cdot d \\ x_{50} &= x_1 - 0.5 \cdot 49 \end{aligned} \tag{2.44}$$

Using Eq. 1.8 for the sum of an arithmetic sequence, we rewrite Eq. 2.42 for $k = 1$ as follows:

$$\begin{aligned} 2x_1 + 1 &= \frac{2x_1 + 99d}{2} \cdot 100; \\ 2x_1 + 1 &= (2x_1 + 99 \cdot (-0.5))50; \\ x_1 &= \frac{1238}{49}. \end{aligned}$$

Substituting the value into Eq. 2.44 we obtain $x_{50} = \frac{75}{98} = \frac{m}{n}$ or $m + n = 75 + 98 = 173$.

Answer. $m/n = 75/98$ and $m + n = 173$.

Let us compare geometric and arithmetic means of two positive numbers a and b. What is greater $\frac{a+b}{2}$ or $\sqrt{ab}$? Because both, a and b are positive, we can raise both sides to the second power: $\frac{(a+b)^2}{4}$ ∧ ab. The symbol ∧ will mean "compare" for us. If an arithmetic mean is greater than a geometric mean, then $\frac{(a+b)^2}{4} - ab > 0$, Thus, $(a+b)^2 - 4ab = a^2 + 2ab + b^2 - 4ab = a^2 - 2ab + b^2 = (a-b)^2 ∧ 0$. Because $(a-b)^2 \geq 0$, then we conclude that the arithmetic mean of two positive numbers is always greater their geometric mean, and is equal to their geometric mean if and only if $a = b$,

$$\frac{a+b}{2} \geq \sqrt{ab}$$
$$a + b \geq 2\sqrt{ab}.$$

Let us solve the following problem:

Problem 84 For how many ordered pairs (x, y) of integers is it true that the arithmetic mean of x and y is exactly 2 more that the geometric mean of x and y?

Solution.

$$\frac{x+y}{2} = 2 + \sqrt{xy}$$
$$x + y = 4 + 2\sqrt{xy}$$
$$(x + y - 4)^2 = 4xy$$
$$x^2 + y^2 + 2xy - 8(x+y) + 16 = 4xy$$
$$x^2 - 2xy + y^2 = 8(x + y - 2)$$
$$(x - y)^2 = 8(x + y - 2)$$
$$(x - y)(x - y) = 2 \cdot 2 \cdot 2 \cdot (x + y - 2)$$

Because x and y are integers from last equation above, we can write only three possible systems:

1. $\begin{cases} x - y = 8 \\ x - y = x + y - 2 \end{cases}$ $\begin{bmatrix} x = 9 \\ y = 1 \end{bmatrix}$

2. $\begin{cases} x - y = 4 \\ x - y = (x + y - 2) \cdot 2 \end{cases}$ $\begin{bmatrix} x = 4 \\ y = 0 \end{bmatrix}$

3. $\begin{cases} x - y = 2 \\ x - y = 4(x + y - 2) \end{cases}$ $\begin{bmatrix} x = \dfrac{9}{4} = 2.25 \\ y = 0.25 \end{bmatrix}$

4. $\begin{cases} x - y = 1 \\ x - y = 8(x + y - 2) \end{cases}$

Systems (3) & (4) do not give us integers. So we have two possible ordered pairs:

Answer. $(9, 1)$ and $(4, 0)$.

2.5   Miscellaneous Problems on Sequences and Series

Problem 85 (Kaganov) Prove that $(a_1 + a_2 + \ldots + a_m)^2 \leq m(a_1^2 + \ldots + a_m^2)$ for any real numbers a_i

Solution. Let us prove it by mathematical induction.
1. $m = 1$. The statement is true for $m = 1$.
2. Assume this statement is true for $(a_1 + a_2 + \ldots + a_{m-1})^2 \leq (m - 1)(a_1^2 + \ldots + a_{m-1}{}^2)$. Denote the left side by α and the right side by β, $\alpha \leq \beta$.
3. Consider that

$$(a_1 + a_2 + \ldots + a_m)^2 = \alpha + a_m{}^2 + 2a_m \cdot (a_1 + a_2 + \ldots + a_m);$$
$$m(a_1{}^2 + \ldots + a_{m-1}{}^2 + a_m{}^2) = (m - 1)(a_1{}^2 + \ldots + a_{m-1}{}^2)$$
$$+ a_1{}^2 + \ldots + a_{m-1}{}^2 + ma_m{}^2$$

Since $(a_1 + a_2 + \ldots + a_{m-1})^2 \leq (m - 1)(a_1^2 + \ldots + a_{m-1}{}^2)$ is true, then

$$(a_1 + a_2 + \ldots + a_{m-1} + a_m)^2$$
$$= \alpha + a_m{}^2 + 2a_m(a_1 + \ldots + a_{m-1}) \leq \beta + \sum_{i=1}^{m-1}(a_i{}^2 + a_m{}^2) + a_m{}^2;$$

From which it follows that

$$\alpha \leq (a_1{}^2 - 2a_1 \cdot a_m + a_m{}^2) + (a_2{}^2 - 2a_2 \cdot a_m + a_m{}^2) + \ldots + (a_{m-1}{}^2 - 2a_{m-1} \cdot a_m + a_m{}^2)$$
$$+ \beta; \alpha \leq \beta + (a_1 - a_m)^2 + (a_2 - a_m)^2 + \ldots + (a_{m-1} - a_m)^2.$$

The proof is complete.

The following problem will demonstrate how the knowledge of sequences helps
us to do calculus problems.

Problem 86 Evaluate $\lim\limits_{x\to 1}\frac{x^{n+1}-1}{x^n-1}$.

Solution. This limit cannot be found directly, because when $x=1$ the denominator
becomes 0. Using formulas for geometric series and applying them for the numerator and denominator:

$$1+x+x^2+\ldots+x^n=\frac{x^{n+1}-1}{x-1}$$
$$1+x+x^2+\ldots+x^{n-1}=\frac{x^n-1}{x-1}$$

We remove discontinuity and evaluate the limit.

$$\lim_{x\to 1}\frac{1+x+x^2+\ldots+x^n}{1+x+x^2+\ldots+x^{n-1}}=\frac{(n+1)\cdot 1}{n\cdot 1}=\frac{n+1}{n},\forall n\in N.$$

Answer. $\lim\limits_{x\to 1}\frac{x^{n+1}-1}{x^n-1}=\frac{n+1}{n}$.

Problem 87 (Rivkin) Given
$$1+a+a^2+\ldots+a^n=(1+a)(1+a^2)(1+a^4)\cdots\left(1+a^{2^k}\right).$$
Find relationship between n and k.

Solution. The left side of the formula can be rewritten as $\frac{a^{n+1}-1}{a-1}$. Multiplying the
both sides by $(a-1)\neq 0$ and because $a\neq 1$, the given relation can be rewritten as
$a^{n+1}-1=(a-1)(1+a)(1+a^2)\ldots(1+a^{2^n})$. Next, using a difference of
squares formula applied several (k) times, the right side can be simplified as
$(a^2-1)(1+a^2)\cdots\left(1+a^{2^k}\right)=(a^4-1)\ldots\left(1+a^{2^k}\right)=a^{2^{k+1}}-1$. Therefore,
$a^{n+1}=a^{2^{k+1}}$. Because by the condition of the problem $a\neq 0,\pm 1$, then the necessary relationship between n and k is $n+1=2^{k+1}$.

Answer. $n=2^{k+1}-1$.

Problem 88 Given a function $g(n) = \frac{f(n+1)}{2-(f(0)+f(1)+f(2)+...+f(n))}$, where $f(x) = \frac{x}{2^x}$. Evaluate $A(m) = \frac{g(m+1)}{g(m)}$ and $B(m) = g(m+1) - g(m)$.

Solution. Let us evaluate several values of function,

$$f(x): \ f(0) = 0, f(1) = \frac{1}{2^1}, \ f(2) = \frac{2}{2^2}, \ f(3) = \frac{3}{2^3}, \ \dots \ ,$$

$$f(n) = \frac{n}{2^n}, \ f(n+1) = \frac{n+1}{2^{n+1}}.$$

Substituting this into formula for $g(n)$ we obtain the following

$$g(n) = \frac{\frac{n+1}{2^{n+1}}}{2 - \left(\frac{1}{2^1} + \frac{2}{2^2} + \frac{3}{2^3} + \frac{4}{2^4} + \dots + \frac{n-1}{2^{n-1}} + \frac{n}{2^n}\right)}. \qquad (2.45)$$

Next, we simplify the sum inside parentheses, by denoting it $S_n = \frac{1}{2^1} + \frac{2}{2^2} + \frac{3}{2^3} + \frac{4}{2^4} + \dots + \frac{n-1}{2^{n-1}} + \frac{n}{2^n}$. Multiplying both sides of the equality by 2 we get $2 \cdot S_n = 1 + \frac{2}{2^1} + \frac{3}{2^2} + \frac{4}{2^3} + \frac{5}{2^4} + \dots + \frac{n-1}{2^{n-2}} + \frac{n}{2^{n-1}}$. Subtracting the left and the right sides of two equations and canceling the same terms, we have

$$S_n = 1 - \frac{n}{2^n} + \frac{1}{2} \cdot \frac{\left(1 - \left(\frac{1}{2}\right)^{n-1}\right)}{\left(1 - \frac{1}{2}\right)}$$

$$= 2 - \frac{n}{2^n} - \frac{1}{2^n} \qquad (2.46)$$

Substituting Eq. 2.46 into Eq. 2.45, we have

$$g(n) = \frac{\frac{n+1}{2^{n+1}}}{2 - \left(2 - \frac{n}{2^n} - \frac{1}{2^{n-1}}\right)} = \frac{(n+1) \cdot 2^n}{2 \cdot 2^n(n+2)}$$

$$= \frac{n+1}{2n+4}$$

Finally,

$$\frac{g(m+1)}{g(m)} = \frac{(m+2)^2}{(m+1)(m+3)}$$

$$g(m+1) - g(m) = \frac{m+2}{2(m+3)} - \frac{m+1}{2(m+2)} = \frac{1}{2(m+2)(m+3)}.$$

Answer. $A(m) = \frac{(m+2)^2}{(m+1)(m+3)}$; $B(m) = \frac{1}{2(m+2)(m+3)}$.

Problem 89 Find all value of r such that all partial sums of the series $\frac{1}{2} + r \cos x + r^2 \cos 2x + r^3 \cos 4x + r^4 \cos 8x + \ldots$ are nonnegative for all real x.

Solution. Consider the second partial sum, $S_2 = \frac{1}{2} + r \cos x \geq 0 \Rightarrow |r| \leq \frac{1}{2}$. Denote

$$\boxed{\psi(y) = r \cos y + r^2 \cos 2y} \Rightarrow$$

$$\psi(2y) = r \cos 2y + r^2 \cos 4y \Rightarrow$$

$$\boxed{\psi(4y) = r \cos 4y + r^2 \cos 8y}$$

$$\psi(8y) = r \cos 8y + r^2 \cos 16y$$

$$\boxed{\psi(16y) = r \cos 16y + r^2 \cos 32y}$$

We can see that the given series can be rewritten as

$$\frac{1}{2} + \left(r \cos x + r^2 \cos 2x\right) + r^2 \left(r \cos 4x + r^2 \cos 8x\right)$$
$$+ r^4 (r \cos 16x + r^2 \cos 32x) + \ldots \tag{2.47}$$
$$= \frac{1}{2} + \psi(x) + r^2 \psi(4x) + r^4 \psi(8x) + \ldots$$

Let us investigate the behavior of $\psi(y)$. Taking the first derivative of it, we obtain that

$$\frac{d\psi}{dy} = -r \sin y - 4r^2 \sin y \cos y$$

$$= -r \sin y (1 + 4r \cos y) = 0$$

$$\frac{d\psi}{dy} = 0 \Leftrightarrow y = \pi n \text{ or } \cos y = -\frac{1}{4r}$$

Case 1. $y = \pi n \Rightarrow \psi(y) = \psi(\pi n) = r \cos \pi n + r^2 2\pi n = (-1)^n r + r^2 \geq -\frac{1}{4}$.

Case 2. $\psi(y) = -\frac{1}{4} + r^2 \left(\frac{1}{8r^2} - 1\right) \geq -\frac{3}{8}$.

Hence we can state that $\psi(y) \geq -\frac{3}{8}$ $\forall y \in \mathbb{R}$. Series (Eq. 2.47) are bounded as follows:

$$S_{2n+1} = \frac{1}{2} + \psi(x) + r^2\psi(4x) + r^4\psi(8x) + \ldots + r^{2(n-1)}\psi\left(4^{n-1}x\right)$$

$$\geq \frac{1}{2} - \frac{3}{8}\left(1 + r^2 + r^4 + \ldots + r^{2(n-1)}\right)$$

$$\geq \frac{1}{2} - \frac{3}{8}\left(1 + \frac{1}{4} + \left(\frac{1}{4}\right)^2 + \ldots + \left(\frac{1}{4}\right)^{n-1}\right)$$

$$= \frac{1}{2} - \frac{3}{8}\left(\frac{1 - \left(\frac{1}{4}\right)^{n-1}}{1 - \frac{1}{4}}\right) = \frac{1}{2 \cdot 4^n} = \frac{1}{2^{2n+1}}$$

Let us find the next partial sum, $S_{2n+2} = S_{2n+1} + r^{2n+1} \cdot \cos\left(2^{2n}x\right) \geq S_{2n+1}$ $-\frac{1}{2^{2n+1}} \geq 0$. Finally, we can conclude that if $|r| \leq \frac{1}{2}$, then all partial sums of the series of Eq. 2.47 are nonnegative.

Problem 90 Given a sequence $S_1 = \sqrt{2}$, $S_{n+1} = \sqrt{2 + S_n}$, prove that this sequence has a limit. Evaluate it.

Proof. Assume that the sequence has a limit S, then $S = \sqrt{S+2} \Rightarrow S^2 = S + 2$ $\Rightarrow S = -1$ or $S = 2$

Answer. 2.

The following problems will make connection between sequences, number theory and geometry.

Problem 91 A side of a square is a. The midpoints of its sides are joined to form an inscribed square. This process is continued as shown in the diagram. Find the sum of the perimeters of the squares if the process is continued without end.

Solution. From the diagram (Figure 2.4), we can see that the sides of the black squares form a geometric progression with the first term of a and common ratio $\frac{1}{2}$: $a, \frac{a}{2}, \frac{a}{4}, \frac{a}{8}, \ldots \frac{a}{2^{n-1}}$. All red squares, in turn, form a geometric progression with the same common ratio but the first term $\frac{a\sqrt{2}}{2}$ (half of the diagonal of the original square): $\frac{a\sqrt{2}}{2}, \frac{a\sqrt{2}}{4}, \frac{a\sqrt{2}}{8}, \ldots$. Because the perimeter of a square with side b is $4b$, we obtain the following expression for the sum of the perimeters of black and red squares:

Figure 2.4 Sketch for
Prob. 91

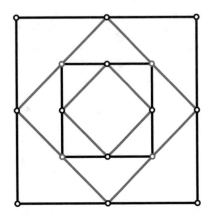

$$P = 4\left(a + \frac{a\sqrt{2}}{2} + a/2 + \frac{a\sqrt{2}}{4} + a/4 + \frac{a\sqrt{2}}{8} + a/8 + \ldots + \frac{a}{2^{n-1}} + \frac{a\sqrt{2}}{2^{n-1}} + \cdots\right)$$

$$= 4a(1 + 1/2 + 1/4 + \ldots) + 4a \cdot \frac{\sqrt{2}}{2}(1 + 1/2 + 1/4 + \ldots)$$

$$= 4a\left(2 + \sqrt{2}\right)\left(1 - \frac{1}{2^n}\right) = 4a\left(2 + \sqrt{2}\right)$$

Answer. $P = 4a\left(2 + \sqrt{2}\right)$

Problem 92 Given a sequence $a_0 = 2$, $a_1 = 5$, $a_n = 5a_{n-1} - 4a_{n-2}$ for $n \geq 2$. Show that $a_n \cdot a_{n+2} - a_{n+1}^2$ is a perfect square for every $n \geq 0$.

Proof. The characteristic polynomial for this recurrent sequence is $r^2 - 5r + 4 = (r-1)(r-4)$, then the general term of the sequence is $a_n = A \cdot 4^n + B \cdot 1^n$. Using the values of the first two terms, the nth term can be written as $a_n = 4^n + 1$. Evaluate

$$a_n \cdot a_{n+2} - a_{n+1}^2 = (4^n + 1)(4^{n+2} + 1) - (4^{n+1} + 1)^2$$
$$= 4^n \cdot 9 = (3 \cdot 2^n)^2 = k^2.$$

The proof is complete.

Problem 93 Prove that there is no infinite arithmetic progression of only prime numbers.

Proof. Consider an arithmetic progression with the first term $a \neq 1$ and common difference d. Then the nth term of this progression can be written as $a_n = a + (n-1)d$. Clearly, if $n = a + 1 \Rightarrow a_n = a + a \cdot d = a(d+1)$. Thus, the first and $(a+1)$ st term, a_{a+1}, of such arithmetic progression are not relatively primes, and this fact does not depend on the value of the common difference. Moreover, in such infinite progression all terms sitting in the positions of $n = a + 1, 2a + 1, 3a + 1, 4a + 1, \ldots$ will be multiples of the first term, a.

For example, in the progression $\underline{3}, 7, 11, \underline{15}, 19, 23, \underline{27}, 31, 35, \underline{39}, \ldots$ there are infinitely many members divisible by 3, we underlined some of them . All of them are in the positions 4, 7, 10, $\ldots (3k + 1), \ldots$. Because our proof was based on an assumption that the first term of a progression is not unit, a reasonable question is what if the first term of an infinite progression equals 1? Can such progression consist of only primes?

The answer is also "no" and the proof of this fact is very similar to the proof above. We just for any given progression start our arguments from the second term. Thus, infinite arithmetic progression $\{a_n\} : 1, 1 + d, 1 + 2d, 1 + 3d, \ldots$ contains progression $\{b_n\} : 1 + d, 1 + 2d, 1 + 3d, \ldots$ the first term of which equals the second term of the first progression, and then again prove that there are infinitely many terms divisible by $(1 + d)$.

Remark. Any infinite arithmetic progression with natural members will have infinitely many multiples of the first, second, third, or any other term and the location of such multiples will depend only on the value of the selected term of a progression. Suppose a number $b \in \mathbb{N}$ is a term of an infinite arithmetic progression, then there are infinitely numbers of terms divisible by b in the relative location $n = b + 1, 2b + 1, 3b + 1, \ldots$. Thus if b is the kth term of the given progression, then all terms divisible by it will have positions of $k, k + b, k + 2b, k + 3b, \ldots$.

For example, since 11 is the third term of the given infinite progression, $3, 7, 11, 15, 19, 23, 27, 31, 35, 39, 43, 47, 51, 55, 59, 63, \ldots$, the terms divisible by 11 will appear at the positions $3, 3 + 11 = 14, 3 + 2 \cdot 11 = 25, 3 + 3 \cdot 11 = 36$, $\ldots, 3 + (m-1) \cdot 11, \ldots$, where m represent the mth consecutive multiple of 11. You can see it yourself, 11 is the first multiple of 11, the second is 55, which is 14th term of the given progression, then the third consecutive multiple of 11 in the progression will correspond to the index $n = 25$ and will be evaluated as $3 + (25 - 1) \cdot 4 = 99$, etc.

We just proved that there is no infinite arithmetic progression that consists of only primes. Is this statement also true for a finite arithmetic progression? The shortest sequence of primes must contain three terms. We can see that the first three terms of the infinite progression discussed above, $\{3, 7, \text{and } 11\}$ are in arithmetic progression given by formula $a_n = 4n - 1$, $n = 1, 2, 3$.

Are there arithmetic progressions with precisely 5, 10 or N prime numbers? The answer is yes, such progressions exist but it is hard to find them.

The previous problem probably gives you some ideas of how to look for such progressions. First, we must select only even numbers as common difference, d. Otherwise, even and odd terms would alternate, which would never be a finite arithmetic progression with only prime terms. Obviously, the first term must be an odd prime. The following theorem formulated and proven by Cantor will help to get us started.

Cantor's Theorem. If N terms of an arithmetic progression are odd primes, then the difference of the progression is divisible by every prime less than N.

The rigorous proof of the existence of an arithmetic sequence with exactly N prime terms was given in 2004 by B. Green and T. Tao. However, their proof does not propose any algorithm of finding such progressions or makes the job of finding it any easy. It is worth to mention that the last longest arithmetic progression of 26 prime numbers was discovered only in 2010.

Here we try to find an arithmetic progression of ten prime terms by solving the following problem.

Problem 94 Propose a finite arithmetic progression formed by ten prime numbers.

Solution. Regarding Cantor's Theorem, the common difference of such progression must be divisible by 2, 3, 5, and 7 (all prime numbers less than $n = 10$). The minimal common difference satisfying this conditions is $d = 2 \cdot 3 \cdot 5 \cdot 7 = 210$. Next, we need to find the starting prime, the first term of the progression. It cannot be 11, because $11 + 210 = 221 = 13 \cdot 17$ is not prime.

Can it be 13? The answer is no because 210 divided by 11 leaves a remainder of 1, $210 = 11 \cdot 19 + 1$. Then the remainder of a term when divided by 11 will increase by one each time as n increases. For example, if the starting prime is 13 which give a remainder of 2 divided by 11, then the nth term has the following form,

$$a_n = 13 + 210 \cdot (n-1) = 11 + 2 + 11 \cdot 19 \cdot (n-1) + n - 1$$
$$= 11k + n + 1 = 11m$$

We can see that if $n = 10$, then the tenth term will be 1903 that divisible by 11 and not prime.

If a starting prime divided by 11 will leave a different remainder, for example, 3,4,5, etc. then a multiple of 11 will be obtained faster each time. Try yourself to select the first term as next prime, 17. Because $17 = 11 + 6$, then the 5th term of the proposed progression will be a multiple of 11...

$$a_n = 17 + 210 \cdot (n-1) = 11 + 11 \cdot 19 \cdot (n-1) + (6+n-1), \quad a_5 = 858$$
$$= 11 \cdot 78.$$

Therefore, the first term must be odd and leave a remainder of one when divided by 11. Let us try $a_1 = 22m + 1$. Consecutive candidates are 23, 67, 89, 199, ... If we try with the first term 23, 67 and 89, we obtain that such progression would have a composite number for the sixth, fourth and second term, respectively. If we set the first term equals 199, then we obtain an arithmetic progression of ten prime numbers, 199, 409, 619, 829, 1039, 1249, 1459, 1669, 1879, and 2089.

Answer. $a_n = 199 + (n-1)210 = 210n - 11, \ 1 \le n \le 10$.

Let us find out other properties of finite arithmetic progressions in integers by solving the following problems.

Problem 95 Is there any arithmetic progression of 50 terms such that any two selected terms are relatively primes? If such progression exists, find it.

Solution. Let us consider an arithmetic progression with the first term of $a_1 = 1 + 49!$ and common difference $d = 49!$. Its nth term can be written as $a_n = 1 + 49! + (n-1)49! = 1 + n \cdot 49!, \ 1 \le n \le 50$.

We can see that any term of the progression is not divisible by any natural number from 1 to 50 because when divided by any such number it gives a remainder of 1. Next, the difference between its kth and mth terms will be $a_k - a_m = (k-m) \cdot 49!$, which means that the difference of any two terms is not divisible by any prime greater than 49. For example, $a_7 - a_4 = 3 \cdot 49! = 3 \cdot 1 \cdot 2 \cdot 3 \cdot 4 \cdot \ldots \cdot 49$. This proves that any two selected terms of the arithmetic progression are relatively prime, because the only common factor they have is one.

Problem 96 Is there an arithmetic progression formed of positive integers such that no term of the progression can be represented as a sum or difference of two primes? If such progression exists, then give an example.

Solution. Consider several arithmetic progressions with the corresponding nth terms ($n \in N$):
a. 6, 10, 14, 18, ..., $4n+2$, ...
b. 11, 19, 27, 35, ..., $8n+3$, ...
c. 47, 89, 131, 173, ..., $42n+5$, ...
d. 37, 67, 97, 127, ..., $30n+7$, ...

The proression of "a" is represented by only even numbers, and clearly many its terms can be written as the sum or difference of two primes, for example,

$10 = 3 + 7$, $14 = 17 - 3$. Any even number is either the sum or difference of two even or two odd numbers; in general prime numbers are odd. One can prove that there exist infinitely many even numbers that can be written both as sums and as difference of two primes. Hence, we further consider only sequences of odd numbers.

Additionally, let us eliminate a sequence of odd numbers b) because its terms $19 = 2 + 17$ and $27 = 29 - 2$. If such arithmetic progression exists, then its terms must be odd numbers. Next, if some of its terms can be represented by a sum or difference of two primes, then one of the primes must be even (2).

Consider progression "c" and assume that one of its terms can be written as sum of two primes:

$$42n + 5 = p_1 + p_2$$
$$42n + 5 = 2 + p_2$$
$$42n + 3 = p_2$$
$$p_2 = 3 \cdot (14n + 1).$$

We can see that p_2 is not prime.

Assume that some term of the progression above can be written as a difference of two primes, i.e., $42n + 5 = p_1 - p_2$. Because each term of the progression is odd, then the second prime must be 2.

$$42n + 5 = p_1 - 2$$
$$42n + 7 = p_1$$
$$p_1 = 7 \cdot (6n + 1)$$

Therefore, no terms of an arithmetic progression $a_n = 42n + 5$ can be represented as sum or difference of two primes. A similar conclusion can be made for progression "d". We leave it for you as a homework (exercise 89).

Problem 97 Find all right triangles with integer sides forming consecutive terms of an arithmetic progression.

Solution. Assume that such triangle exists and that its sides are $a = a$, $b = a + d$, $c = a + 2d$. then they must satisfy Pythagorean Theorem:

$$a^2 + (a + d)^2 = (a + 2d)^2$$
$$a^2 + a^2 + 2ad + d^2 = a^2 + 4ad + 4d^2$$
$$a^2 = 3d^2 + 2ad$$
$$(a - d)^2 = (2d)^2$$
$$a - d = 2d$$
$$a = 3d, \ b = 4d, \ c = 5d, \ d \in \mathbb{N}.$$

Therefore, there are infinitely many such right triangles. For example, the sides of the following right triangles form an arithmetic progression and are Pythagorean triples: (3, 4, 5), (6, 8, 10), (9, 12, 15), (12, 16, 20)...

Answer. $(a, b, c) = (3d, 4d, 5d)$, $d \in \mathbb{N}$.

Problem 98   It is known that the numbers $x(x+1)$, $y(y+1)$, $z(z+1)$ are in increasing arithmetic progression. Find integer numbers x, y and z.

Solution. Assume that such numbers exist and that
$$\begin{cases} x = x \\ y = ax + b, \text{ where integer coefficients } a, b, c, d \text{ are to be determined.} \\ z = cx + d \end{cases}$$
Because $x(x+1)$, $y(y+1)$, $z(z+1)$ form an arithmetic progression, then $y(y+1) - x(x+1) = z(z+1) - y(y+1)$. Substituting here the expressions from the system above, we obtain the following chain of true equalities:

$$(ax+b)(ax+b+1) - x(x+1) = (cx+d)(cx+d+1) - (ax+b)(ax+b+1)$$
$$2(a^2x^2 + 2abx + b^2) + 2ax + 3b - x^2 - x = c^2x^2 + 2cdx + d^2 + cx + d$$

By equating the constant terms, the coefficients of linear and quadratic terms, respectively, we obtain the system of three equations in four undetermined integer parameters:

$$\begin{cases} 2b(b+1) = d(d+1) \\ 2a + 4ab - 1 = 2cd + c \\ 2a^2 - 1 = c^2 \end{cases}$$

Consider the last equation of the system, $1 + c^2 = 2a^2$. In order to have any solutions in integers, we know that parameter c must be an odd number, then $c = 2n + 1$. Substituting this back into the equation we have

$$1 + (2n+1)^2 = 2a^2$$
$$1 + 4a^2 + 4a + 1 = 2a^2$$
$$2n^2 + 2n + 1 = a^2$$
$$2n^2 + 2n = a^2 - 1$$
$$2n(n+1) = (a-1)(a+1)$$

The right hand side is represented by the product of two numbers that differ by 2, hence they either both odd or both even. Because the left side is even then $(a-1)$ and $(a+1)$ must be even, for example, $a-1 = 2m$, $a+1 = 2m+2$. Substituting this into the discussed equation, we obtain

$$2n(n + 1) = 2m(2m + 2)$$
$$n(n + 1) = 2m(m + 1).$$

The last equation has solution only if its variables satisfy the system

$$\begin{cases} n = m + 1 \\ n + 1 = 2m \end{cases} \Rightarrow m + 2 = 2m \Rightarrow a = 2m + 1 = 2 \cdot 2 + 1 = 5.$$

Knowing a, we can evaluate the corresponding positive c, $1 + c^2 = 2 \cdot 5^2 = 50$ $\Rightarrow c^2 = 49$, $c = 7$. Similarly to the solution of the underlined equation above, we can find positive solution to the first equation of the system.

$$d(d + 1) = 2b(b + 1)$$
$$\begin{cases} d = b + 1 \\ 2b = d + 1 \end{cases} \Rightarrow b = 2, \ d = 3.$$

Note that we found all four parameters using only solutions of the first and the last equations. This is very typical when solving equations in integers. The second equation can be used for checking. Thus, $1 + 2 \cdot 7 \cdot 3 + 7 = 4 \cdot 5 \cdot 2 + 2 \cdot 5 = 50$. Finally, we found that if $x = x$, $y = 5x + 2$, $z = 7x + 3$, then $x(x + 1) = x^2 + x$, $y(y + 1) = 25x^2 + 25x + 6$, $z(z + 1) = 49x^2 + 49x + 12$ are in the increasing arithmetic progression with common difference $24x^2 + 24x + 6$.

Answer. $x = x$, $y = 5x + 2$, $z = 7x + 3$, $x \in \mathbb{N}$.

Problem 99 A sequence is defined by $a_n = \frac{1}{n+n^2}$, $n \geq 1$. Given $a_m + a_{m+1}$ $+ \ldots + a_{n-1} = \frac{1}{17}$, $m < n$, evaluate $n - m$.

Solution. Factoring the denominator of the nth term, we notice that it can be written as $a_n = \frac{1}{n+n^2} = \frac{1}{n(n+1)} = \frac{1}{n} - \frac{1}{n+1}$. Replacing each term on the left of the given condition, we have $\frac{1}{m} - \frac{1}{m+1} + \frac{1}{m+1} - \frac{1}{m+2} + \ldots + \frac{1}{n} - \frac{1}{n+1} = \frac{1}{17}$. After cancellation of the opposite terms we obtain

$$\frac{1}{m} - \frac{1}{n} = \frac{n - m}{mn} = \frac{1}{17}$$
$$17(n - m) = mn$$

The last equation must be solved in integers and can be written as $17m = n(17 - m)$. Because from the condition of the problem $17 - m < 17$, and 17 is prime, we know that n must be a multiple of 17. Let $n = 17k$. After substitution we have

$$17m = 17k(17 - m)$$
$$m = k(17 - m).$$

This equation has integer solutions if and only if the second factor on the right hand side equals one, i.e.

$$17 - m = 1$$
$$m = 16, \ k = 17, \ n = 16 \cdot 17.$$
$$n - m = 16 \cdot 17 - 16 = 16^2 = 256.$$

Answer. $n - m = 256$.

Problem 100 Given a sequence
$u_1 = 2, \ u_2 = 8, \ \ldots, \ u_n = 4u_{n-1} - u_{n-2}, \ n = 3, 4, 5, \ \ldots,$ Prove   that
$u_n^2 - u_{n+1} \cdot u_{n-1} = 4$.

Proof. We   can   evaluate   some   terms   of   the   recurrence   as
$u_3 = 4u_2 - u_1 = 4 \cdot 8 - 2 = 30,$ $u_4 = 4u_3 - u_2 = 4 \cdot 30 - 8 = 112$. It is clear
that $u_4^2 - u_3 u_2 = 30^2 - 112 \cdot 8 = 4$. Because $ab = ba$, $u_n \cdot 4u_{n-1} = u_{n-1} \cdot 4u_n$.
Using the recurrent relationship for the left and right hand sides, we obtain the
following chain of true equations:

$$u_n \cdot (u_n + u_{n-2}) = u_{n-1}(u_{n+1} + u_{n-1})$$
$$u_n^2 - u_{n+1}u_{n-1} = u_{n-1}^2 - u_n u_{n-2} = u_{n-2}^2 - u_{n-1}u_{n-3} = \ldots$$
$$= u_2^2 - u_3 u_1 = 8^2 - 30 \cdot 2 = 4.$$

The proof is complete.

Chapter 3
Series Convergence Theorems and Applications

Usually convergence of series is taught in mathematical analysis. If we can find a partial sum for a given numerical series, then we take the limit of this sum to infinity and can easily decide on the series convergence. If series is functional, for example, power series or trigonometric series, then finding their partial sum would also benefit further investigation, but the question is not whether or not it converges but for which value of x it converges. In the previous two chapters we discuss several methods of finding partial and infinite sums of series. In this chapter, we look at series from a different angle and you will learn new methods of finding finite or infinite sums for a given numerical or functional series. As you would learn in the previous chapters, finding a formula for a partial sum is usually a challenging task and sometimes it is simply impossible.

Let us consider the following infinite series:

$$a)\ 1 + 2 + 3 + \ldots + n + \ldots$$

$$b)\ \frac{1}{2} + \frac{1}{6} + \frac{1}{12} + \ldots + \frac{1}{n(n+1)} + \ldots$$

$$c)\ 1 + \frac{1}{2} + \frac{1}{3} + \ldots + \frac{1}{n} + \ldots$$

$$d)\ 1 - \frac{1}{2} + \frac{1}{3} - \frac{1}{4} + \ldots + (-1)^{n+1} \cdot \frac{1}{n} + \ldots$$

We can evaluate the nth partial sum for the first two series and make a conclusion that the first series diverges and the second one converges to 1 (Problem 50). It is hard to find n^{th} partial sum for the last two series; but in this chapter, you will learn that series c) is called a harmonic series and that it diverges and that series d) is called a Leibniz alternating series and it converges to ln2. While it is almost obvious for the first series to diverge even without calculating its partial sum, because each term of the series is increasing without bound as n is increasing, so as partial sums, the convergent and divergent behavior of the second and third series, respectively,

© Springer International Publishing Switzerland 2016
E. Grigorieva, *Methods of Solving Sequence and Series Problems*,
DOI 10.1007/978-3-319-45686-7_3

cannot be explained only by the behavior of their n^{th} term. Thus, for both the second and third series, the nth term will decrease as n increases, but only the second series converges.

It looks that in order to converge, it is necessary but not sufficient that the limit of the nth term of an infinite series would approach zero. Thus, series b) converges to 1 and series d) converges to ln2. On one hand, in order to find the infinite sum for b) we do not need any additional knowledge, just the ability to evaluate the nth partial sum. On the other hand, for the last series, the infinite sum can be found using power series. This is why in this chapter you not only review and learn the necessary and sufficient convergence theorems for numerical series but additionally you are introduced convergence of functional series, such as power series and trigonometric series.

Other methods of finding infinite sums for numerical series are taught, such as method of power series, method of integration, and differentiation and Abel's method. Additionally, we demonstrate how convergence (divergence) of some numerical series can be established using Maclaurin or Taylor expansions of some known functions and vice versa.

You will learn that an infinite sum of conditionally convergent alternating series depends on the order of terms, while the sum of absolutely convergent series does not depend on how we group and add its terms. Application of power series to solving differential equations, finding integrals or for approximation is also discussed. You will understand why this or that series are better for estimation of a given irrational number and that series convergence can be fast and slow. This is why, for example, the famous alternating series $1 - 1/3 + 1/5 - 1/7 + \ldots$ converges to $\frac{\pi}{4}$ but is never used for approximation of π. Moreover, this chapter introduces you to generating functions and how generating functions can be used to find the nth term of a sequence given by recursion. Hence this chapter will be useful for students studying calculus and their teachers.

Since many contest problems require knowledge of the topics mentioned above, especially when you have to decide whether a series will have a limit to infinity, we summarize important facts from mathematical analysis that will help you to handle some challenging problems.

3.1 Numerical Series

Consider the series $\sum_{n=1}^{\infty} u_n = u_1 + u_2 + u_3 + \ldots$ with its numerical terms and the associated sequence of partial sums $\{S_n\}$. Expressions of the type $u_{n+1} + u_{n+2} + \ldots = \sum_{k=n+1}^{\infty} u_k$, representing numerical series are called the nth remainder of the series $\sum_{k=1}^{\infty} u_k$ and are denoted by $r_n = \sum_{k=1}^{\infty} u_{n+k}$ or $r_n = \sum_{n=k+1}^{\infty} u_n$.

For any convergent series, $S = \sum_{k=1}^{\infty} u_k = \sum_{k=1}^{n} u_k + \sum_{k=n+1}^{\infty} u_k = S_n + r_n$. Here S_n is the nth partial sum of the series. In order series $\sum_{k=1}^{\infty} u_k$ to be convergent, it is necessary and sufficient that any its remainder r_n be convergent. Obviously, if numerical series $\sum_{k=1}^{\infty} u_k$ converges, the limit of partial sums exists, i.e., $\lim_{n \to \infty} S_n = S$, then $\lim_{n \to \infty} r_n = 0$. Thus, removing any finite number of terms does not influence the series convergence.

We say that the series $\sum_{n=1}^{\infty} u_n$ is convergent if and only if, a sequence $\{S_n\}$ is convergent. The total sum of the series is the limit of the sequence $\{S_n\}$, which we denote by $\lim_{n \to \infty} S_n = \sum_{n=1}^{\infty} u_n$. Hence, the series convergence is related to the convergence of a sequence. Many make mistakes by confusing the convergence of a sequence of partial sums with the convergence of the sequence of numbers. There are some steps you need to take:

1. For a series to be convergent, first, it must pass a necessary condition. The limit of the nth term must go to zero as n goes to infinity. Otherwise, the series diverges.
2. If the limit of the nth term is zero so the necessary condition is fulfilled, then the series might converge, but not necessarily, and further investigation is important.

3.1.1 Necessary and Sufficient Convergence Conditions

Theorem 3.1 (Necessary condition for convergence of numerical series) If the series $\sum_{k=1}^{\infty} u_k$ converges, then the limit of its common term at infinity equals zero, i.e., $\lim_{k \to \infty} u_k = 0$.

The following statement is also true.

Corollary 3.1 If series $\sum_{n=1}^{\infty} u_n$ is convergent, then $\lim_{n \to \infty} u_n = 0$. If $\lim_{n \to \infty} u_n \neq 0$ or if this limit is undefined, then the series cannot converge, and it diverges.

Example. Consider the following series: $\sum\limits_{n=1}^{\infty} \frac{5n^2-3n}{4-2n^2}$. By checking that $\lim\limits_{n\to\infty} u_n = \frac{5n^2-3n}{4-2n^2} = -\frac{5}{2} \neq 0$, we can be sure that the series is divergent.

However, even if the necessary condition is fulfilled, it is not sufficient for us to know that the series is convergent. We must further investigate using one of a number of different theorems and principles.

Example. The well-known harmonic series $1 + \frac{1}{2} + \frac{1}{3} + \frac{1}{4} + \ldots + \frac{1}{n} + \ldots$ is divergent, even though it passes the necessary condition: $\lim\limits_{n\to\infty} u_n = \lim\limits_{n\to\infty} \frac{1}{n} = 0$.

3.1.2   Nonnegative Numerical Series

The series $\sum\limits_{k=1}^{\infty} a_k$, $a_k \geq 0$ $\forall k \in \mathbb{N}$, are called nonnegative numerical series. For nonnegative series, the following properties are valid:

- Exchanging the terms, removal or adding of the finite number of terms of the series does not influence its convergence or divergence;
- If series $\sum\limits_{k=1}^{\infty} a_k$ and $\sum\limits_{k=1}^{\infty} b_k$ converge and their sums equal S_a and S_b respectively,

  then the series $\sum\limits_{k=1}^{\infty} (a_k + b_k)$ also converges and $\sum\limits_{k=1}^{\infty} (a_k + b_k) = S_a + S_b$.

- Series $\sum\limits_{k=1}^{\infty} (a_k + b_k)$ is called the sum of the series $\sum\limits_{k=1}^{\infty} a_k$ and $\sum\limits_{k=1}^{\infty} b_k$;
- If series $\sum\limits_{k=1}^{\infty} a_k$ converges and its sum equals S, then $\sum\limits_{k=1}^{\infty} \alpha \cdot a_k$ also converges and

  $\sum\limits_{k=1}^{\infty} \alpha \cdot a_k = \alpha \cdot S$. Series $\alpha \sum\limits_{k=1}^{\infty} a_k$ is called the product of series $\sum\limits_{k=1}^{\infty} a_k$ by a real number α;

- If series $\sum\limits_{k=1}^{\infty} a_k$ converges, then any series obtained by grouping of its terms without changing their order also converges and has the same sum as the original series.

Let us apply these rules by considering the following problem now.

Problem 101 Is the series $\sum\limits_{n=1}^{\infty} \frac{2^n+5^n}{10^n}$ convergent or divergent? Find its sum if exists.

Solution. Because the nth term of the series can be written as $a_n = \frac{1}{5^n} + \frac{1}{2^n}$, then the given series is the sum of two convergent infinite geometric series that is convergent. The sum can be found as $\frac{\frac{1}{5}}{\left(1 - \frac{1}{5}\right)} + \frac{\frac{1}{2}}{\left(1 - \frac{1}{2}\right)} = \frac{1}{4} + 1 = \frac{5}{4}$.

Answer. $\sum_{n=1}^{\infty} \frac{2^n + 5^n}{10^n} = \frac{5}{4}$.

The following theorem was stated and proved by Cauchy (Baron Augustine-Louis Cauchy, 1789–1857, French mathematician) and this theorem allows us to establish whether a series converges or diverges.

Theorem 3.2 (Cauchy Necessary and Sufficient Series Convergence Theorem) In order the series $\sum_{n=1}^{\infty} u_n$ to be convergent, it is necessary and sufficient that for any positive $\varepsilon > 0$, there exists a number $N = N(\varepsilon)$ such that for any $\forall n > N$, $\forall p \in \mathbb{N}$ the following inequality is valid: $|u_{n+1} + u_{n+2} + \ldots + u_{n+p}| < \varepsilon$. This last sum of the terms of the series is called the segment of length p.

Consider the harmonic series: $1 + \frac{1}{2} + \frac{1}{3} + \frac{1}{4} + \ldots + \frac{1}{n} + \ldots = \sum_{n=1}^{\infty} \frac{1}{n}$. Each term, c, is the **harmonic mean** of its neighbors, a and b because $c = \frac{2ab}{a+b}$. You can see that $c = \frac{1}{3}$ is between $a = \frac{1}{2}$ and $b = \frac{1}{4}$ and the following is true $\frac{1}{3} = \frac{2 \cdot \frac{1}{2} \cdot \frac{1}{4}}{\frac{1}{2} + \frac{1}{4}}$.

Let us prove that the series is divergent. Consider a segment of length p, such that $n = p$

$$|u_{n+1} + u_{n+2} + \ldots + u_{n+p}| = \frac{1}{n+1} + \frac{1}{n+2} + \frac{1}{n+3} + \ldots + \frac{1}{n+p-1} + \frac{1}{n+p}$$

$$= \frac{1}{n+1} + \frac{1}{n+2} + \frac{1}{n+3} + \ldots + \frac{1}{n+n-1} + \frac{1}{n+n}$$

It is obvious that $\frac{1}{n+k} > \frac{1}{n+n}$, $k = 1, 2, 3, \ldots n - 1$. Therefore, $\frac{1}{n+1} + \frac{1}{n+2} + \frac{1}{n+3} + \ldots + \frac{1}{n+n-1} + \frac{1}{n+n} > \frac{1}{n+n} + \frac{1}{n+n} + \frac{1}{n+n} + \ldots + \frac{1}{n+n} = n \cdot \frac{1}{2n} = \frac{1}{2}$ so $|u_{n+1} + u_{n+2} + \ldots + u_{n+p}| > \frac{1}{2}$. Hence for $\varepsilon = \frac{1}{2}$, the segment of length $p = n$ is appeared to be bigger than ε. Therefore, harmonic series is divergent. Later we prove divergence of harmonic series using comparison theorems.

Let us solve the following problem.

Problem 102 Is the series $\sum_{n=1}^{\infty} \frac{2^n + n^2}{n \cdot 2^n}$ convergent or divergent?

Solution. Let us rewrite the nth term of the series as the sum of two other terms as $a_n = \frac{1}{n} + \frac{n}{2^n}$. Now the given series is $\sum\limits_{n=1}^{\infty} \frac{2^n + n^2}{n \cdot 2^n} = \sum\limits_{n=1}^{\infty} \frac{1}{n} + \sum\limits_{n=1}^{\infty} \frac{n}{2^n}$. Even the second series is convergent (see Problem 31 solved earlier) the first series (harmonic) is divergent, therefore the entire given series will diverge.

3.1.2.1 Comparison Theorems. Criteria of Series Convergence

Comparison tests are something that comes naturally. Based on my own experience, some students are able to compare the terms of the given series with the other series, convergence or divergence of which they know. For example, $\frac{1}{n^2} < \frac{1}{n(n-1)}$, $\forall n \in \mathbb{N}$, $n > 1$. Given $1 + \sum\limits_{n=2}^{\infty} \frac{1}{n(n-1)} = S$, then students make a conclusion that $\sum\limits_{n=1}^{\infty} \frac{1}{n^2} < S$. Similar ideas we have already used in solving Problem 67.

In the following theorems we use two series: $\sum\limits_{n=1}^{\infty} u_n$ and $\sum\limits_{n=1}^{\infty} v_n$.

Theorem 3.3 (Comparison Criterion) Let N be some integer $N \geq 1$.

If $0 \leq u_n \leq v_n$ for all $n \geq N$ and $\sum\limits_{n=1}^{\infty} v_n$ converges, then $\sum\limits_{n=1}^{\infty} u_n$ converges.

If $0 \leq v_n \leq u_n$ for all $n \geq N$ and $\sum\limits_{n=1}^{\infty} v_n$ diverges, then $\sum\limits_{n=1}^{\infty} u_n$ diverges.

Theorem 3.4 (Quotient Comparison Theorem) Let N be some integer $N \geq 1$.

If for the two series $\sum\limits_{n=1}^{\infty} u_n$ and $\sum\limits_{n=1}^{\infty} v_n$, where $u_n, v_n > 0$ the inequality $k \leq \frac{u_n}{v_n} \leq K$ holds $\forall k > 0, K > 0$, then both series behave the same way: Either both are convergent or both are divergent.

Corollary 3.2 (Limit Comparison Corollary) If $\lim\limits_{n \to \infty} \frac{u_n}{v_n} = A$, where $A \neq 0$ or $A \neq \infty$, then both series behave the same way: either both convergent or both divergent.

In order to investigate whether or not series converge or diverge we use the following standard series with known behavior:

- Geometric series: $\sum\limits_{k=1}^{\infty} ar^{k-1}, (a \neq 0)$, converges if $|r| < 1$ and diverges if $|r| \geq 1$;

- Dirichlet generalized harmonic series: $\sum\limits_{k=1}^{\infty} \frac{1}{k^p}$, converges for $p > 1$ and diverges if $p \leq 1$.

The generalized series, called Dirichlet series, named later after Peter Gustav Dirichlet (German mathematician, 1805–1859), got a lot of interest hundreds years ago. Thus, Jakob Bernoulli (Swiss mathematician, 1654–1705) and many other mathematicians of that time including his own brother, Johann Bernoulli (1667–1748) knew that the series at $p = 2$ must converge, so that the infinite sum $\sum\limits_{n=1}^{\infty} \frac{1}{n^2} = 1 + \frac{1}{4} + \frac{1}{9} + \frac{1}{16} + \ldots$ must have a limit. Leonhart Euler (Swiss mathematician, 1707–1786) who was a student of Johann Bernoulli, became immediately famous when he solved this problem, 40 years after it was stated and found that the sum of this infinite series is $\frac{\pi^2}{6}$. Much of Euler's genius lay in his ability to apply new and creative approaches to existing problems. This was how he solved what became known as the Basel Problem. The problem challenged mathematicians to determine the exact value of the infinite series above. The question was originally proposed by mathematician Pietro Mengoli in 1644.

The development of infinite series was a topic of much interest in the seventeenth and eighteenth centuries. Neither Leibniz nor Jakob Bernoulli was able to determine the sum of this series, although the Bernoulli brothers were able to determine that the upper bound of the summation was 2. Jacob did this by comparing the series $1 + \frac{1}{4} + \frac{1}{9} + \frac{1}{16} + \ldots$ with a second series with a slightly different denominator. This second series, $1 + \frac{1}{1 \cdot 2} + \frac{1}{2 \cdot 3} + \frac{1}{3 \cdot 4} + \ldots$ can be rewritten as: $1 + \sum\limits_{n=2}^{\infty} \frac{1}{n(n-1)}$, a telescoping series, which Bernoulli knew to be convergent, with the sum equal to 2. (Please show it yourself or look at the solution of Problem 50.) Noticing that, after the first term, every term in the first series is less than every term in the second series, Jakob Bernoulli concluded that the top series must converge and its sum is less than 2.

Although the upper bound of the series was determined, Jakob Bernoulli still could not find the sum and in 1689 he brought the unsolved problem to the attention of the larger mathematical community. Forty five years later, in 1734, Euler determined that $\sum\limits_{n=1}^{\infty} \frac{1}{n^2} = \frac{\pi^2}{6}$ using a typically creative approach and his solution to this problem made the 28 year old mathematician famous (see Dunham [16]). By analogy, Euler also evaluated exact sums of an infinite Dirichlet series at $p = 4$ and $p = 6$ and found the following:

$$\sum_{n=1}^{\infty} \frac{1}{n^4} = 1 + \frac{1}{2^4} + \frac{1}{3^4} + \cdots = \frac{\pi^4}{90},$$

$$\sum_{n=1}^{\infty} \frac{1}{n^6} = 1 + \frac{1}{2^6} + \frac{1}{3^6} + \cdots = \frac{\pi^6}{945}.$$

Finding exact sums of convergent infinite series is a challenging problem in many cases. For example, it is established that Dirichlet series, $\sum_{n=1}^{\infty} \frac{1}{n^p}$, is convergent for $p > 1$, however, there is no much success in finding infinite sums for odd values of p. There is an obvious question: Can we find such infinite sums? Leonard Euler attempted to find such sums but finally simply said, "The problem is very complex," which discouraged many mathematicians. Moreover, even now such sums are not found.

Some methods of finding exact infinite sums for even powers of p will be explained and demonstrated in Section 3.3. Let us see how the Comparison Criterion, (Theorem 3.3) can be applied for a familiar harmonic series,

$$1 + \frac{1}{2} + \frac{1}{3} + \frac{1}{4} + \ldots + \frac{1}{n} + \ldots = \sum_{n=1}^{\infty} \frac{1}{n}. \tag{3.1}$$

The limit of the nth term is zero, so the series passes the Necessary Condition (Theorem 3.1). However, the series is divergent. Let us prove that the limit of the partial sums does not exist. First, we combine some terms together and then we try to find the lower boundary for each group inside parentheses.

$$1 + \left(\frac{1}{2}\right) + \left(\frac{1}{3} + \frac{1}{4}\right) + \left(\frac{1}{5} + \frac{1}{6} + \frac{1}{7} + \frac{1}{8}\right)$$

$$+ \left(\frac{1}{9} + \frac{1}{10} + \frac{1}{11} + \frac{1}{12} + \frac{1}{13} + \frac{1}{14} + \frac{1}{15} + \frac{1}{16}\right) + \cdots$$

$$\frac{1}{3} + \frac{1}{4} > \frac{1}{4} + \frac{1}{4} = \frac{1}{2}$$

$$\frac{1}{5} + \frac{1}{6} + \frac{1}{7} + \frac{1}{8} > 4 \cdot \frac{1}{8} = \frac{1}{2}$$

$$\frac{1}{9} + \frac{1}{10} + \frac{1}{11} + \frac{1}{12} + \frac{1}{13} + \frac{1}{14} + \frac{1}{15} + \frac{1}{16} > 8 \cdot \frac{1}{16} = \frac{1}{2}, \text{ etc.}$$

Therefore, the sums inside each parentheses are greater than $\frac{1}{2}$.

Next, consider the infinite series,

$$\sum_{n=1}^{\infty} v_k = 1 + \frac{1}{2} + \frac{1}{4} + \frac{1}{4} + \frac{1}{8} + \frac{1}{8} + \frac{1}{8} + \frac{1}{8}$$

$$+ \frac{1}{16} + \frac{1}{16} + \frac{1}{16} + \frac{1}{16} + \frac{1}{16} + \frac{1}{16} + \frac{1}{16} + \frac{1}{16} + \cdots \tag{3.2}$$

If we group the terms of this series in the same pattern as we used above, each group after the first term will be equal to ½. Therefore, we can state that the sum of Eq. 3.1 is greater than the sum of Eq. 3.2. For (3.2) we can record the following partial sums:

$$n = 2^k : S_2 = 1 + \frac{1}{2}$$

$$S_{2^2} = S_4 = 1 + 2 \cdot \frac{1}{2}$$

$$S_{2^3} = S_8 = 1 + 3 \cdot \frac{1}{2}$$

$$\cdots$$

$$S_{2^k} = S_n = 1 + k \cdot \frac{1}{2}$$

When $k \to \infty$, $\lim S_{2^k} = \infty$, the series of Eq. 3.2 is divergent, so by Theorem 3.3, the series of Eq. 3.1 is divergent.

3.1.2.2 Sufficient Convergence Theorems

In order series $\sum_{k=1}^{\infty} u_k$ with nonnegative terms to converge, it is necessary and sufficient that the sequence of the partial sums $\{S_n\}$ for this series be bounded. There are several series convergence theorems, the most important are D'Alembert Sufficient Convergence Theorem (Jean D'Alembert, French mathematician, 1717–1783) and the Cauchy Sufficient Convergence Test.

Theorem 3.5 (D'Alembert Sufficient Convergence Theorem) Let the series $\sum_{n=1}^{\infty} u_n$, where $u_n > 0$ be given, and if starting from some n^{th} term of the series, the inequality $\frac{u_{n+1}}{u_n} \leq M < 1$ holds, then the series is convergent. If $\frac{u_{n+1}}{u_n} \geq M > 1$, then the series is divergent.

Corollary 3.3 (D'Alembert Sufficient Ratio Test) Let for series $\sum\limits_{k=1}^{\infty} u_k$

$(u_k > 0)$ there exists the limit $\lim\limits_{k\to\infty} \frac{u_{k+1}}{u_k} = L$.

If $\lim\limits_{n\to\infty} \frac{u_{n+1}}{u_n} = L < 1$, then $\sum\limits_{n=1}^{\infty} u_n$ is convergent

If $\lim\limits_{n\to\infty} \frac{u_{n+1}}{u_n} = L > 1$, then $\sum\limits_{n=1}^{\infty} u_n$ is divergent

If $\lim\limits_{n\to\infty} \frac{u_{n+1}}{u_n} = 1$, then it is inconclusive and additional investigation of the

series would be needed.

Theorem 3.6 (Cauchy Convergence Sufficient Root Test) If for the series

$\sum\limits_{k=1}^{\infty} u_k$ $(u_k > 0)$ there exists $\lim\limits_{k\to\infty} \sqrt[k]{u_k} = L$, then:

If $L < 1$, then the series is convergent.
If $L > 1$, then it is divergent.

It is inconclusive if $L = 1$. (Additional investigation of the series will be needed.)

From the existence of the limit $\lim\limits_{k\to\infty} \frac{u_{k+1}}{u_k}$, it follows that there also exists the limit

$\lim\limits_{k\to\infty} \sqrt[k]{u_k}$.

Remark. The converse statement is not always true because the Cauchy Root Test is "stronger" than the D'Alembert Ratio Test.

For infinite series where the antiderivative of the terms can be found, it is useful to apply the Cauchy Integral Convergence Test.

Theorem 3.7 (Cauchy Integral Convergence Test)

If the terms of $\sum\limits_{k=1}^{\infty} u_k$ satisfy the inequality $u_1 > u_2 > u_3 > \ldots > u_n > \ldots$ and

have type $u_k = f(k)$ where $f(x)$ is nonnegative integrable function that

is monotonically decreasing on the interval $[1; +\infty)$, then series $\sum\limits_{k=1}^{\infty} u_k$

and improper integral $\int\limits_{1}^{\infty} f(x)dx$ converge or diverge simultaneously.

In the case of convergence, the following inequality holds:
$\int\limits_{1}^{\infty} f(x)dx \le \sum\limits_{n=1}^{\infty} u_n \le \int\limits_{1}^{\infty} f(x)dx + u_1$.

Note. Remember that if $\int\limits_{1}^{\infty} f(x)dx = \lim\limits_{n\to\infty} \int\limits_{1}^{n} f(x)dx$ exists, then an improper integral is convergent and if the limit at $n \to \infty$ does not exist, then the integral is divergent.

If the Cauchy Integral Convergence Test is applicable, then it allows us to find for convergent series the lower and upper boundary for the infinite sum. Let us see an application of Theorem 3.7 by solving Problem 103.

> **Problem 103** For the following series: $\frac{1}{3} + \frac{1}{8} + \frac{1}{15} + \frac{1}{24} + \frac{1}{35} + \ldots$. Find the nth term of the series. Investigate whether or not the given infinite series is convergent or divergent. When you are making your statement, provide the corresponding theorem that you used.

Solution. Notice that each denominator is a product of two natural numbers that differ by two. Thus, the sum can be written as $\sum\limits_{n=1}^{\infty} \frac{1}{n^2+2n}$. The necessary condition is satisfied, because $\lim\limits_{n\to\infty} \frac{1}{n^2+2n} = 0$. Applying the Cauchy Integral Test we can find that $f(x) = \frac{1}{x(x+2)} = \frac{1}{2}\left(\frac{1}{x} - \frac{1}{x+2}\right)$ and that $\int\limits_{1}^{\infty} f(x)dx = \frac{\ln 3}{2}$. Therefore, the given infinite series is convergent and their sum is bounded, $\frac{\ln 3}{2} \leq \sum\limits_{n=1}^{\infty} \frac{1}{n^2+2n} \leq \frac{\ln 3}{2} + \frac{1}{3}$.

Since the numbers within each denominator differ by two, we can evaluate this sum exactly as $\sum\limits_{n=1}^{\infty} \frac{1}{n^2+2n} = \frac{1}{2}\left(\sum\limits_{n=1}^{\infty} \frac{1}{n} - \sum\limits_{n=1}^{\infty} \frac{1}{n+2}\right)$ so

$$S_n = \frac{1}{2}\left(1 + \frac{1}{2} + \frac{1}{3} + \frac{1}{4} + \ldots + \frac{1}{n} - \frac{1}{3} - \frac{1}{4} - \ldots - \frac{1}{n} - \frac{1}{n+1} - \frac{1}{n+2}\right)$$

$$S_n = \frac{1}{2}\left(1 + \frac{1}{2} - \frac{1}{n+1} - \frac{1}{n+2}\right)$$

and $\lim\limits_{n\to\infty} S_n = \frac{3}{4}$. The series converges. Moreover, its sum is precisely between the lower and upper bounds given by Theorem 3.7.

$$\frac{\ln 3}{2} \leq \frac{3}{4} \leq \frac{\ln 3}{2} + \frac{1}{3}.$$

Answer. The series converges to ¾.

Here are some problems for practice. You need to decide which theorem to use.

Problem 104 Given $\sum\limits_{n=1}^{\infty} u_n = \frac{3}{2} + \frac{1}{1} + \frac{5}{4} + \frac{9}{7} + \frac{13}{10} + \frac{17}{13} + \cdots$.

a) Find the nth term of the series.
b) Investigate whether or not the given infinite series is convergent or divergent. When you are making your statement, provide the corresponding theorem that you used.

Solution. Ignoring the first term, we notice that the numerator' numbers differ by 4 and the denominator's numbers differ by 3. This series will have the following n^{th} term: $u_n = \frac{4n-7}{3n-5}$. Note that the first term is also described by this formula.

a) The necessary condition does not hold, because $\lim\limits_{n \to \infty} \frac{4n-7}{3n-5} = \frac{4}{3} \neq 0$.

b) Therefore, the series diverges.

Problem 105 $\cos\frac{1}{3} + \cos\frac{1}{6} + \cos\frac{1}{9} + \cos\frac{1}{12} + \cos\frac{1}{15} + \cdots$.

a) Find the nth term of the series.
b) Investigate whether or not the given infinite series is convergent or divergent. When you are making your statement, provide the corresponding theorem that you used.

Solution. This series can be written as $\cos\frac{1}{3} + \cos\frac{1}{6} + \cos\frac{1}{9} + \cos\frac{1}{12} + \cos\frac{1}{15} + \cdots = \sum\limits_{n=1}^{\infty} u_n = \sum\limits_{n=1}^{\infty} \cos\frac{1}{3n}$.

a. The necessary condition does not hold, because
$$\lim\limits_{n \to \infty} \cos\frac{1}{3n} = \cos\left(\lim\limits_{n \to \infty} \frac{1}{3n}\right) = \cos 0 = 1 \neq 0.$$
b. Therefore, the series diverges.

Problem 106 Investigate whether the series $\sum\limits_{n=1}^{\infty} \frac{2n-1}{\left(\sqrt{2}\right)^n}$ is convergent or divergent.

Solution. Consider the D'Alembert Sufficient Ratio Test (Corollary 3.3)

$$\lim_{n\to\infty}\frac{u_{n+1}}{u_n}=\lim_{n\to\infty}\frac{(2(n+1)-1)}{(\sqrt{2})^{n+1}}\cdot\frac{(\sqrt{2})^n}{(2n-1)}=\lim_{n\to\infty}\left\{\frac{2n+1}{(\sqrt{2})}\cdot\frac{1}{(2n-1)}\right\}=\frac{1}{\sqrt{2}}$$

Since $\frac{1}{\sqrt{2}}<1$, then $\sum\limits_{n=1}^{\infty}\frac{2n-1}{(\sqrt{2})^n}$ converges.

Problem 107 Investigate whether the series is convergent or divergent

$$\sum_{n=1}^{\infty}\left(\frac{n}{3n-1}\right)^{2n}.$$

Solution. Let us use Cauchy Convergence Sufficient Root Test (Theorem 3.6) by evaluating $\lim\limits_{n\to\infty}\sqrt[n]{\left(\frac{n}{3n-1}\right)^{2n}}=\lim\limits_{n\to\infty}\left(\frac{n}{3n-1}\right)^2=\frac{1}{9}<1$. The series is convergent.

Problem 108 For the series $\sum\limits_{n=3}^{\infty}\frac{1}{n^2-4}$. a) Investigate whether or not it is convergent or divergent. b) If it is convergent then evaluate the partial sum and its limit when n increases without bound.

Solution.

a. Let us use the Cauchy Integral Convergence Test (Theorem 3.7) to demonstrate that the series is convergent. First, remember from our earlier work that

$$\frac{1}{x^2-4}=\frac{1}{(x-2)(x+2)}=\frac{1}{4}\left(\frac{1}{x-2}-\frac{1}{x+2}\right)\quad\text{so that,}\quad\int\limits_3^{\infty}\frac{dx}{x^2-4}=\frac{1}{4}\left[\int\limits_3^{\infty}\frac{dx}{x-2}-\int\limits_3^{\infty}\frac{dx}{x+2}\right]=\frac{\ln 5}{4}.$$

Since the limit exists, the integral is convergent and the series is convergent and its infinite sum is bounded as $\frac{\ln 5}{4}\leq\sum\limits_{n=3}^{\infty}\frac{1}{n^2-4}\leq\frac{\ln 5}{4}+\frac{1}{5}$.

b. Now let us look at the sum,

$$\sum_{n=3}^{\infty}\frac{1}{n^2-4}=\frac{1}{4}\left(\sum_{n=3}^{\infty}\frac{1}{n-2}-\sum_{n=3}^{\infty}\frac{1}{n+2}\right)$$

$$\sum_{n=3}^{\infty}\frac{1}{n-2}=1+\frac{1}{2}+\frac{1}{3}+\frac{1}{4}+\frac{1}{5}+\ldots+\frac{1}{n-2}+\ldots$$

$$\sum_{n=3}^{\infty}-\frac{1}{n+2}=-\frac{1}{5}-\frac{1}{6}-\frac{1}{7}-\ldots-\frac{1}{n-2}-\frac{1}{n-1}-\frac{1}{n}-\frac{1}{n+1}-\frac{1}{n+2}-\ldots$$

Therefore, with many of the terms cancelling, we are left with

$$S_n = \frac{1}{4}\left[1 + \frac{1}{2} + \frac{1}{3} + \frac{1}{4} - \frac{1}{n-1} - \frac{1}{n} - \frac{1}{n+1} - \frac{1}{n+2}\right]$$

$$\lim_{n\to\infty} S_n = \frac{1}{4}\left(1 + \frac{1}{2} + \frac{1}{3} + \frac{1}{4}\right) = \frac{25}{48}.$$

It is true that

$$\frac{\ln 5}{4} \le \frac{25}{48} \le \frac{\ln 5}{4} + \frac{1}{5}$$

$$0.4022359 \le 0.520833 \le 0.602359$$

Answer. 25/48.

Remark. In order to prove that the series $\sum\limits_{n=3}^{\infty} \frac{1}{n^2-4}$ is convergent, we can compare it

with convergent series $\sum\limits_{n=3}^{\infty} \frac{1}{n^2}$. Considering Corollary 3.2, we obtain that

$\lim\limits_{n\to\infty} \frac{a_n}{b_n} = \lim\limits_{n\to\infty} \frac{n^2}{n^2-4} = 1$. Therefore, the series behaves the same, i.e., converge.

Problem 109 Investigate whether or not the series $\sum\limits_{n=1}^{\infty} u_n = \sum\limits_{n=1}^{\infty} \frac{1}{n^2+2}$ is

convergent or divergent.

Solution. Consider an auxiliary series $\sum\limits_{n=1}^{\infty} v_n = \sum\limits_{n=1}^{\infty} \frac{1}{n^2}$. We can see that

$0 < u_n = \frac{1}{n^2+2} < \frac{1}{n^2} = v_n$. Since $\sum\limits_{n=1}^{\infty} \frac{1}{n^2}$ converges then $\sum\limits_{n=1}^{\infty} \frac{1}{n^2+2}$ converges. We have

used the Comparison Criterion (Theorem 3.3) to prove convergence.

In order to investigate convergence of numerical series, we usually have to evaluate limits at infinity, so often the well-known conditions of equivalence of infinitesimals will be used:

$$\sin\left(\frac{1}{n}\right) \sim \frac{1}{n}, \quad \tan\left(\frac{1}{n}\right) \sim \frac{1}{n}, \quad \ln\left(1+\frac{1}{n}\right) \sim \frac{1}{n}, \tag{3.3}$$

$$\arcsin\frac{1}{n} \sim \frac{1}{n}, \quad \arctan\frac{1}{n} \sim \frac{1}{n}, \quad e^{\frac{1}{n}} - 1 \sim \frac{1}{n}. \tag{3.4}$$

The following formula is also very important to know:

$$\text{Stirling's Formula: } n! \sim \sqrt{2\pi n} \cdot \left(\frac{n}{e}\right)^n \tag{3.5}$$

Additionally, we use the following limits:

$$\lim_{n\to\infty} \frac{\ln n}{n^p} = 0 \; (p > 0), \quad \lim_{n\to\infty}\left(1+\frac{1}{n}\right)^n = e, \quad \lim_{n\to\infty} \sqrt[n]{n} = 1. \tag{3.6}$$

Problem 110 Is $\arcsin 1 + \arcsin\frac{1}{2} + \arcsin\frac{1}{3} + \dots + \arcsin\frac{1}{n} + \dots$ convergent or divergent?

Solution. Let $\sum\limits_{n=1}^{\infty} u_n = \sum\limits_{n=1}^{\infty} \arcsin\frac{1}{n}$, $\sum\limits_{n=1}^{\infty} v_n = \sum\limits_{n=1}^{\infty}\frac{1}{n}$. By Corollary 3.2 and using the first formula of Eq. 3.4, we have that $\lim\limits_{n\to\infty}\frac{u_n}{v_n} = 1$. Therefore, both series behave the same, and hence it is divergent as harmonic series.

Problem 111 Using the Cauchy Integral Convergence Test prove that the series $\sum\limits_{n=1}^{\infty}\frac{1}{n^2+1}$ is convergent.

Proof. Common term of this series is $u_n = \frac{1}{n^2+1} = f(n)$. Replacing n by x, we obtain function $f(x) = \frac{1}{x^2+1}$. This function satisfies the condition of the Cauchy Integral Convergence Test (Theorem 3.7) because it is positive and monotonically decreasing as x increasing. Evaluate the improper integral,

$$\int\limits_{1}^{+\infty} \frac{dx}{x^2+1} = \arctan x \Big|_{1}^{+\infty} = \lim_{x\to+\infty} \arctan x - \arctan 1 = \frac{\pi}{2} - \frac{\pi}{4} = \frac{\pi}{4}.$$

Since it converges, the series also converges. Additionally, we could obtain the lower and upper boundary for the infinite sum, $\frac{\pi}{4} \le \sum\limits_{n=1}^{\infty}\frac{1}{n^2+1} \le \frac{\pi}{4} + \frac{1}{2}$, and to obtain its approximation as $0.785398 \le \sum\limits_{n=1}^{\infty}\frac{1}{n^2+1} \le 1.2854$.

Remark. Convergence of this series can be proven with the use of Comparison Criterion (Theorem 3.3) and by comparing this series with Dirichlet series $\sum\limits_{n=1}^{\infty} \frac{1}{n^2}$, that is convergent at $p = 2 > 1$. Since $u_n = \frac{1}{n^2+1} < v_n = \frac{1}{n^2}$, then the given series is convergent. Further, using the Comparison Criterion, we also can obtain

$$\sum_{n=1}^{\infty} \frac{1}{n^2+1} < \sum_{n=1}^{\infty} \frac{1}{n^2} = \frac{\pi^2}{6} \approx 1.6449.$$

Obviously, the Cauchy Integral Convergence Test gives us a better upper bound approximation for the infinite sum.

Problem 112 Is the series $\sum\limits_{n=1}^{\infty} \frac{n^n}{2^n \cdot n!}$ convergent or divergent?

Solution. In order to apply the D'Alembert Sufficient Ratio Test, we find that the common term of the series is $u_n = \frac{n^n}{2^n \cdot n!}$. Replacing n by $(n+1)$ we obtain $u_{n+1} = \frac{(n+1)^{n+1}}{2^{n+1} \cdot (n+1)!}$. Next, we find: $\frac{u_{n+1}}{u_n} = \frac{(n+1)^{n+1} \cdot 2^n \cdot n!}{2^{n+1} \cdot (n+1)! \cdot n^n} = \frac{(n+1)^n}{2 \cdot n^n} = \frac{1}{2} \cdot \left(1 + \frac{1}{n}\right)^n$. Using the second formula of Eq. 3.6, we obtain $\lim\limits_{n\to\infty} \frac{u_{n+1}}{u_n} = \frac{1}{2} \lim\limits_{n\to\infty} \left(1 + \frac{1}{n}\right)^n = \frac{e}{2}$. However, $e > 2$, then $\frac{e}{2} > 1$, so by the D'Alembert Sufficient Ratio Test the series diverges.

3.1.2.3 Gauss and Dirichlet Convergence Theorems

Next, we discuss what can be done if both D'Alembert Ratio and Cauchy Convergence Sufficient Root Test are inconclusive and the Cauchy Integral test cannot be applied. How can we determine if a given series converges or diverges? Suppose that the limit as n approaches infinity of the ratio of the $(n+1)^{st}$ to the nth terms equals one. Then the ratio test is inconclusive. In these cases Gauss proposed a ratio test of preceding to following consecutive terms.

Theorem 3.8 (Gauss's Convergence Theorem). Consider series

$$a_1 + a_2 + \ldots + a_n + a_{n+1} + \ldots,$$

$$\frac{a_n}{a_{n+1}} = \lambda + \frac{\mu}{n} + \frac{v_n}{n^2}, \qquad \text{where } \lambda, \mu \text{ are constants and } |v_n| \leq M$$

(bounded value).

The series $\sum\limits_{n=1}^{\infty} a_n$ is convergent if $\lambda > 1$ or $\lambda = 1$, $\mu > 1$.

The series $\sum\limits_{n=1}^{\infty} a_n$ is divergent if $\lambda < 1$ or $\lambda = 1$, $\mu \leq 1$.

Gauss's Convergence Theorem is very useful if other tests do not allow us to make a conclusion regarding series convergence. Let us demonstrate its application by solving the following problem.

Problem 113 The series $\sum_{n=1}^{\infty} a_n$ is defined by $a_n = \left(\frac{1 \cdot 3 \cdot 5 \cdot 7 \cdot \ldots \cdot (2n-1)}{2 \cdot 4 \cdot 6 \cdot 8 \cdot \ldots \cdot 2n}\right)^2$. Is it convergent?

Solution. First, we try applying the ratio test (Corollary 3.3): $\lim_{n \to \infty} \frac{a_{n+1}}{a_n} = \lim_{n \to \infty} \left(\frac{2n+1}{2n+2}\right)^2 = \lim_{n \to \infty} \frac{4n^2 + 4n + 1}{4n^2 + 8n + 4} = 1$. The limit is one, so the ratio test is inconclusive. Let us consider the reciprocal of this ratio and apply Gauss principle, $\frac{a_n}{a_{n+1}} = \frac{4n^2 + 8n + 4}{4n^2 + 4n + 1} = \frac{1 + \frac{2}{n} + \frac{1}{n^2}}{1 + \frac{1}{n} + \frac{1}{4n^2}} = 1 + \frac{1}{n} + \frac{v_n}{n^2}$. We can see that $\lambda = 1$, $\mu = 1 \leq 1$. Therefore, the series diverges.

If $a_n > 0$ then it is often useful to use a different form of Gauss's Convergence Theorem (Corollary 3.4).

Corollary 3.4 (Gauss) If $a_n > 0$ and there exists a number $\varepsilon > 0$, such that

$$\frac{a_{n+1}}{a_n} = 1 + \frac{\mu}{n} + O\left(\frac{1}{n^{1+\varepsilon}}\right), \quad n \to \infty, \tag{3.7}$$

then the series $\sum_{n=1}^{\infty} a_n$ converges if $\mu < -1$ and it diverges if $\mu \geq -1$.

Gauss's Convergence Theorem considers the case where $\lim_{n \to \infty} \frac{a_{n+1}}{a_n} = 1$ (D'Alembert's ratio test is inconclusive) and compares a sequence $\{a_n\}$ with the sequence $\{n^{-\mu}\}$. Basically, this test compares the given series with a geometric progression. ($\mu = 1$).

Corollary 3.5 (Dirichlet) If $a_n > 0$ and there exist two numbers p and $C > 0$, such that

$$a_n \sim \frac{C}{n^p}, \quad n \to \infty, \tag{3.8}$$

then $\sum_{n=1}^{\infty} a_n$ converges if $p > 1$, and diverges if $p \leq 1$.

We apply this property when we can compare series with Dirichlet series $\sum\limits_{n=1}^{\infty} \frac{1}{n^p}$.

The question is when and how we use one of these two tests. The answer depends on which of two comparisons, Eq. 3.7 or Eq. 3.8 is easier to obtain.

Let us practice by solving the following problems.

Problem 114 Is the series $\sum\limits_{n=1}^{\infty} \frac{\sin\left(\frac{\pi}{n}\right)}{n^p}$ convergent or divergent?

Solution. Using the first formula of Eq. 3.3 for a sine function with $\sin\left(\frac{\pi}{n}\right) \sim \frac{\pi}{n} \Rightarrow a_n \sim \frac{\pi}{n^{p+1}}$, $n \to \infty$. Then using Eq. 3.8, we can state that the series will converge if $p > 0$ $(p+1 > 1)$ and diverge if $p \le 0$ $(p+1 \le 1)$.

Problem 115 Is the series $\sum\limits_{n=1}^{\infty} \frac{n!e^n}{n^{n+p}}$ convergent or divergent?

Solution. Let us apply Corollary 3.4 for the ratio of two consecutive terms:

$$\frac{a_{n+1}}{a_n} = \frac{(n+1)!e^{n+1}n^{n+p}}{(n+1)^{n+p+1}n!e^n}$$

$$= e\frac{n^{n+p}}{(n+1)^{n+p}} = e\left(\frac{1}{1+\frac{1}{n}}\right)^{n+p}$$

$$= e\left(1+\frac{1}{n}\right)^{-(n+p)} \sim 1 + \frac{1}{n}\left(\frac{1}{2}-p\right) + O\left(\frac{1}{n^2}\right), \quad n \to \infty.$$

Above we used a binomial distribution. Since $\mu = \frac{1}{2} - p$, then if $p > \frac{3}{2}$ then the given series is convergent and if $p \le \frac{3}{2}$ by Gauss's Convergence Theorem.

In particular, necessary representation can be obtained by using Eq. 3.6 and applying Taylor's formula, which we practice in Section 3.2.2.

3.1.3 Alternating Series

In this Section, the following important topics will be discussed:

- Alternating Series. Leibniz Convergence Theorem
- Absolutely and conditionally convergent series
- Abel's and Dirichlet's Convergence Theorems
- The error of estimation of an infinite sum by a finite sum

The alternating series is a series in which all terms alternate by sign, $\sum_{k=1}^{\infty} (-1)^{k-1} a_k = a_1 - a_2 + a_3 - a_4 + \ldots + (-1)^{k-1} a_k + \ldots$, where a_k are numbers of the same sign. For example, $1 - \frac{1}{2} + \frac{1}{3} - \frac{1}{4} + \ldots + \frac{(-1)^{n+1}}{n} + \ldots$ is an alternating Leibniz series.

3.1.3.1 Leibniz Criteria for Alternating Series

Theorem 3.9 (Leibniz Theorem) Let all terms of the alternating series $\sum_{k=1}^{\infty} (-1)^{k-1} a_k$, where $a_k \geq 0$ satisfy the conditions:

1) $a_k \geq a_{k+1} \ \forall k \in \mathbb{N}$;
2) $\lim\limits_{k \to \infty} a_k = 0$,

then the series $\sum_{k=1}^{\infty} (-1)^{k-1} a_k$ converges and its sum S is less than or equal to the first term, i.e., $S \leq a_1$.

Series, satisfying the condition of Theorem 3.9 are called Leibniz series. Leibniz was a German mathematician (1646–1716), together with Issac Newton, he is often called a founder of mathematical analysis.

The remainder $r_n = (-1)^n (a_{n+1} - a_{n+2} + \ldots)$ of the Leibniz series satisfies the inequality $|r_n| \leq a_{n+1}$. Therefore, if we approximate the sum of convergent alternating series by a partial sum, then the value of the error will not exceed the absolute value of the first dropped term, $|a_{n+1}|$ and will have its sign.

If all terms of an alternating series

$$\sum_{n=1}^{\infty} (-1)^n \cdot a_n = a_1 - a_2 + \ldots + (-1)^n \cdot a_n + \ldots \quad (a_n > 0) \quad (3.9)$$

1. Are monotonically decreasing by absolute value, $a_{n+1} < a_n (n = 1, 2, 3, \ldots)$
2. Approach zero, i.e., $\lim\limits_{n \to \infty} a_n = 0$, the series of Eq. 3.9 is convergent, its sum S is positive and does not exceed the first term of the series, $0 < S < a_1$.

Alternatively, if the alternating series starts from a negative term, $-a_1 + a_2 - a_3 + \ldots (a_n > 0)$, and for this series the conditions 1) and 2) of the Leibniz Theorem are valid, then this series is convergent, its sum S is negative and it satisfies the inequality, $-a_1 < S < 0$.

Let us solve the following problem.

Problem 116 Investigate whether or not the following series is convergent or divergent $\sum_{n=1}^{\infty} (-1)^{n+1} \cdot \left(1 + \frac{1}{n}\right)^n$.

Solution. This series is alternating. The value of the common term is $a_n = \left(1 + \frac{1}{n}\right)^n$. Let us find the limit of the nth term to infinity (see Eq. 3.6): $\lim_{n \to \infty} a_n = \lim_{n \to \infty} \left(1 + \frac{1}{n}\right)^n = e$. Because the limit of the common term does not approach zero, then the series diverges (the necessary condition is not fulfilled).

3.1.3.2 Absolutely and Conditionally Convergent Alternating Series

Convergent alternating series can be either absolutely or conditionally convergent.

Definition. The series $\sum_{k=1}^{\infty} a_k$ is **absolutely convergent**, if the series with nonnegative terms $\sum_{k=1}^{\infty} |a_k|$ is convergent. If the series absolutely convergent, then it is convergent.

The converse statement, in general, is not true. Absolutely convergent series have the following properties:

- if the series $\sum_{k=1}^{\infty} a_k$ is absolutely convergent and $\sum_{k=1}^{\infty} a_k = S$, $\sum_{k=1}^{\infty} |a_k| = \Sigma$, then $|S| \leq \Sigma$;

- if the series $\sum_{k=1}^{\infty} a_k$ and $\sum_{k=1}^{\infty} b_k$ are absolutely convergent, then for any α and β, the series $\sum_{k=1}^{\infty} (\alpha a_k + \beta b_k)$ is absolutely convergent;

- if the series $\sum_{k=1}^{\infty} a_k$ is absolutely convergent, the series made from the same terms taken in a different order is also absolutely convergent and its sum equals the sum of the original series.

Example 3. Consider the following alternating series:

$$1 - \frac{1}{2} + \frac{1}{2^2} - \frac{1}{2^3} + \cdots \tag{3.10}$$

Let make a series from this one by interchanging neighboring members of the series:

$$-\frac{1}{2} + 1 - \frac{1}{2^3} + \frac{1}{2^2} - \cdots \tag{3.11}$$

Because the series made of the absolute values of the original series, $\sum\limits_{n=0}^{\infty} \frac{1}{2^n} = 1 + \frac{1}{2} + \frac{1}{2^2} + \frac{1}{2^3} + \cdots = \frac{1}{1-\frac{1}{2}} = 2$ converges as a geometric series, then the original series of Eq. 3.10 is absolutely convergent and its sum equals $S = \frac{1}{1-\left(-\frac{1}{2}\right)} = \frac{2}{3}$. Hence, the series of Eq. 3.11 obtained by interchanging terms of Eq. 3.10 will converge as well to the same sum, 2/3.

- If the series $\sum\limits_{k=1}^{\infty} a_k$ and $\sum\limits_{k=1}^{\infty} b_k$ absolutely convergent, the series made from all possible paired product $a_k b_m$ of terms of these series, in any order, is also absolutely convergent.

The last statement can be reformulated as follows:

Theorem 3.10 Let series $\sum\limits_{k=1}^{\infty} a_k$ and $\sum\limits_{k=1}^{\infty} b_k$ be absolutely convergent and converge to the sums, S_a and S_b, respectively. Then a new series $\sum\limits_{k=1}^{\infty} a_k \cdot \sum\limits_{k=1}^{\infty} b_k$ containing all possible pair products $a_k b_m$ is also absolutely convergent and its sum equals $S = S_a \cdot S_b$.

This theorem is a very important theorem and it means that absolute convergent series can be multiplied term by term. Because in an absolute convergent series any terms can be added and combined in any way, then we can rewrite the product of two series as

$$\sum_{k=1}^{\infty} a_k \cdot \sum_{k=1}^{\infty} b_k = a_1 \cdot b_1 + (a_1 b_2 + a_2 b_1) + (a_1 b_3 + a_2 b_2 + a_3 b_1) + \cdots$$
$$+ (a_1 b_n + a_2 b_{n-1} + \cdots + a_n b_1) + \cdots \tag{3.12}$$
$$= S = S_a \cdot S_b.$$

As it follows from other properties of alternating series that only absolute convergent series have all properties of finite sums.

Problem 117 Given $\sum\limits_{n=1}^{\infty} a_n = \sum\limits_{n=1}^{\infty} \frac{1}{3^{n-1}}$, $\sum\limits_{n=1}^{\infty} b_n = \sum\limits_{n=1}^{\infty} \frac{(-1)^{n-1}}{3^{n-1}}$. Find the product of two series. It is convergent or divergent?

Solution. Because both series are absolutely convergent, then the product of two series will be also absolutely convergent and Theorem 3.10 can be applied. We multiply the series using Eq. 3.12. Here $a_1 = 1$, $a_2 = \frac{1}{3}$, ..., $a_{n-1} = \frac{1}{3^{n-2}}$ are terms of the first series and $b_1 = 1$, $b_2 = -\frac{1}{3}$, ..., $b_{n-1} = \frac{(-1)^{n-2}}{3^{n-2}}$. Let us find several products needed for Eq. 3.12:

$$a_1 b = 1$$

$$a_1 b_2 + a_2 b_1 = 1 \cdot \left(-\frac{1}{3}\right) + \frac{1}{3} \cdot 1 = 0$$

$$a_1 b_3 + a_2 b_2 + a_3 b_1 = 1 \cdot \frac{1}{3^2} + \frac{1}{3} \cdot \left(-\frac{1}{3}\right) + \frac{1}{3^2} \cdot 1 = \frac{1}{3^2}$$

$$a_1 b_4 + a_2 b_3 + a_3 b_2 + a_4 b_1 = 1\left(-\frac{1}{3^3}\right) + \frac{1}{3} \cdot \frac{1}{3^2} + \frac{1}{3^2} \cdot \left(-\frac{1}{3}\right) + \frac{1}{3^3} \cdot 1 = 0$$

Next, we calculate

$$(a_1 b_n + a_2 b_{n-1} + \cdots + a_{n-1} b_2 + a_n b_1)$$

$$= 1 \cdot (-1)^{n-1} \frac{1}{3^{n-1}} + \frac{1}{3} \cdot (-1)^{n-2} \frac{1}{3^{n-2}} + \cdots + \frac{1}{3^{n-2}} \cdot \left(-\frac{1}{3}\right) + \frac{1}{3^{n-1}} \cdot 1$$

$$= \frac{(-1)^{n-1}}{3^{n-1}} + \frac{(-1)^{n-2}}{3^{n-1}} + \cdots - \frac{1}{3^{n-1}} + \frac{1}{3^{n-1}}$$

$$= \begin{cases} 0, & \text{if } n = 2k \\ \frac{1}{3^{n-1}}, & \text{if } n = 2k - 1 \end{cases} \quad k \in \mathbb{N}.$$

$$\sum_{k=1}^{\infty} a_k \cdot \sum_{k=1}^{\infty} b_k == 1 + 0 + \frac{1}{3^2} + 0 + \ldots + \frac{1}{3^{n-1}} + 0$$

$$= 1 + \frac{1}{3^2} + \frac{1}{3^4} + \frac{1}{3^6} + \ldots + \frac{1}{3^{2m-2}} + \ldots = \sum_{m=1}^{\infty} \frac{1}{3^{2m-2}}.$$

$$= \frac{1}{1 - \frac{1}{9}} = \frac{9}{8}.$$

On the other hand, $S_a = \sum_{k=1}^{\infty} a_k = 1 + \frac{1}{3} + \frac{1}{9} + \ldots = \frac{1}{1-\frac{1}{3}} = \frac{3}{2}$, $S_b = \frac{1}{1+\frac{1}{3}} = \frac{3}{4}$ so $S_a \cdot S_b = \frac{3}{2} \cdot \frac{3}{4} = \frac{9}{8}$.

Answer. 9/8.

Problem 118 Prove that the series $\sum_{n=1}^{\infty} \frac{\sin n}{n^3}$ is absolutely convergent.

Proof. Consider the series:

$$\sum_{n=1}^{\infty} \frac{|\sin n|}{n^3} = \frac{|\sin 1|}{1^3} + \frac{|\sin 2|}{2^3} + \ldots + \frac{|\sin n|}{n^3} + \ldots \quad (3.13)$$

Since $|\sin n| \leq 1$, then each term of Eq. 3.13 does not exceed the corresponding term of the following series

$$\sum_{n=1}^{\infty} \frac{1}{n^3} = \frac{1}{1^3} + \frac{1}{2^3} + \ldots + \frac{1}{n^3} + \ldots \quad (3.14)$$

The series of Eq. 3.14 is a Dirichlet series of type $\sum_{n=1}^{\infty} \frac{1}{n^p}$, where $p = 3$. Since $p > 1$, the series of Eq. 3.14 is convergent. By Theorem 3.3, series of Eq. 3.13 is also convergent. Then, by the absolute convergence theorem, the given alternating series is absolutely convergent.

Definition. Series $\sum_{k=1}^{\infty} a_k$ is called **conditionally convergent**, if series $\sum_{k=1}^{\infty} a_k$ converges, but the series $\sum_{k=1}^{\infty} |a_k|$ diverges.

Example. The series $\sum_{n=1}^{\infty} \frac{(-1)^{n+1}}{n} = 1 - \frac{1}{2} + \frac{1}{3} - \ldots + (-1)^{n+1} \cdot \frac{1}{n} + \ldots$ is called Leibniz and is convergent by the Leibniz Theorem. On the other hand, its corresponding absolute value series, $\sum_{n=1}^{\infty} \frac{1}{n} = 1 + \frac{1}{2} + \frac{1}{3} + \ldots + \frac{1}{n} + \ldots$ is divergent (harmonic series). Therefore, the Leibniz series is conditionally convergent series.

Example. The series $\sum\limits_{n=1}^{\infty} \frac{(-1)^{n+1}}{n^p}$ $(p > 0)$ is an alternating series. If $p > 0$ it satisfies the Leibniz Theorem, i.e.

1) $\frac{1}{(n+1)^p} < \frac{1}{n^p}$ $(n = 1, 2, 3, \ldots)$

2) $\lim\limits_{n\to\infty} \frac{1}{n^p} = 0$

and, hence, it is convergent. If we replace all its terms by their absolute values, we obtain the Dirichlet series,

$$\sum_{n=1}^{\infty} \frac{1}{n^p}, \tag{3.15}$$

which is convergent for $p > 1$ and divergent for $p \le 1$. Therefore, the series of Eq. 3.15 is absolutely convergent for $p > 1$ and conditionally convergent for $0 < p \le 1$.

For the series $\sum\limits_{k=1}^{\infty} a_k$ denote by $a_1^+, a_2^+, \ldots, a_k^+, \ldots$ and $a_1^-, a_2^-, \ldots, a_k^-, \ldots$ respectively, its nonnegative and negative terms taken in the same order, in which they are in the series $\sum\limits_{k=1}^{\infty} a_k$. Consider the series $\sum\limits_{k=1}^{\infty} a_k^+$ and $\sum\limits_{k=1}^{\infty} a_k^-$. If the series $\sum\limits_{k=1}^{\infty} a_k$ conditionally converges, then both series $\sum\limits_{k=1}^{\infty} a_k^+$ and $\sum\limits_{k=1}^{\infty} a_k^-$ are divergent.

Example. Consider a familiar Leibniz series, $\sum\limits_{n=1}^{\infty} \frac{(-1)^{n+1}}{n} = 1 - \frac{1}{2} + \frac{1}{3} - \ldots +$ $(-1)^{n+1} \cdot \frac{1}{n} + \ldots$. For this conditionally convergent series, we have $\sum\limits_{k=1}^{\infty} a_k^+ = 1 +$ $\frac{1}{3} + \frac{1}{5} + \frac{1}{7} + \ldots + \frac{1}{2n-1} + \ldots$ and $\sum\limits_{k=1}^{\infty} a_k^- = -\frac{1}{2} - \frac{1}{4} - \frac{1}{6} - \ldots - \frac{1}{2n} - \ldots$, which are both divergent.

Earlier we have learned that interchanging the order of the terms of an absolutely convergent alternating series does not change the series sum. This fact is not true for a conditionally convergent series and the sum of the series will depend on the order of its terms!

Example. Consider conditionally convergent Leibniz series again and denote its sum by S,

$$S = 1 - \frac{1}{2} + \frac{1}{3} - \frac{1}{4} + \frac{1}{5} - \frac{1}{6} + \ldots + (-1)^{n-1}\frac{1}{n} + \ldots \tag{3.16}$$

Next, let us change the order of the terms in the infinite sum, in such a way that a positive term will follow by two negative terms:

$$\left(1 - \frac{1}{2} - \frac{1}{4}\right) + \left(\frac{1}{3} - \frac{1}{6} - \frac{1}{8}\right) + \left(\frac{1}{5} - \frac{1}{10} - \frac{1}{12}\right) + \dots$$

This can be further regrouped as

$$\left(1 - \frac{1}{2}\right) - \frac{1}{4} + \left(\frac{1}{3} - \frac{1}{6}\right) - \frac{1}{8} + \left(\frac{1}{5} - \frac{1}{10}\right) - \frac{1}{12} + \dots \tag{3.17}$$

and evaluated as

$$\frac{1}{2} - \frac{1}{4} + \frac{1}{6} - \frac{1}{8} + \frac{1}{10} - \frac{1}{12} + \dots$$
$$= \frac{1}{2} \cdot \left(1 - \frac{1}{2} + \frac{1}{3} - \frac{1}{4} + \frac{1}{5} - \frac{1}{6} + \dots\right) = \frac{1}{2} \cdot S.$$

We can see that after interchanging the order of terms in the series of Eq. 3.16, we obtained the series of Eq. 3.17, the sum of which decreased by half!

Can we increase the sum of infinite series of Eq. 3.16 by rearranging its term differently? The answer is "Yes." For example, we can rearrange the terms of Eq. 3.16 in such a way that two positive terms will follow one negative and so on and this series will have a larger sum, $\frac{3S}{2}$, than the original series of Eq. 3.16, i.e., $\left(1 + \frac{1}{3} - \frac{1}{2}\right) + \left(\frac{1}{5} + \frac{1}{7} - \frac{1}{4}\right) + \left(\frac{1}{9} + \frac{1}{11} - \frac{1}{6}\right) + \dots = \frac{3}{2} S.$

Let us prove that the sum on the left is 3/2 of the Leibniz sum (Eq. 3.16). The left-hand side can be rewritten as

$$\left(1 - \frac{1}{2} + \frac{1}{3} - \frac{1}{4} + \left\{\frac{1}{2} + \frac{1}{4} - \frac{1}{2}\right\}\right) + \left(\frac{1}{5} - \frac{1}{6} + \frac{1}{7} - \frac{1}{8} + \left\{\frac{1}{6} + \frac{1}{8} - \frac{1}{4}\right\}\right)$$
$$+ \left(\frac{1}{9} - \frac{1}{10} + \frac{1}{11} - \frac{1}{12} + \left\{\frac{1}{10} + \frac{1}{12} - \frac{1}{6}\right\}\right) + \dots$$

Next, we separately add all the terms inside all parentheses and all the terms inside all braces and obtain:

$$\left(1 - \frac{1}{2} + \frac{1}{3} - \frac{1}{4} + \frac{1}{5} - \frac{1}{6} + \frac{1}{7} - \frac{1}{8} + \frac{1}{9} - \frac{1}{10} + \dots\right)$$
$$+ \left\{\frac{1}{4} + \frac{1}{6} - \frac{1}{8} + \frac{1}{10} - \frac{1}{12} + \dots\right\}$$

Denoting the Leibniz infinite sum inside parentheses by S, let us express the sum inside braces in terms of S:

$$S + \frac{1}{2}\left(\frac{1}{2} + \frac{1}{3} - \frac{1}{4} + \frac{1}{5} - \frac{1}{6} + \ldots\right)$$
$$= S + \frac{1}{2}\left(1 - \frac{1}{2} + \frac{1}{3} - \frac{1}{4} + \frac{1}{5} - \frac{1}{6} + \ldots\right)$$
$$= S + \frac{1}{2}S = \frac{3}{2} \cdot S.$$

This completes the proof.

The German mathematician Bernhard Riemann (1826–1866) proved that the sum of conditionally convergent series depends on the order of the terms and that by rearranging terms of a conditionally convergent series, the sum of the new series can become equal to any a priory given number. Changing the order of the terms can change the sum. Moreover, some rearrangements will lead to a divergent alternating series.

The following statement is valid.

Theorem 3.11 (Riemann) If the series $\sum\limits_{k=1}^{\infty} a_k$ is conditionally convergent, then for any real number S, its terms can be rearranged in such a way that the sum of the obtained series will be S.

3.1.3.3 Convergence of Series Formed by the Product of Terms

If the nth term of a series can be represented as a product of the common terms of two other series, i.e., $\sum\limits_{k=1}^{\infty} a_k b_k$ then the following statements are true:

Theorem 3.12 (Dirichlet Theorem) Let for series $\sum\limits_{k=1}^{\infty} a_k b_k$ the following be true:

1. The sequence $\{a_k\}$ is monotonic and $\lim\limits_{k \to \infty} a_k = 0$
2. The sequence of the sums $\{B_n\}$, $B_n = b_1 + b_2 + \ldots + b_n$, is bounded, i.e.,

$$\exists M > 0 \; \forall n \in N \left(|B_n| = \left|\sum_{k=1}^{n} b_k\right| \le M\right),$$

then the series $\sum\limits_{k=1}^{\infty} a_k b_k$ converges.

Let us see how Theorem 3.12 can be used by solving the next problem.

Problem 119 Investigate convergence of the series: $\sum\limits_{n=1}^{\infty} \frac{\sin n \cdot \sin n^2}{n}$.

Solution. Let $a_n = \frac{1}{n}$, $b_n = \sin n \cdot \sin n^2$. It is clear that the first sequence $\{a_n\}$ is monotonically decreasing and approaching zero. Applying formula of the product of two sines to the common term of the second sequence, we obtain
$\sin k \cdot \sin k^2 = \frac{1}{2}\left(\cos\left(k^2 - k\right) - \cos\left(k^2 + k\right) \right)$, then

$$B_n = \sum_{k=1}^{n} \sin k \cdot \sin k^2 = \frac{1}{2}\left(\sum_{k=1}^{n} \cos\left(k^2 - k\right) - \cos\left(k^2 + k\right) \right)$$

$$= \frac{1}{2}\left(\cos 0 - \cos 2 + \cos 2 - \cos 6 + \ldots + \cos\left(n^2 - n\right) - \cos\left(n^2 + n\right) \right)$$

$$= \frac{1}{2}\left(1 - \cos\left(n^2 + n\right)\right).$$

Because $|B_n| = \left|\frac{1}{2}(1 - \cos\left(n^2 + n\right))\right| \leq 1 \quad \forall n \in N$, then the series $\sum\limits_{n=1}^{\infty} \frac{\sin n \cdot \sin n^2}{n}$ is convergent.

Theorem 3.13 (Abel's Theorem) For series $\sum\limits_{k=1}^{\infty} a_k b_k$, let the following be true:

1. The sequence $\{a_k\}$ is bounded and monotonic,
2. The series $\sum\limits_{k=1}^{\infty} b_k$ converges.

   Then the series $\sum\limits_{k=1}^{\infty} a_k b_k$ converges.

Using Abel's Theorem, let us solve the following problem.

Problem 120 Investigate convergence of the series $\sum\limits_{n=1}^{\infty} \frac{\left(3 + 2\cos\frac{\pi}{2n}\right)}{\sqrt[9]{n^{11}} + 10}$.

Solution. Note that sequence $a_n = \{3 + 2\cos\frac{\pi}{2n}\}$ is bounded and monotonic and the series $\sum\limits_{n=1}^{\infty} b_n = \sum\limits_{n=1}^{\infty} \frac{1}{\sqrt[9]{n^{11}} + 10} < \sum\limits_{n=1}^{\infty} \frac{1}{n^{\frac{11}{6}}}$ converges. Hence, by Abel's Theorem the given series is convergent.

In order to solve next problem, we need to recall the inequality between an arithmetic and geometric means.

Problem 121 Given $\sum\limits_{n=1}^{\infty} a_n^2$ and $\sum\limits_{n=1}^{\infty} b_n^2$ are convergent series. Prove that the

series, $\sum\limits_{n=1}^{\infty} a_n b_n$ is absolutely convergent.

Proof. Obviously, $\sum\limits_{n=1}^{\infty} \frac{\left(a_n^2+b_n^2\right)}{2}$ is convergent. Using the inequality between arith-

metic and geometric mean, $|a_n b_n| \leq \frac{a_n^2+b_n^2}{2} \ \forall n \in N$, we can see that an absolute

value of each term of the series $\sum\limits_{n=1}^{\infty} |a_n b_n|$ does not exceed the corresponding term of

$\sum\limits_{n=1}^{\infty} \frac{\left(a_n^2+b_n^2\right)}{2}$. Therefore, by the Comparison Criterion (Theorem 3.3), the series

$\sum\limits_{n=1}^{\infty} a_n b_n$ is absolutely convergent.

3.1.3.4 An Error of Estimation of an Infinite Sum by a Partial Sum

The following corollary from Leibniz's Theorem (Theorem 3.9) allows us to estimate an infinite sum of alternating series by the sum of its first several terms.

Corollary 3.6 If the sum of the Leibniz series, S, is replaced by the sum of the first n terms (S_n) of the series, then absolute error $|r_n|$ does not exceed the absolute value of the first of the dropped terms: $|r_n| \leq |a_{n+1}|$.

The sign of the error (r_n) coincides with the sign of the first dropped term. Here $r_n = S - S_n$.

Let us apply this corollary by solving the problems below.

Problem 122 How many terms of the series $\sum\limits_{n=1}^{\infty} (-1)^{n+1} \cdot \frac{1}{n^3} =$

$1 - \frac{1}{2^3} + \frac{1}{3^3} - \frac{1}{4^3} + \ldots + \frac{(-1)^n}{n^3} + \ldots$ are needed in order to estimate the sum of the series with accuracy of 0.001?

Solution. This is an alternating series satisfying the Leibniz Theorem. Moreover, the corresponding positive series is convergent as Dirichlet series with $p = 3$, i.e.,

Table 3.1 Numerical series convergence criteria

Necessary Condition	
Positive numerical series	**Alternating numerical series**
Comparison Theorems	Leibniz Theorem
D'Alembert Ratio Test	Absolute Convergence Theorem
Cauchy Root Test	Riemann Theorem
Cauchy Integral Test	Dirichlet Theorem
Gauss or Dirichlet Test	Abel's Theorem

$1 > \frac{1}{2^3} > \frac{1}{3^3} > \frac{1}{4^3} > \ldots;$ $\lim\limits_{n \to \infty} \frac{1}{n^3} = 0.$ Hence, the given series is absolutely convergent. In order to calculate the sum of this series with the specified accuracy, it is necessary to find such a term, the absolute value of which is less than 0.001, i.e., $\frac{1}{n^3} < 0.001$ or $n^3 > 1000$, or $n > 10$. Hence, we need to add 10 the first terms of the series. Since $a_{11} = \frac{1}{11^3} < 0.001$, we obtain the following error estimate: $|r_{10}| \leq a_{11} < 0.001.$

Answer. We need to add the first ten terms of the series.

Problem 123 Find an error of estimation the sum of series $\sum\limits_{n=1}^{\infty} \frac{(-1)^{n+1}}{n} = 1$

$-\frac{1}{2} + \frac{1}{3} - \ldots + (-1)^{n+1} \cdot \frac{1}{n} + \ldots$ by the sum of its first three terms.

Solution. The Leibniz series is conditionally convergent because $|a_{n+1}| < |a_n|$, $\lim\limits_{n \to \infty} a_n = 0.$ We have that

$S_3 = 1 - \frac{1}{2} + \frac{1}{3} = \frac{5}{6}.$ The error of such estimation must satisfy $|error| \leq |-\frac{1}{4}|$ $= \frac{1}{4} = 0.25$ and the sign of the error is negative so as its fourth term $a_4 = -\frac{1}{4}$ (the first term of the series that was dropped).

Finally, let us summarize the most important convergence theorems for numerical series in Table 3.1.

3.2 Functional Series

The series

$$\sum_{n=1}^{\infty} u_n(x) = u_1(x) + u_2(x) + \ldots + u_n(x) + \ldots \qquad (3.18)$$

is called functional series if all its terms are functions of independent variable x. At each fixed value of $x = x_0$ the functional series of Eq. 3.18 becomes numerical series

$$\sum_{n=1}^{\infty} u_n(x_0) = u_1(x_0) + u_2(x_0) + \ldots + u_n(x_0) + \ldots \tag{3.19}$$

The question usually is not whether the series converges but for which value of x it converges. If series of Eq. 3.19 converge, then x_0 is the convergence point of the series of Eq. 3.18. The set of all convergence points x of functional series of Eq. 3.18 is called the convergence interval and the function $S(x) = \lim_{n \to \infty} S_n(x) = \lim_{n \to \infty} \sum_{k=1}^{n} u_k(x)$ is called the sum of the given series. The function $r_n(x) = S(x) - S_n(x)$ is called the remainder of the series of Eq. 3.19. If the series of Eq. 3.19 diverges, then the value x_0 is called the point of divergence of the series.

There are two types of functional series:

1. Power series represented by $\sum_{n=0}^{\infty} a_n x^n$, where each term is a multiple of x.

2. Trigonometric series that can be represented by $a_0 + \sum_{n=1}^{\infty} (a_n \sin nx + b_n \cos nx)$,

   where each term is a constant multiple of sine and cosine.

In the simplest cases, to determine the region of convergence of the series of Eq. 3.19, the known criteria of convergence of numerical series, assuming that x is fixed can be applied. While power series are known to be well-behaved, trigonometric series behave unpredictably most of the time. Please recall Problem 70 of Chapter 2. By proving the relationship, we actually find the nth partial sum of the series, that is,

$$S_n = \sin x + \sin 2x + \sin 3x + \ldots + \sin nx = \frac{\sin \frac{nx}{2} \cdot \sin \frac{(n+1)x}{2}}{\sin \frac{x}{2}}$$

Assume now that we have the corresponding infinite sum $\sum_{n=1}^{\infty} \sin nx$. Does the following limit exist?

$$\lim_{n \to \infty} S_n(x) = \lim_{n \to \infty} \frac{\sin \frac{nx}{2} \cdot \sin \frac{(n+1)x}{2}}{\sin \frac{x}{2}}.$$

Applying the formula for the product of two sines, which also can be found in the solution of Problem 70, we can rewrite the right hand side as

$$\lim_{n \to \infty} S_n(x) = \lim_{n \to \infty} \frac{\cos \frac{x}{2} - \cos \frac{(2n+1)x}{2}}{\sin \frac{x}{2}} = \cot \frac{x}{2} - \frac{1}{\sin \frac{x}{2}} \cdot \lim_{n \to \infty} \cos \left(\frac{2n+1}{2} \cdot x \right)$$

The infinite series diverges. Below we focus only on power series.

3.2.1 Power Series

In this section, we discuss the following important topics:

- Definition of power series, Abel's theorem
- Convergence of a power series
- Properties of convergent power series

Definition. Series of type, $a_0 + a_1(x - x_0) + \ldots + a_k(x - x_0)^k + \ldots = \sum_{k=0}^{\infty} a_k(x - x_0)^k$, are called power series of $(x - x_0)$. Here a_k, x_0 – are given real numbers, x – independent variable. Numbers a_k are called coefficients of power series.

At $x_0 = 0$ we have power series in x: $a_0 + a_1 x + \ldots + a_k x^k + \ldots = \sum_{k=0}^{\infty} a_k x^k$. Because the replacement of $x - x_0 = \xi$ for series $\sum_{k=0}^{\infty} a_k(x - x_0)^k$ can reduce it to the series $\sum_{k=0}^{\infty} a_k \xi^k$, then without loss of generality we can consider only the series $\sum_{k=0}^{\infty} a_k x^k$. The power series $\sum_{k=0}^{\infty} a_k x^k$ is always convergent at point $x = 0$; at $x \neq 0$ the power series can either converge or diverge.

Theorem 3.14 (Abel's Theorem) If power series $\sum_{k=0}^{\infty} a_k x^k$ converges at point $x_0 \neq 0$, then it is absolutely convergent on the interval $-|x_0| < x < |x_0|$. If at point $x_1 \neq 0$, the power series $\sum_{k=0}^{\infty} a_k x^k$ diverges, then it diverges at all points x, such that $|x| > |x_1|$.

Proof. Abel, Niels Henrik, Norwegian mathematician, 1802–1829.

Part 1. Assume that the series $\sum_{k=0}^{\infty} a_k x^k$ converges at point $x_0 = \beta \neq 0$, i.e.,

$\sum_{k=0}^{\infty} a_k \beta^k = a_0 + a_1 \beta + a_2 \beta^2 + \ldots + a_n \beta^n + \ldots$ converges, then by Theorem 3.1, $\lim_{n \to \infty} a_n \beta^n = 0$, and there exists a number M, such that $|a_n \beta^n| < M$, $n = 0, 1, 2, \ldots$ because the sequence is bounded. Consider now the following series

$$|a_0| + |a_1x| + \left|a_2x^2\right| + \left|a_3x^3\right| + \ldots + \left|a_nx^n\right| + \ldots$$

Clearly, if $|x| < \beta$, then for the nth term of the series the following is true

$$\left|a_nx^n\right| < \left|a_n\beta^n\right| \cdot \left|\frac{x}{\beta}\right|^n < M \cdot \left|\frac{x}{\beta}\right|^n, \quad \left|\frac{x}{\beta}\right| < 1.$$

Then all the terms of the series of the absolute values of the power series are also similarly bounded and we have the following true inequlities:

$$|a_0| < M$$

$$|a_1x| < M\left|\frac{x}{\beta}\right|$$

$$\left|a_2x^2\right| < M\left|\frac{x}{\beta}\right|^2$$

$$\ldots$$

$$\left|a_nx^n\right| < M\left|\frac{x}{\beta}\right|^n$$

$$\ldots$$

Adding   the   left   and   right   sides   of   the   inequalities,   we   obtain
$\sum_{n=0}^{\infty}\left|a_nx^n\right| < M\left(1 + \left|\frac{x}{\beta}\right| + \left|\frac{x}{\beta}\right|^2 + \cdots + \left|\frac{x}{\beta}\right|^n + \ldots\right) = \frac{M}{1 - \left|\frac{x}{\beta}\right|}$, which converges by Theorem 3.2 because the expression inside parentheses is decreasing infinite geometric series with the known sum. Hence, $\sum_{k=0}^{\infty} a_kx^k$ will absolutely converge.

Part 2. Assume that the numerical series $\sum_{k=0}^{\infty} a_k\beta^k = a_0 + a_1\beta + a_2\beta^2 + \ldots + a_n\beta^n$
$+\ldots$ diverges at $x = \beta \neq 0$, but the power series $\sum_{k=0}^{\infty} a_kx^k$ converges for $|x| > \beta$.
Convergence of the power series would lead to the convergence of the numerical series, which is a contradiction. The proof is complete.

It follows from Abel's Theorem that if the power series $\sum_{k=0}^{\infty} a_kx^k$ diverges at least at one point $x \neq 0$, then there always exists a number $R > 0$, such that the power series   converges   absolutely   for   all $x \in (-R; R)$ and   diverges   for   all $x \in (-\infty; -R) \cup (R; +\infty)$.

The number $R \geq 0$ is called the radius of convergence of the power series $\sum_{k=0}^{\infty} a_kx^k$, if the power series converges at each point of the interval $(-R; R)$ and diverges for $|x| > R$. The interval $(-R; R)$ is called the convergence interval. At the end of the interval, $x = \pm R$, the series $\sum_{k=0}^{\infty} a_kx^k$ can be as convergent as divergent. If the series converges at least at one point $x_1 = R$ or $x_2 = -R$, then these points along (together) with the convergence interval form convergence range.

If the series $\sum\limits_{k=0}^{\infty} a_k x^k$ converges only at one point $x = 0$, then $R = 0$; if it converges

for all $x \in \mathbb{R}$, then $R = \infty$. Let for the coefficients of the series $\sum\limits_{k=0}^{\infty} a_k x^k$, the limit

$\overline{\lim\limits_{k \to \infty}} \sqrt[k]{|a_k|} \neq 0$ exists, then the convergence radius can be found by the Cauchy-

Hadamard formula: $R = \dfrac{1}{\overline{\lim\limits_{k \to \infty}} \sqrt[k]{|a_k|}}$. By analogy, if the limit, $\lim\limits_{k \to \infty} \left| \dfrac{a_{k+1}}{a_k} \right| = L$, exists,

then $R = \dfrac{1}{L} = \lim\limits_{k \to \infty} \left| \dfrac{a_k}{a_{k+1}} \right|$ is the radius of convergence.

For the power series of a general type $\sum\limits_{k=0}^{\infty} a_k (x - x_0)^k$ there exists $R \in \mathbb{R}, R \geq 0$,
such that this series converges absolutely for $|x - x_0| < R$ and diverges for
$|x - x_0| > R$. Here the number $R \geq 0$ is called the convergence radius, and the interval $(x_0 - R; x_0 + R)$ is the convergence interval of the power series.

Power Series $\sum\limits_{k=0}^{\infty} a_k x^k$ have the following properties:

- if the radius of convergence of a power series is different from zero, then its sum
  is continuous on the interval of convergence $(-R; R)$;
- operation of term by term differentiation and integration on any interval $[x_0; x]$
  $\subset (-R; R)$ of power series does not change its radius of convergence;
- if the radius of convergence of a power series is different from zero, then the
  power series can be differentiated term by term on the interval of convergence;
- power series can be integrated term by term on any interval $[x_0; x]$ contained in
  the interval of convergence.
- If limits of integration α, β belong to the convergence interval of the power
  series, then definite integral of the series with these limits equal the sum of
  integrals of the terms of this series. The interval of convergence is the same.

Let $S(x) = a_0 + a_1 x + a_2 x^2 + \ldots + a_n x^n + \ldots$ be power series with interval of
convergence $(-R, R)$. The derivative series $\phi(x) = a_1 + 2a_2 x + 3a_3 x^2 + \ldots + na_n$
$x^{n-1} + \ldots$ converges on the same interval $|x| < R$ and its sum $\phi(x) = S'(x) = \frac{dS(x)}{dx}$.
A simplest power series is a geometric series, $1 + x + x^2 + \ldots + x^n + \ldots$. It is
convergent at $|x| < 1$. Hence, for the given series the radius of convergence is $R = 1$,
and the convergence interval is $(-1, 1)$. The sum of this series is $S(x) = \frac{1}{1-x}$. For
function $\qquad S(x) = \frac{1}{1-x} \qquad$ we $\qquad$ have $\qquad$ the $\qquad$ following $\qquad$ expansion,
$\frac{1}{1-x} = 1 + x + x^2 + \ldots + x^n + \ldots \quad (|x| < 1)$. Differentiating both sides of it, we
obtain $\frac{1}{(1-x)^2} = 1 + 2x + 3x^2 + 4x^3 + \ldots \ (|x| < 1)$.

Problem 124 Find the radius and convergence interval of the power series
$\sum\limits_{n=1}^{\infty} \frac{x^n}{n!}$.

Solution. Applying the formula for the convergence radius, $\lim\limits_{n\to\infty}\left|\dfrac{a_n}{a_{n+1}}\right| =$

$\lim\limits_{n\to\infty}\left|\dfrac{1}{n!}:\dfrac{1}{(n+1)!}\right| = \lim\limits_{n\to\infty}(n+1) = \infty.$ $R = \infty$, it means that the series is convergent for all x, in $(-\infty, +\infty)$. Further, notice that from it follows from the series convergence that $\lim\limits_{n\to\infty}\dfrac{x^n}{n!} = 0$ for all real x.

Problem 125 Find convergence interval of power series: $\sum\limits_{n=1}^{\infty}\dfrac{n^n}{n!}x^n$.

Solution. We find the radius of convergence using the D'Alembert Ratio Test,

$R = \lim\limits_{n\to\infty}\dfrac{n^n\cdot(n+1)!}{n!\cdot(n+1)^{n+1}} = \lim\limits_{n\to\infty}\dfrac{n+1}{\left(1+\frac{1}{n}\right)^n\cdot(n+1)} = \dfrac{1}{e}.$ Therefore, this series is convergent on

the interval $-\frac{1}{e} < x < \frac{1}{e}$. Let us investigate convergence at the ends of the interval:

1. At the left end, $x = -\frac{1}{e}$, the series $\sum\limits_{n=1}^{\infty}\dfrac{n^n}{n!}\cdot\left(-\dfrac{1}{e}\right)^n$ becomes an alternating series.

   The absolute value of its common term equals $\dfrac{n^n}{n!\cdot e^n}$ and using Stirling's Formula

   (Eq. 3.5), $n! \sim \sqrt{2\pi n}\cdot\left(\dfrac{n}{e}\right)^n$, it is equivalent at $n\to\infty$ to $\dfrac{n^n}{\sqrt{2\pi n}\cdot\left(\frac{n}{e}\right)^n\cdot e^n} = \dfrac{1}{\sqrt{2\pi n}} \to 0.$

   By the Leibniz Theorem, this series is convergent at the left end of the convergence interval.

2. Let us check for convergence of this series at the other end. At $x = \frac{1}{e}$ this series

   becomes $\sum\limits_{n=1}^{\infty}\dfrac{n^n}{n!\cdot e^n}$; using Stirling's formula and the Comparison Criterion (The-

   orem 3.3), we have $a_n = \dfrac{n^n}{n!\cdot e^n} \sim \dfrac{n^n}{\sqrt{2\pi n}\cdot\left(\frac{n}{e}\right)^n\cdot e^n} = \dfrac{1}{\sqrt{2\pi\cdot n^{1/2}}} = \dfrac{1}{\sqrt{2\pi}}\cdot\dfrac{1}{n^{1/2}}.$ Since this is a

   Dirichlet type series with $p = \frac{1}{2}$, then the given series is divergent at $x = \frac{1}{e}$.

   Therefore, the interval $\left[-\frac{1}{e}; \frac{1}{e}\right)$ is the maximal convergence interval for the given series.

Remark. We can find the interval of absolute convergence of a functional series by directly applying either the D'Alembert Ratio Test or the Cauchy Root Test.

Consider the series, $\sum\limits_{n=1}^{\infty}|u_n(x)| = |u_1(x)| + |u_2(x)| + \dots$. Let $\lim\limits_{n\to\infty}\left|\dfrac{u_{n+1}(x)}{u_n(x)}\right| = L(x)$ or

$\lim\limits_{n\to\infty}\sqrt[n]{|u_n(x)|} = L(x)$. By solving the inequality $L(x) < 1$ or $L(x) > 1$, we can find the convergence or divergence interval for the series, respectively.

Problem 126 Find convergence radius of the series $\sum\limits_{n=1}^{\infty}\dfrac{(x-4)^n}{n\cdot3^n}$.

Solution. For the series $\sum\limits_{n=1}^{\infty} \frac{(x-4)^n}{n \cdot 3^n}$ we evaluate $\lim\limits_{n\to\infty} \left| \frac{(x-4)^{n+1} n \cdot 3^n}{(n+1) \cdot 3^{n+1} (x-4)^n} \right| =$ $\lim\limits_{n\to\infty} \left| \frac{(x-4)n}{3(n+1)} \right| = \frac{|x-4|}{3} < 1$. Solving the inequality $|x-4| < 3$, we find the convergence interval as $1 < x < 7$. Next, we check the series' behavior at the ends of the interval. At $x = 1$ the series becomes $\sum\limits_{n=1}^{\infty} \frac{(-1)^n}{n}$ Leibniz series and it is conditionally convergent. At $x = 7$ the series becomes $\sum\limits_{n=1}^{\infty} \frac{1}{n}$ i.e., the harmonic series and is divergent.

Answer. The series is convergent for all $x \in [1, 7)$.

3.2.2 Taylor and Maclaurin Series

In this section, you will learn or review the following topics:

- Expansion of functions in power series
- Expansion of elementary functions in Maclaurin series
- Applications of power series

Many functions can be written as infinite power series using the expansion of Taylor or Maclaurin. The Taylor series represents a function as an infinite sum calculated from the values of its derivatives at a point. The Taylor series was introduced by the British mathematician Brook Taylor in 1715. A Taylor series that is based at $x = 0$ is called a Maclaurin series, named for the Scottish mathematician Colin Maclaurin. In order to approximate a function one can use a finite number of terms of the series.

Let the function $f(x)$ have derivatives of any order in the neighborhood of a. Then the series, $f(a) + f'(a)(x-a) + \frac{f'(a)}{2!}(x-a)^2 + \ldots + \frac{f^{(k)}(a)}{k!}(x-a)^k + \ldots$ is called the Taylor series of the function $f(x)$ at $x = a$. The convergence radius R of the series $\sum\limits_{k=0}^{\infty} \frac{f^{(k)}(a)}{k!}(x-a)^k$ can be either zero, or different from zero; moreover, in the latter case the sum $S(x)$ of Taylor Series can be different from $f(x)$. It is important to determine when in the formula, $f(x) \sim \sum\limits_{k=0}^{\infty} \frac{f^{(k)}(a)}{k!}(x-a)^k$ an equal sign is allowed, i.e., when the Taylor series converges to a function $f(x)$. If $S(x) = f(x)$ on $(a - R; a + R)$, then we say that function $f(x)$ can be expanded in a Taylor series in the neighborhood of a. Partial sums of Taylor series, $S_n(x) = f(a) + f'(a)(x-a) + \frac{f''(a)}{2!}(x-a) + \ldots + \frac{f^{(n)}(a)}{n!}(x-a)^n$ are Taylor's polynomials of $f(x)$ at a.

Theorem 3.15 (Taylor) If a function $f(x)$ has in the neighborhood $(x_0 - R; x_0 + R)$ of point x_0 derivatives of any order and $\forall x \in (x_0 - R; x_0 + R)$ the condition $\left|f^{(k)}(x)\right| \leq M \frac{k!}{R^n}$, $k = 0, 1, 2, \ldots$ is valid, then on $(x_0 - R; x_0 + R)$ function $f(x)$ can be expanded uniquely:
$$f(x) = f(x_0) + f'(x_0)(x - x_0) + \ldots + \frac{f^{(k)}(x_0)}{k!}(x - x_0)^k + \ldots.$$

A function that can be differentiated infinitely many times is called C^∞ differentiable function.

Corollary 3.7 In order C^∞ differentiable function $f(x)$ in the neighborhood x_0 can be represented by Taylor series, it is necessary and sufficient that the remainder of the expansion would approach zero, $\lim_{n \to \infty} R_n(x) = 0$ $\forall x \in (x_0 - R; x_0 + R)$.

Corollary 3.8 If for any $x \in (x_0 - R; x_0 + R)$ all derivatives of $f(x)$ are bounded by the same constant M, the Taylor series $\sum_{k=0}^{\infty} \frac{f^{(k)}(x_0)}{k!}(x - x_0)^k$ converges to this function $f(x)$ on the interval $|x - x_0| < R$.

At $x_0 = 0$, Taylor's formula has type, $f(x) = f(0) + f'(0)x + \frac{f'(0)}{2!}x^2 + \ldots + \frac{f^{(k)}(0)}{k!}x^k + \ldots$ and is called a Maclaurin series.

Basic Maclaurin series expansions:

$$\frac{1}{1-x} = 1 + x + x^2 + \ldots + x^n + \ldots \quad (|x| < 1) \tag{3.20}$$

$$e^x = 1 + \frac{x}{1!} + \frac{x^2}{2!} + \ldots + \frac{x^k}{k!} + \ldots = \sum_{k=0}^{\infty} \frac{x^k}{k!}, x \in \mathbb{R}, \tag{3.21}$$

$$\cosh x = 1 + \frac{x^2}{2!} + \frac{x^4}{4!} + \ldots + \frac{x^{2k}}{(2k)!} + \ldots = \sum_{k=0}^{\infty} \frac{x^{2k}}{(2k)!}, x \in \mathbb{R} \tag{3.22}$$

$$\sinh x = x + \frac{x^3}{3!} + \frac{x^5}{5!} + \ldots + \frac{x^{2k+1}}{(2k+1)!} + \ldots = \sum_{k=0}^{\infty} \frac{x^{2k+1}}{(2k+1)!}, x \in \mathbb{R}, \tag{3.23}$$

$$\sin x = x - \frac{x^3}{3!} + \ldots + (-1)^n \frac{x^{2k+1}}{(2k+1)!} + \ldots = \sum_{k=0}^{\infty} (-1)^k \frac{x^{2k+1}}{(2k+1)!}, x \in \mathbb{R} \tag{3.24}$$

$$\cos x = 1 - \frac{x^2}{2!} + \frac{x^4}{4!} - \ldots + (-1)^k \frac{x^{2k}}{(2k)!} + \ldots = \sum_{k=0}^{\infty} (-1)^k \frac{x^{2k}}{(2k)!} \qquad (3.25)$$

$$\ln(1+x) = x - \frac{x^2}{2} + \frac{x^3}{3} - \frac{x^4}{4} + \ldots = \sum_{k=0}^{\infty} (-1)^k \frac{x^{k+1}}{k+1}, \quad -1 < x \le 1 \qquad (3.26)$$

$$(1+x)^m = 1 + mx + \frac{m(m-1)}{2!}x^2 + \ldots + \frac{m(m-1)\ldots(m-k+1)}{k!}x^k + \ldots$$

$$= 1 + \sum_{k=1}^{\infty} \frac{m(m-1)(m-2)\ldots(m-k+1)}{k!}x^k, \quad |x| < 1. \qquad (3.27)$$

Note that power expansions for hyperbolic cosine of Eqs. 3.22–3.23 (cosh(x) and sinh(x)) are obtained using an exponential power series (Eq. 3.21) applied to the equations, $\cosh x = \frac{e^x + e^{-x}}{2}$ and $\sinh x = \frac{e^x - e^{-x}}{2}$. Using Eqs. 3.20–3.27, many other power series can be derived.

Series $\sum_{n=1}^{\infty} \frac{m(m-1)(m-2)\ldots(m-n+1)}{n!}x^n$ are binomial because at $m = n \in \mathbb{N}$ all its coefficients starting from the $(n+1)^{th}$ term become zero and power series becomes Newton's binomial distribution, $(1+x)^n = 1 + nx + \frac{n(n-1)}{2!}x^2 + \ldots + x^n = \sum_{k=0}^{n} C_n^k x^k$.

Proof. Consider the function $f(x) = (1+x)^m$ and find its several derivatives:

$$f'(x) = m(1+x)^{m-1}$$
$$f''(x) = m(m-1)(1+x)^{m-2}$$
$$\ldots$$
$$f^{(n)}(x) = m(m-1)(m-2)\ldots(m-n+1)(1+x)^{m-n}$$
at $x = 0$
$$f(0) = 1$$
$$f'(0) = m$$
$$f''(0) = m(m-1)$$
$$\ldots$$

$$(1+x)^m = 1 + mx + \frac{m(m-1)}{1 \cdot 2}x^2 + \ldots + x^n = \sum_{k=0}^{m} C_m^k x^k, \quad |x| < 1.$$

Let us demonstrate how by using binomial distribution of Eq. 3.27, we can derive other infinite series. For example, substituting x by $-x$ into Eq. 3.27, we obtain

$\frac{1}{1+x} = (1-(-x))^{-1} = 1 - x + x^2 - x^3 + \ldots + (-1)^n x^n + \ldots \quad |x| < 1, \quad m = -1.$

For $m = -\frac{1}{2}$ Eq. 3.27 takes the form,

$$\frac{1}{\sqrt{1+x}} = 1 - \frac{1}{2}x - \frac{1}{2}\cdot\left(-\frac{3}{2}\right)\cdot\frac{1}{2!}x^2 - \frac{1}{2}\cdot\left(-\frac{3}{2}\right)\cdot\left(-\frac{5}{2}\right)\cdot\frac{1}{3!}x^3 + \ldots$$

$$= 1 - \frac{1}{2}x + \frac{3}{8}x^2 - \frac{5}{16}x^3 + \frac{35}{128}x^4 + \cdots + \frac{(-1)^n(2n-1)!!}{(2n)!!}x^n + \ldots$$

$$\frac{1}{\sqrt{1-x}} = 1 + \frac{1}{2}x + \frac{1}{2}\cdot\frac{3}{4}\cdot x^2 + \frac{1}{2}\cdot\frac{3}{4}\cdot\frac{5}{6}\cdot x^3 + \ldots$$

$$+ \frac{1\cdot3\cdot5\cdot7\cdots(2n-1)}{2\cdot4\cdot6\cdot8\cdots(2n)}\cdot x^n + \ldots, |x| < 1.$$

where the !! notation is defined by

$$(2m)!! = 2\cdot4\cdot6\cdot\ldots\cdot2m$$
$$(2m-1)!! = 1\cdot3\cdot5\cdot\ldots\cdot(2m-1).$$

For $m = \frac{1}{2}$ we can obtain power expansion for two other similar functions,

$$\sqrt{1+x} = (1-(-x))^{\frac{1}{2}} = 1 + \frac{1}{2}\cdot x - \frac{1\cdot1}{2\cdot4}\cdot x^2 + \frac{1\cdot1\cdot3}{2\cdot4\cdot6}\cdot x^3 + \ldots$$

$$+ (-1)^n\frac{(2n-1)!!}{(2n+2)!!}\cdot x^{n+1} + \ldots$$

$$\sqrt{1-x} = (1-x)^{\frac{1}{2}} = 1 - \frac{1}{2}\cdot x - \frac{1\cdot1}{2\cdot4}\cdot x^2 - \frac{1\cdot1\cdot3}{2\cdot4\cdot6}\cdot x^3 - \ldots$$

$$- \frac{(2n-1)!!}{(2n+2)!!}\cdot x^{n+1} - \ldots, \quad |x| < 1$$

The Taylor and Maclaurin series are used in calculating approximate values of functions, integrals, solution of differential equations and for finding sums of infinite numerical series.

3.2.2.1 Finding the Power Expansion of Functions

Let us solve some of the problems.

Problem 127 Find the Maclaurin series for $f(x) = e^{1-4x^3}$.

Solution. The given function can be factored as $f(x) = e \cdot e^{-4x^3}$. The second factor can be expanded in to power series using Eq. 3.21 by substituting $x \to -4x^3$:

$$e^{1-4x^3} = e \cdot \left(1 - 4x^3 + \frac{(-4x^3)^2}{2!} + \ldots + \frac{(-4x^3)^n}{n!} + \ldots\right)$$

$$= e - 4ex^3 + \frac{4^2 \cdot ex^6}{2!} + \ldots + \frac{(-1)^n 4^n ex^{3n}}{n!} + \ldots$$

Answer. $f(x) = e^{1-4x^3} = \sum\limits_{k=0}^{\infty} \frac{(-1)^k 4^k ex^{3k}}{k!}$.

Problem 128 Find the Taylor expansion as powers of x for $f(x) = \ln(3 + x)$.

Solution. $\ln(3 + x) = \ln\left(3 \cdot \left(1 + \frac{x}{3}\right)\right) = \ln 3 + \ln\left(1 + \frac{x}{3}\right)$, $y = \frac{x}{3}$.
Next, using Eq. 3.26, we obtain:

$$\ln(1 + y) = y - \frac{y^2}{2} + \frac{y^3}{3} - \frac{y^4}{4} + \ldots = \sum_{k=0}^{\infty} (-1)^k \frac{y^{k+1}}{k+1}, \quad y \in (-1, 1],$$

$$\ln(3 + x) = \ln 3 + \frac{x}{3} - \frac{x^2}{2 \cdot 3^2} + \frac{x^3}{3 \cdot 3^3} - \frac{x^4}{4 \cdot 3^4} + \ldots = \ln 3 + \sum_{k=0}^{\infty} (-1)^k \frac{x^{k+1}}{(k+1)3^{k+1}}$$

Answer. $\ln(3 + x) = \ln 3 + \sum\limits_{k=0}^{\infty} (-1)^k \frac{x^{k+1}}{(k+1)3^{k+1}}$.

Problem 129 Find the Maclaurin series in x for $f(x) = \frac{1}{5-x}$.

Solution. We use Eq. 3.20,

$$\frac{1}{1 - y} = 1 + y + y^2 + \cdots + y^n + \ldots$$

$$f(x) = \frac{1}{5 - x} = \frac{1}{5} \cdot \frac{1}{1 - \frac{x}{5}};$$

$$f(x) = \frac{1}{5}\left(1 + \frac{x}{5} + \left(\frac{x}{5}\right)^2 + \ldots\right) = \frac{1}{5} + \frac{x}{5^2} + \ldots \frac{x^n}{5^{n+1}} + \ldots,$$

so $\left|\frac{x}{5}\right| < 1$ or $x \in (-5, 5)$.

The next problem is an example of how fraction decomposition, learned earlier in this book, can help us with finding Maclaurin series for a rational function.

Problem 130 Find the Maclaurin series for $g(x) = \frac{1}{x^2-3x+4}$.

Solution. Let us factor a quadratic binomial and then rewrite the function as a difference of two fractions:

$$\frac{1}{(x-4)(x+1)} = \frac{1}{5}\left(\frac{1}{x-4} - \frac{1}{x+1}\right) = -\frac{1}{5\cdot4} \cdot \frac{1}{\left(\frac{4-x}{4}\right)} - \frac{1}{5}(1+x)^{-1}$$

$$= -\frac{1}{20} \cdot \left(\frac{1}{1-\left(\frac{x}{4}\right)}\right) - \frac{1}{5} \cdot \left(\frac{1}{1+x}\right)$$

We used the fact that two quantities inside the denominator (factors) differ by 5. For each fraction we use own power expansions:

$$\frac{1}{1-\frac{x}{4}} = 1 + \frac{x}{4} + \left(\frac{x}{4}\right)^2 + \ldots + \left(\frac{x}{4}\right)^n + \ldots, \quad \left|\frac{x}{4}\right| < 1$$

$$\frac{1}{1+x} = 1 - x + x^2 - x^3 + \ldots + (-1)^n x^n + \ldots, \quad |x| < 1$$

The given function is a linear combination of two distributions, hence we obtain

$$g(x) = -\frac{1}{20} - \frac{1}{5} + \left(-\frac{1}{80} + \frac{1}{5}\right)x + \ldots + \left(\frac{(-1)^{n+1}}{5} - \frac{1}{20} \cdot \left(\frac{1}{4}\right)^n\right)x^n + \ldots$$

Answer. $g(x) = \sum_{n=0}^{\infty} \left(\frac{(-1)^{n+1}}{5} - \frac{1}{20} \cdot \left(\frac{1}{4}\right)^n\right)x^n$.

Problem 131 Find the Maclaurin series of function $f(x) = \arcsin x$.

Solution. For the expansion of $\frac{1}{\sqrt{1-x^2}}$ in a Maclaurin series, we use Eq. 3.27, replacing in this formula x by $-x^2$ and using $m = -\frac{1}{2}$. We obtain

$$\frac{1}{\sqrt{1-x^2}} = 1 + \frac{1}{2}x^2 + \frac{1\cdot3}{2\cdot4}x^4 + \ldots + \frac{1\cdot3\cdot5\cdot\ldots\cdot(2n-1)}{2\cdot4\cdot6\cdot\ldots\cdot2n}x^{2n} + \ldots$$

This series is convergent for $|x| < 1$. Integrating this series, we find, $\int\limits_0^x \frac{dx}{\sqrt{1-x^2}} =$

$$\int\limits_0^x \left(1 + \tfrac{1}{2}x^2 + \tfrac{1\cdot3}{2\cdot4}x^4 + \ldots\right) dx = x + \tfrac{1}{2}\cdot\tfrac{x^3}{3} + \tfrac{1\cdot3}{2\cdot4}\cdot\tfrac{x^5}{5} + \ldots + \tfrac{1\cdot3\cdot5\cdot\ldots\cdot(2n-1)}{2\cdot4\cdot6\cdot\ldots\cdot2n}\cdot\tfrac{x^{2n+1}}{2n+1} + \ldots$$

Because $\int\limits_0^x \frac{dx}{\sqrt{1-x^2}} = \arcsin\ x$, then $\arcsin\ x = x + \tfrac{1}{6}\cdot x^3 + \ldots + \tfrac{(2n-1)!!}{(2n)!!}\cdot\tfrac{x^{2n+1}}{2n+1} + \ldots$

This series is convergent for $|x| < 1$.

Problem 132 Prove that $\ln\left(\frac{n+1}{n}\right) = 2\left(\frac{1}{2n+1} + \frac{1}{3(2n+1)^3} + \frac{1}{5(2n+1)^5} + \ldots\right)$.

Proof. Let us rewrite the logarithmic function as follows:

$$\ln\left(\frac{n+1}{n}\right) = \ln\frac{1 + \frac{1}{2n+1}}{1 - \frac{1}{2n+1}} = \ln\left(1 + \frac{1}{2n+1}\right) - \ln\left(1 - \frac{1}{2n+1}\right).$$

Applying Eq. 3.26 using $x \to \frac{1}{2n+1}$ and $x \to -\frac{1}{2n+1}$ respectively:

$$\ln\left(1 + \frac{1}{2n+1}\right) = \frac{1}{2n+1} - \frac{1}{2(2n+1)^2} + \frac{1}{3(2n+1)^3} - \ldots$$

$$\ln\left(1 - \frac{1}{2n+1}\right) = -\frac{1}{2n+1} - \frac{1}{2(2n+1)^2} - \frac{1}{3(2n+1)^3} - \ldots$$

and subtracting two expansions, we obtain the requested formula.

3.2.2.2 Method of Undetermined Coefficients

Problem 133 Find the Maclaurin series of $f(x) = x \cdot \cot x$.

Solution. Assume that $f(x) = x \cot x = a_0 + a_1 x + a_2 x^2 + \ldots + a_n x^n + \ldots$, where the coefficients are to be determined. By definition of a cotangent, we can state that $\cot x = \frac{\cos x}{\sin x} \Rightarrow \cos x = \cot x \cdot \sin x$. Multiplying both sides of the second relationship by x, we have $x \cdot \cos x = (x \cot x) \cdot \sin x$. Then we can multiply the power expansion of $\cos x$ by x on the left and undetermined series and series for $\sin x$:

$$x\left(1 - \frac{x^2}{2!} + \frac{x^4}{4!} - \frac{x^6}{6!} + \ldots + (-1)^n \cdot \frac{x^{2n}}{(2n)!} + \ldots\right)$$

$$= \left(x - \frac{x^3}{3!} + \frac{x^5}{5!} - \ldots (-1)^{n-1} \cdot \frac{x^{2n-1}}{(2n-1)!} \cdot\right) \cdot (a_0 + a_1 x + a_2 x^2 + \ldots + a_n x^n + \ldots)$$

Equating the coefficients of the same powers of x on both sides, we obtain

$$a_0 = 1$$

$$a_1 = 0$$

$$a_2 - \frac{a_0}{6} = -\frac{1}{2} \Rightarrow \boxed{a_2 = -\frac{1}{3}}$$

$$a_3 = 0$$

$$\frac{1}{24} = \frac{a_0}{120} - \frac{a_2}{6} + a_4$$

$$\boxed{a_4 = -\frac{1}{45}}$$

Therefore,

$$x \cot x = 1 - \frac{1}{3}x^2 - \frac{1}{45}x^4 + \ldots \tag{3.28}$$

On the other hand, it can be shown (See for example, "Mathematical Analysis" by Vinogradova, Olehnik, and Sadovnichii [14], page 330) that the same function on the left has the following representation:

$$x \cot x = 1 + 2 \cdot \sum_{m=1}^{\infty} \frac{x^2}{x^2 - \pi^2 m^2}, \quad x \neq \pi k, \ k = 0, 1, 2, \ldots \tag{3.29}$$

If $|x| < 1$, let us manipulate the mth term of the series of Eq. 3.29,

$$\frac{x^2}{x^2 - \pi^2 m^2} = \frac{\frac{x^2}{\pi^2 m^2}}{\frac{x^2}{\pi^2 m^2} - 1} = -\left(\frac{\frac{x^2}{\pi^2 m^2}}{1 - \frac{x^2}{\pi^2 m^2}}\right) = -\sum_{n=1}^{\infty} \left(\frac{x^2}{\pi^2 m^2}\right)^n \tag{3.30}$$

It is easy to see why Eq. 3.30 is true. Consider a derivative, $\left(\frac{y}{1-y}\right)' = \frac{1}{(1-y)^2} = (1-y)^{-2} = 1 + 2y + \ldots$. By integrating both sides, we have $\frac{y}{1-y} = y + y^2 + y^3 + \ldots$. Substituting $y = \frac{x^2}{\pi^2 m^2}$, we get Eq. 3.30. If we substitute the mth term given by Eq. 3.30 into Eq. 3.29 we obtain the power series for the requested function,

$$x\cot x = 1 - 2 \cdot \sum_{m=1}^{\infty} \sum_{n=1}^{\infty} \left(\frac{x^2}{\pi^2 m^2}\right)^n = 1 - 2 \cdot \sum_{n=1}^{\infty} \left(\frac{x^2}{\pi^2}\right)^n \cdot \sum_{m=1}^{\infty} \left(\frac{1}{m^2}\right)^n. \quad (3.31)$$

The right hand sides of Eqs. 3.29 and 3.31 are the power expansions of the same function. By equating coefficients for x^2 and x^4, respectively, we can obtain two very famous formulas:

Let $n = 1$, $-\frac{1}{3}x^2 = -\frac{2}{\pi^2}\left(\sum_{m=1}^{\infty}\frac{1}{m^2}\right) \cdot x^2 \implies \boxed{\sum_{m=1}^{\infty}\frac{1}{m^2} = \frac{\pi^2}{6}}$

Let $n = 2$, $-\frac{1}{45}x^4 = -\frac{2x^4}{\pi^4}\left(\sum_{m=1}^{\infty}\frac{1}{m^4}\right) \implies \boxed{\sum_{m=1}^{\infty}\frac{1}{m^4} = \frac{\pi^4}{90}}$

Both formulas were first proven by Euler. The first formula is called the Basel Problem. Its solution is demonstrated in Section 3.3.

3.2.2.3 Using Complex Numbers

Problem 134 demonstrates how complex numbers help in the evaluation of the Maclaurin Series of some functions.

Problem 134 Find the Maclaurin series of $f(x) = e^x \sin x$.

Solution. Using the Euler formula $\sin x = \frac{e^{ix} - e^{-ix}}{2i}$ we can rewrite our function as

$f(x) = e^x\left(\frac{e^{ix} - e^{-ix}}{2i}\right) = \frac{e^{(1+i)x} - e^{(1-i)x}}{2i}$ and its n^{th}derivative as

$$f^{(n)}(x) = \frac{(1+i)^n e^{(1+i)x} - (1-i)^n e^{(1-i)x}}{2i} =$$
$$= \frac{e^x((1+i)^n e^{ix} - (1-i)^n e^{-ix})}{2i}$$

Using De Moivre's Formula, we can simplify it as

$$(1+i)^n = (\sqrt{2})^n\left(\cos\frac{\pi}{4} \cdot n + i\sin\frac{\pi}{4} \cdot n\right) = (\sqrt{2})^n e^{i\frac{\pi}{4}n}$$
$$(1-i)^n = (\sqrt{2})^n\left(\cos\frac{7\pi}{4} \cdot n + i\sin\frac{7\pi}{4} \cdot n\right) = (\sqrt{2})^n\left(\cos\frac{-\pi}{4} \cdot n + i\sin\frac{-\pi}{4} \cdot n\right)$$
$$= (\sqrt{2})^n e^{-i\cdot\frac{\pi}{4}n}.$$

Substituting these into formula for the nth derivative, we get

$$f^{(n)}(x) = \frac{e^x\left(\left(\sqrt{2}\right)^n e^{\frac{\pi}{4}n}e^{ix} - \left(\sqrt{2}\right)^n e^{-i\frac{\pi}{4}n}e^{-ix}\right)}{2i} = \frac{e^x \cdot \left(\sqrt{2}\right)^n}{2i} \cdot \left(e^{\left(\frac{\pi}{4}n+x\right)i} - e^{-\left(\frac{\pi}{4}n+x\right)i}\right)$$

$$= \left(\sqrt{2}\right)^n e^x \sin\left(x + \frac{\pi}{4}n\right).$$

At zero it has value $f^{(n)}(0) = \left(\sqrt{2}\right)^n \sin\frac{\pi n}{4}$. The Maclaurin series is
$$f(x) = e^x \sin x = \sum_{n=0}^{\infty} \frac{\left(\sqrt{2}\right)^n \sin\frac{\pi n}{4}}{n!} \cdot x^n.$$

3.3 Methods of Finding Sums for Infinite Series

Finding an infinite sum for convergent series is not always easy. However, in this section you will learn several methods that are worthwhile to mention.

3.3.1 Using Method of Partial Sums

Problem 135 Evaluate the sum, $S_n = x + 4x^3 + 7x^5 + 10x^7 + \ldots + (3n-2)x^{2n-1}$

Solution. Multiply   the   sum   by x^2,   i.e., $S_n \cdot x^2 = x^3 + 4x^5 + 7x^7 + \ldots + (3n-2)x^{2n+1}$ and subtract this product to get $S_n - S_n x^2 = (1-x^2)S_n = x + 3x^3 + 3x^5 + 3x^7 + \ldots + 3x^{2n-1} - (3n-2)x^{2n+1}$ or $S_n(1-x^2) = x - (3n-2)x^{2n+1} + 3 \cdot \frac{x^3\left(1-x^{2(n-1)}\right)}{1-x^2}$. Dividing both sides by $(1-x^2)$, we obtain $S_n = \frac{x-(3n-2)x^{2n+1}}{(1-x^2)} + \frac{3x^3\left(1-x^{2(n-1)}\right)}{(1-x^2)^2}$. It is clear that convergence (divergence) of a functional series depends on the value of the independent variable x, i.e., $\lim_{n\to\infty}\frac{u_{n+1}}{u_n} = x^2 \cdot \lim_{n\to\infty}\frac{3n+1}{3n-2} = x^2 < 1 \Rightarrow |x| < 1$.   If $-1 < x < 1$, then   the corresponding infinite series converges to $\lim_{n\to\infty} S_n = \frac{x}{(1-x^2)} + \frac{3x^3}{(1-x^2)^2}$. If $x = 1$, then the given series becomes an arithmetic series with the first term 1 and the common difference of 3, i.e., $S_n = 1 + 4 + 7 + \ldots + 3n - 2 = \frac{1+3n-2}{2} \cdot n = \frac{n(3n-1)}{2}$,

Hence the corresponding infinite series is divergent. A similar result can be obtained for $x = -1$.

Answer. $S_n = \frac{x-(3n-2)x^{2n+1}}{(1-x^2)} + \frac{3x^3\left(1-x^{2(n-1)}\right)}{(1-x^2)^2}.$

Problem 136 Evaluate the infinite sum $\sum_{n=1}^{\infty} \frac{2n-1}{\left(\sqrt{2}\right)^n}.$

Solution. In Problem 106 we establish that this series converges. Let us evaluate the partial sum of the series. First, we write several terms of the series:

$$S_n = \frac{1}{\sqrt{2}} + \frac{3}{\left(\sqrt{2}\right)^2} + \frac{5}{\left(\sqrt{2}\right)^3} + \frac{7}{\left(\sqrt{2}\right)^4} + \ldots + \frac{2n-3}{\left(\sqrt{2}\right)^{n-1}} + \frac{2n-1}{\left(\sqrt{2}\right)^n}$$

Next, we multiply this sum by $\sqrt{2}$,

$$\sqrt{2}S_n = 1 + \frac{3}{\sqrt{2}} + \frac{5}{\left(\sqrt{2}\right)^2} + \frac{7}{\left(\sqrt{2}\right)^3} + \frac{9}{\left(\sqrt{2}\right)^4} + \ldots + \frac{2n-3}{\left(\sqrt{2}\right)^{n-2}} + \frac{2n-1}{\left(\sqrt{2}\right)^{n-1}}.$$

Now we rewrite it in a different form by extracting 2 within each numerator:

$$\sqrt{2}S_n = 1 + \frac{2+1}{\sqrt{2}} + \frac{2+3}{\left(\sqrt{2}\right)^2} + \frac{2+5}{\left(\sqrt{2}\right)^3} + \frac{2+7}{\left(\sqrt{2}\right)^4} + \ldots + \frac{2+2n-5}{\left(\sqrt{2}\right)^{n-2}} + \frac{2+2n-3}{\left(\sqrt{2}\right)^{n-1}}$$

Break this into two series, add the last term to the second series, then subtract it outside the braces to keep things in balance.

$$\sqrt{2}S_n = 1 + \left(\frac{2}{\sqrt{2}} + \frac{2}{\left(\sqrt{2}\right)^2} + \ldots + \frac{2}{\left(\sqrt{2}\right)^{n-2}} + \frac{2}{\left(\sqrt{2}\right)^{n-1}}\right)$$
$$+ \left\{\frac{1}{\sqrt{2}} + \frac{3}{\left(\sqrt{2}\right)^2} + \frac{5}{\left(\sqrt{2}\right)^3} + \ldots + \frac{2n-1}{\left(\sqrt{2}\right)^n}\right\} - \frac{2n-1}{\left(\sqrt{2}\right)^n}$$

Notice that the expression inside parentheses is $(n-1)$ terms of the geometric series with the first term $\frac{2}{\sqrt{2}}$ and common ratio $\frac{1}{\sqrt{2}}$. The expression inside the braces is the unknown partial sum, S_n. Collecting like terms and factoring S_n on the left hand side, and after evaluating the sum of the geometric series on the right, we obtain the following:

$$(\sqrt{2}-1)S_n = 1 + \frac{\sqrt{2}\left(1-\dfrac{1}{\sqrt{2^{n-1}}}\right)}{1-\dfrac{1}{\sqrt{2}}} - \frac{2n-1}{(\sqrt{2})^n}$$

$$(\sqrt{2}-1)S_n = 1 + \frac{2\left(1-\dfrac{1}{\sqrt{2^{n-1}}}\right)}{\sqrt{2}-1} - \frac{2n-1}{(\sqrt{2})^n}$$

$$(\sqrt{2}-1)S_n = 1 + 2(1+\sqrt{2})\left(1-\frac{1}{(\sqrt{2})^{n-1}}\right) - \frac{2n-1}{(\sqrt{2})^n}$$

Dividing the last equation by $(\sqrt{2}-1)$, solving it for the partial sum and simplifying, we have

$$S_n = \frac{1+2(\sqrt{2}+1)\left(1-\dfrac{1}{\sqrt{2^{n-1}}}\right) - \dfrac{2n-1}{\sqrt{2^n}}}{\sqrt{2}-1}$$

$$= 7+5\sqrt{2} - \frac{2(\sqrt{2}+1)^2}{\sqrt{2^{n-1}}} - \frac{(2n-1)(\sqrt{2}+1)}{\sqrt{2^n}}$$

Finally, the partial sum is given by $S_n = 7+5\sqrt{2} - \frac{2(\sqrt{2}+1)^2}{\sqrt{2^{n-1}}} - \frac{(2n-1)(\sqrt{2}+1)}{\sqrt{2^n}}$ and $\lim\limits_{n\to\infty} S_n = 7+5\sqrt{2}$.

It is easy to see that this formula is correct. Let $n=2$, then from the given series we obtain $S_2 = \frac{1}{\sqrt{2}} + \frac{3}{2} = \frac{3+\sqrt{2}}{2}$. Next, if we substitute $n=2$ into the formula, we have

$$S_2 = \frac{(7+5\sqrt{2})\cdot 2 - 2\sqrt{2}(3+2\sqrt{2}) - 3\sqrt{2} - 3}{2}$$

$$= \frac{14+10\sqrt{2} - 6\sqrt{2} - 8 - 3\sqrt{2} - 3}{2} = \frac{\sqrt{2}+3}{2}.$$

Answer. $\lim\limits_{n\to\infty} S_n = 7+5\sqrt{2}$.

3.3.2 Using Power Series of Elementary Functions

Knowledge of the Taylor or Maclaurin series of common functions helps us to find sums of convergent series.

Example. Consider the sum $\sum\limits_{n=0}^{\infty} \frac{2^n}{n!} = 1 + \frac{2}{1!} + \frac{2^2}{2!} + \frac{2^3}{3!} + \frac{2^4}{4!} + \ldots + \frac{2^n}{n!} + \ldots$.

By D'Alembert (Corollary 3.3), this series is convergent. If we compare this series to the Maclaurin series for $y = e^x$ (Eq. 3.21), it is clear that the given series is that one for $x = 2$. Therefore, its sum is e^2.

Let us see the following problem.

Problem 137 Find the sum of an infinite series $1 - \frac{100^2}{2!} + \frac{100^4}{4!} - \frac{100^6}{6!} + \ldots + \frac{(-1)^n \cdot 100^{2n}}{(2n)!} + \ldots$.

Solution. This series coincides with the series for $y = \cos x$ at $x = 100$ (Eq. 3.28). Hence, the sum equals $\cos(100)$.

Answer. $\cos(100)$.

Let us consider the following problem.

Problem 138 Evaluate the sum of an infinite series $\left(\frac{1}{2}\right)^3 - \left(\frac{1}{2}\right)^5 \cdot \frac{1}{3!} + \left(\frac{1}{2}\right)^7 \cdot \frac{1}{5!} - \left(\frac{1}{2}\right)^9 \cdot \frac{1}{7!} + \ldots + \frac{(-1)^{n-1}}{2^{2n+1} \cdot (2n-1)!} + \ldots$.

Solution. We can factor out $\left(\frac{1}{2}\right)^2$ and obtain the quantity inside parentheses that looks familiar:

$$\left(\frac{1}{2}\right)^2 \cdot \left(1 - \left(\frac{1}{2}\right)^3 \cdot \frac{1}{3!} + \left(\frac{1}{2}\right)^5 \cdot \frac{1}{5!} - \left(\frac{1}{2}\right)^7 \cdot \frac{1}{7!} + \ldots + \frac{(-1)^{n-1}}{2^{2n-1} \cdot (2n-1)!} + \ldots \right).$$

We can see that the expression inside parenthesis is the infinite series for $y = \sin x$ at $x = \frac{1}{2}$ (Eq. 3.13). Therefore, the given sum is $\frac{\sin(0.5)}{4}$.

Answer. $\frac{\sin(0.5)}{4}$.

The next problem is also interesting.

Problem 139 Find $1 + \frac{1}{2} \cdot \frac{1}{3} + \frac{1}{2} \cdot \frac{3}{4} \cdot \frac{1}{5} + \frac{1 \cdot 3 \cdot 5 \cdot 1}{2 \cdot 4 \cdot 6 \cdot 7} + \ldots + \frac{(2n-1)!!}{(2n)!!} \cdot \frac{1}{2n+1} + \ldots$.

Solution. In Problem 131 we established that for all $|x| < 1$ that $x + \sum_{n=1}^{\infty} \frac{(2n-1)!!}{(2n)!!} \cdot \frac{x^{2n+1}}{2n+1} = \arcsin x$. By comparing two formulas, we find that they are equal at $x = 1$. Hence $1 + \sum_{n=1}^{\infty} \frac{(2n-1)!!}{(2n)!!} \cdot \frac{1}{2n+1} = \arcsin 1 = \frac{\pi}{2}$.

Answer. $\frac{\pi}{2}$.

There are famous series with the known sums:

$$\sum_{n=1}^{\infty} \frac{(-1)^{n+1}}{n} = \ln 2 \quad \text{(Leibniz series)} \tag{3.32}$$

$$\sum_{n=1}^{\infty} \frac{1}{n^2} = \frac{\pi^2}{6} \quad \text{(Dirichlet series)} \tag{3.33}$$

$$\sum_{n=1}^{\infty} \frac{1}{n^4} = \frac{\pi^4}{90} \tag{3.34}$$

$$\sum_{n=1}^{\infty} \frac{1}{n(n+m)} = \frac{1}{m} \cdot \sum_{k=1}^{\infty} \frac{1}{k} \tag{3.35}$$

Let us solve the problems using these series and their sums.

Problem 140 Find the sum of an infinite series $\frac{1}{4} + \frac{1}{36} + \frac{1}{144} + \frac{1}{400} + \frac{1}{900} + \frac{1}{1764} + \dots$.

Solution. The series can be rewritten as

$$\left(\frac{1}{1 \cdot 2}\right)^2 + \left(\frac{1}{2 \cdot 3}\right)^2 + \left(\frac{1}{3 \cdot 4}\right)^2 + \left(\frac{1}{4 \cdot 5}\right)^2 + \dots + = \sum_{n=1}^{\infty} \frac{1}{n^2 \cdot (n+1)^2}$$

$$= \sum_{n=1}^{\infty} \left(\frac{1}{n} - \frac{1}{n+1}\right)^2 = \sum_{n=1}^{\infty} \frac{1}{n^2} - 2 \cdot \sum_{n=1}^{\infty} \frac{1}{n(n+1)} + \sum_{n=1}^{\infty} \frac{1}{(n+1)^2}$$

The first series is Dirichlet series (Eq. 3.33); it converges to $\frac{\pi^2}{6}$. The second series of Eq. 3.35 converges to 2, and the last series can be written as $\sum_{n=1}^{\infty} \frac{1}{(n+1)^2} = \frac{1}{2^2} + \frac{1}{3^2} + \frac{1}{4^2} + \dots = \sum_{n=1}^{\infty} \frac{1}{n^2} - 1$. Here we again extracted the series of Eq. 3.33.

Finally, $\sum_{n=1}^{\infty} \frac{1}{n^2 \cdot (n+1)^2} = 2 \cdot \sum_{n=1}^{\infty} \frac{1}{n^2} - 2 \cdot \sum_{n=1}^{\infty} \frac{1}{n(n+1)} - 1 = 2 \cdot \frac{\pi^2}{6} - 2 - 1 = \frac{\pi^2}{3} - 3$.

Answer. $\frac{\pi^2}{3} - 3$.

Problem 141 Prove that $\sum\limits_{n=1}^{\infty} \frac{1}{n^2} = \frac{\pi^2}{6}$.

Proof. This problem is also known as Basel Problem. We summarize the ideas of Euler's proof. Euler knew that any polynomial having n zeros, $x_1, x_2, \ldots x_n$ can be factored as $P(x) = 0 \Leftrightarrow \left(1 - \frac{x}{x_1}\right)\left(1 - \frac{x}{x_2}\right)\left(1 - \frac{x}{x_3}\right) \cdot \ldots \cdot \left(1 \frac{x}{x_n}\right) = 0$. Next, Euler considered the function $f(x) = \frac{\sin x}{x}$, $x \neq 0$, $\lim\limits_{x \to 0} \frac{\sin x}{x} = 1$. On one hand, because $\sin x = 0 \Leftrightarrow x = \pi n, n = 0, \pm 1, \pm 2, \ldots$, then $\frac{\sin x}{x} = 0 \Leftrightarrow x = \pi n, n = \pm 1, \pm 2, \pm 3, \ldots$ he represented as a function with infinitely many zeros:

$$\frac{\sin x}{x} = \left(1 - \frac{x}{\pi}\right)\left(1 + \frac{x}{\pi}\right)\left(1 - \frac{x}{2\pi}\right)\left(1 + \frac{x}{2\pi}\right) \ldots \left(1 - \frac{x}{\pi n}\right)\left(1 + \frac{x}{\pi n}\right) \ldots$$

Applying the difference of squares formula for each pair of factors, this was rewritten as infinite product:

$$\frac{\sin x}{x} = \left(1 - \frac{x^2}{\pi^2}\right)\left(1 - \frac{x^2}{4\pi^2}\right)\left(1 - \frac{x^2}{9\pi^2}\right) \ldots$$

On the other hand, for the same function we can find infinite Maclaurin series by dividing Eq. 3.24 by x:

$$\frac{\sin x}{x} = 1 - \frac{x^2}{3!} + \frac{x^4}{5!} - \frac{x^6}{7!} + \ldots = \sum_{k=0}^{\infty} \frac{(-1)^k x^{2k}}{(2k+1)!}$$

The right hand sides of two polynomial representations of the same function must be equal, then the coefficients at each power of x must be also equated. The constant term in both polynomials is 1.

Let us equate the coefficients of x^2:

$$x^2 \cdot \left(-\frac{1}{6}\right) = x^2 \left(-\frac{1}{\pi^2} - \frac{1}{4\pi^2} - \frac{1}{9\pi^2} - \ldots\right)$$

$$1 + \frac{1}{4} + \frac{1}{9} + \ldots = \frac{\pi^2}{6}$$

This proves the first infinite sum: $1 + \frac{1}{2^2} + \frac{1}{3^2} + \ldots = \sum\limits_{n=1}^{\infty} \frac{1}{n^2} = \frac{\pi^2}{6}$.

Remark. I noticed that sometimes students do not understand right away how the coefficients of x^2 in the polynomial written as an infinite product and the corresponding Maclaurin series were equated.

For example, a finite product can be expanded as follows:

$$(A + B)(C + D)(E + F) = A \cdot C \cdot E + A \cdot C \cdot F + A \cdot C \cdot E + A \cdot D \cdot F + B \cdot C \cdot E$$
$$+ B \cdot C \cdot F + B \cdot D \cdot E + B \cdot D \cdot F.$$

We had three parentheses and each term of the expansion on the right hand side, e.g., $A \cdot C \cdot E$ equals the product of the terms taken from each parentheses but only one term from each parenthesis is present in each such a product.

If we multiply only three factors on the Euler's infinite product, we obtain:

$$\left(1 - \frac{x^2}{\pi^2}\right)\left(1 - \frac{x^2}{4\pi^2}\right)\left(1 - \frac{x^2}{9\pi^2}\right) = 1 \cdot 1 \cdot 1 + 1 \cdot 1 \cdot \left(-\frac{x^2}{9\pi^2}\right) + 1 \cdot \left(-\frac{x^2}{4\pi^2}\right) \cdot 1$$

$$+ 1 \cdot \left(-\frac{x^2}{4\pi^2}\right) \cdot \left(-\frac{x^2}{9\pi^2}\right) + \left(-\frac{x^2}{\pi^2}\right) \cdot 1 \cdot 1 + \left(-\frac{x^2}{\pi^2}\right) \cdot 1 \cdot \left(-\frac{x^2}{9\pi^2}\right)$$

$$+ \left(-\frac{x^2}{\pi^2}\right)\left(-\frac{x^2}{4\pi^2}\right) \cdot 1 + \left(-\frac{x^2}{\pi^2}\right)\left(-\frac{x^2}{4\pi^2}\right)\left(-\frac{x^2}{9\pi^2}\right).$$

From this formula, one can see that the constant term is 1, and that the coefficient of x^2 is $-\frac{1}{\pi^2} - \frac{1}{4\pi^2} - \frac{1}{9\pi^2}$. For the infinite product considered in the Basel Problem, this coefficient will be represented by the corresponding infinite series of similar type.

Problem 142 Evaluate $\sum_{n=2}^{\infty} \frac{(-1)^n}{n^2+n-2}$.

Solution. The absolute value of the common term of the series can be represented as $\frac{1}{n^2+n-2} = \frac{1}{(n-1)(n+2)} = \frac{1}{3} \cdot \frac{1}{n-1} - \frac{1}{3} \cdot \frac{1}{n+2}$ (because two quantities inside the denominator differ by 3).

Using sigma notation's properties and shifting index of the summation, $\sum_{n=2}^{\infty} \frac{(-1)^n}{n-1}$

$$= \sum_{n=1}^{\infty} \frac{(-1)^{n+1}}{n} \text{ (Leibniz)}$$

$$-\sum_{n=2}^{\infty} \frac{(-1)^n}{n+2} = \sum_{n=2}^{\infty} \frac{(-1)^{n+1}}{n+2} = \sum_{n=1}^{\infty} \frac{(-1)^{n+1}}{n} - 1 + \frac{1}{2} - \frac{1}{3}.$$

Finally, using Eq. 3.32, we obtain the answer,

$$\sum_{n=2}^{\infty} \frac{(-1)^n}{n^2+n-2} = \frac{1}{3}\sum_{n=1}^{\infty}\frac{(-1)^{n+1}}{n} + \frac{1}{3}\sum_{n=2}^{\infty}\frac{(-1)^{n+1}}{n+2}$$

$$= \frac{1}{3}\sum_{n=1}^{\infty}\frac{(-1)^{n+1}}{n} + \frac{1}{3}\left(\sum_{n=1}^{\infty}\frac{(-1)^{n+1}}{n} - 1 + \frac{1}{2} - \frac{1}{3}\right)$$

$$= \frac{2}{3}\sum_{n=1}^{\infty}\frac{(-1)^{n+1}}{n} - \frac{5}{18} = \frac{2}{3}\ln 2 - \frac{5}{18}.$$

Answer. $\frac{2}{3}\ln 2 - \frac{5}{18}$.

Problem 143 Evaluate the sum of infinite series $\sum_{n=1}^{\infty}\frac{1}{n(2n+1)}$.

Solution. Rewriting this series in an equivalent form by multiplying it and dividing by 2, we obtain two quantities within the denominator that differ by two, and hence the series can be evaluated as the difference of two series:

$$\sum_{n=1}^{\infty}\frac{2}{(2n)(2n+1)} = 2\cdot\left(\sum_{n=1}^{\infty}\frac{1}{2n} - \sum_{n=1}^{\infty}\frac{1}{2n+1}\right)$$

$$= 2\left(\left(\frac{1}{2}+\frac{1}{4}+\frac{1}{6}+\ldots\right) - \left\{\frac{1}{3}+\frac{1}{5}+\frac{1}{7}+\ldots\right\}\right)$$

$$= -2(\ln 2 - 1).$$

Answer. $\sum_{n=1}^{\infty}\frac{1}{n(2n+1)} = -2(\ln 2 - 1)$.

3.3.3   *Method of Differentiation and Integration of Series*

By differentiating and integrating well-known Maclaurin series Eqs. 3.20–3.27, new power series can be obtained and used for finding sums of convergent infinite numerical series. In this section we use the following rules of differentiation:

$$(e^x)' = e^x, \frac{d}{dx}(e^u) = e^u \frac{du}{dx}$$

$$(x^n)' = nx^{n-1}, \quad \frac{d}{dx}(u^n) = nu^{n-1} \frac{du}{dx}$$

$$(\ln(x))' = \frac{1}{x}, \quad \frac{d}{dx}(\ln u) = \frac{1}{u} \frac{du}{dx}$$

$$(u \cdot v)' = u'v + uv' \text{ (derivative of a product)}$$

$$\left(\frac{u}{v}\right)' = \frac{u'v - uv'}{v^2} \text{ (derivative of a quotient)}$$

By applying Maclaurin's formula and sigma notation, some well-known functions can be written as infinite series.

$$\ln\frac{1}{1-x} = \sum_{n=1}^{\infty} \frac{x^n}{n}, \quad |x| < 1 \tag{3.36}$$

$$\frac{1-x^m}{1-x} = 1 + x + x^2 + x^3 + \ldots + x^{m-1} \tag{3.37}$$

How can we prove Eq. 3.36? Equation 3.20 for an infinite geometric series after integration of both sides will give us Eq. 3.36 by expansion of a natural logarithm. Also we applied properties of logarithms to the left hand side:

$$\int\limits_0^x \frac{dx}{1-x} = -\ln(1-x) = \ln(1-x)^{-1} = \ln\frac{1}{1-x}, \quad (0 < x < 1)$$

From these formulas new ones can be obtained. For example, if we take a derivative of both sides of Eq. 3.37, we obtain $\left(\frac{1-x^m}{1-x}\right)' = 1 + 2x + 3x^2 + \ldots + (m-1)x^{m-2}$ so

$$\frac{-mx^{m-1}(1-x) - (1-x^m)(-1)}{(1-x)^2} = 1 + 2x + 3x^2 + 4x^3 + \ldots + (m-1)x^{m-2}.$$

Finally, we obtain the formula for finding a partial sum such as

$$1 + 2x + 3x^2 + 4x^3 + \ldots + (m-1)x^{m-2} = \frac{x^m(m-1) - mx^{m-1} + 1}{(1-x)^2} \tag{3.38}$$

Let us see how Eq. 3.28 can be applied to the Problem 144.

Problem 144 Find the sum: $1 + 2x + 3x^2 + 4x^3 + \ldots + (n+1)x^n = \sum_{k=0}^{n}(k+1)x^k$.

Solution. We consider two different ways of solving this problem. Method 1 is based on no knowledge of differentiation. In Method 2 we present the use of Eq. 3.28 obtained earlier.

Method 1. The given sum is a power series. You can notice that each term is neither an arithmetic nor geometric sequence term.

Let us denote $S_n = 1 + 2x + 3x^2 + 4x^3 + \ldots + (n+1)x^n$. Multiplying both sides of it by x we obtain, $xS_n = x + 2x^2 + 3x^3 + 4x^4 + \ldots + nx^n + (n+1)x^{n+1}$. Subtracting second expression from the first, $S_n - xS_n = 1 + x + x^2 + x^3 + \ldots + x^n - (n+1)x^{n+1}$. The first part of this formula can be evaluated as the sum of a geometric sequence with $b_1 = 1$, $r = x$. Thus, $(1-x)S_n = \frac{1 \cdot (1-x^{n+1})}{1-x} - (n+1)x^{n+1}$, if $x \neq 1$ and then $S_n = \frac{1-x^{n+1}}{(1-x)^2} - \frac{(n+1)x^{n+1}}{1-x}$.

If $x = 1$, then $S_n = 1 + 2 + 3 + 4 + \ldots + (n+1) = \frac{(n+1)(n+2)}{2}$.

We of course notice that this is an arithmetic series.

Method 2. If we compare Eq. 3.28 and the given problem, we can see that they are equal if $m = n + 2$. Therefore, $S_n = 1 + 2x + 3x^2 + 4x^3 + \ldots + (n+1)x^n = \frac{x^{n+2}(n+1)-(n+2)x^{n+1}+1}{(1-x)^2}$, which is identical to what we found using Method 1, if you put fractions over the common denominator and simplify the expression.

Can we find the value of the corresponding infinite sum? $\sum_{n=0}^{\infty} (n+1)x^n = \lim_{n \to \infty} S_n = \frac{1}{(1-x)^2}$. Because $|x| < 1 \Rightarrow x^{n+1} \to 0$, $(n+1)x^{n+1} \to 0$ as $n \to \infty$. A similar problem is offered to you in the homework section.

Answer. If $x \neq 1$, $S_n = \frac{1-x^{n+1}}{(1-x)^2} - \frac{(n+1)x^{n+1}}{1-x}$, if $x = 1$, $S_n = \frac{(n+1)(n+2)}{2}$

Knowledge of power series allows us to find sums of convergent infinite numerical series.

Problem 145 Evaluate the infinite sum: $\frac{1}{2 \cdot 3^2} + \frac{1}{3 \cdot 3^3} + \frac{1}{4 \cdot 3^4} + \ldots + \frac{1}{n \cdot 3^n} + \ldots$.

Solution. The given series can be rewritten as $S = \frac{1}{2} \cdot \left(\frac{1}{3}\right)^2 + \frac{1}{3} \cdot \left(\frac{1}{3}\right)^3 + \frac{1}{4} \cdot \left(\frac{1}{3}\right)^4 + \ldots + \frac{1}{n} \cdot \left(\frac{1}{3}\right)^n + \ldots = \sum_{k=2}^{\infty} \frac{\left(\frac{1}{3}\right)^k}{k}$. Then if we replace $x = \frac{1}{3}$ in the series of Eq. 3.26 for the expansion of $\ln(1-x)$, then the obtained expression above will be that sum minus its first term, $\ln\frac{1}{1-\frac{1}{3}} = \frac{1}{3} + S$ so $S = \ln\frac{3}{2} - \frac{1}{3}$.

Answer. $\ln\frac{3}{2} - \frac{1}{3}$.

Problem 146 Find the sum of series $S = 1 - \frac{1}{3} + \frac{1}{5} - \frac{1}{7} + \frac{1}{9} - \ldots$.

Solution. First, we replace in the Maclaurin series for $\frac{1}{1-x}$ (Eq. 3.20) $x \to -z^2$, $\frac{1}{1+z^2} = 1 - z^2 + z^4 - \ldots + (-1)^n z^{2n} + \ldots$. Multiplying both sides by dz and integrating, we obtain

$$\int_0^x \frac{dz}{1+z^2} = \int_0^x dz - \int_0^x z^2 dz + \int_0^x z^4 dz - \cdots + (-1)^n \int_0^x z^{2n} dz + \cdots$$

$$\arctan x = x - \frac{x^3}{3} + \frac{x^5}{5} - \cdots + (-1)^n \frac{x^{2n+1}}{2n+1} + \ldots, \quad |x| < 1.$$

If $x = 1$ then $\arctan 1 = \pi/4$. Therefore, $1 - \frac{1}{3} + \frac{1}{5} - \frac{1}{7} + \frac{1}{9} - \ldots = \frac{\pi}{4}$.

Answer. $\frac{\pi}{4}$.

Problem 147 Find the sum of the series $1 + \frac{x^4}{4!} + \frac{x^8}{8!} + \frac{x^{12}}{12!} + \ldots + \frac{x^{4k}}{(4k)!} + \ldots$.

Solution. The radius of convergence for this series is 1. Moreover, because this series is convergent at $x = -1$ and at $x = 1$ it is convergent on the closed interval $[-1, 1]$. Let $S(x)$ be the sum of this series and let us find several successive derivatives of this function:

$$S'(x) = \frac{x^3}{3!} + \frac{x^7}{7!} + \ldots + \frac{x^{4k-1}}{(4k-1)!} + \cdots$$

$$S''(x) = \frac{x^2}{2!} + \frac{x^6}{6!} + \ldots + \frac{x^{4k-2}}{(4k-2)!} + \cdots$$

$$S'''(x) = \frac{x}{1} + \frac{x^5}{5!} + \ldots + \frac{x^{4k-3}}{(4k-3)!} + \cdots$$

$$S^{(4)}(x) = 1 + \frac{x^4}{4!} + \ldots + \frac{x^{4k-4}}{(4k-4)!} + \cdots$$

It is not hard to see that the fourth derivative of the function equals the function itself. Hence, the unknown sum satisfies the following differential equation with the corresponding initial conditions,

$$\frac{d^4 S(x)}{dx^4} = S(x),$$

$$S(0) = 1, \quad S'(0) = S''(0) = S'''(0) = 0.$$

The solution to this differential equation is the sum, $S(x) = \frac{1}{4}e^x + \frac{1}{4}e^{-x} + \frac{1}{2}\cos x = \frac{\cosh x + \cos x}{2}$.

Answer. $S(x) = \sum_{k=0}^{\infty} \frac{x^{4k}}{(4k)!} = \frac{\cosh x + \cos x}{2}$.

3.3.4 Abel's Method

Abel's Theorem 3.14 can be reformulated as follows: Let the powers series $\sum_{k=0}^{\infty} a_k x^k$ converge on the interval $x \in (-1, 1)$ and $S(x)$ be its sum. Moreover, if this series also converges at $x = 1$, then the sum $S(x)$ is continuous as x goes to 1 from the left.

> **Abel's Method.** If $\sum_{n=1}^{\infty} a_n$ converges, then its sum can be found as
>
> $$S = \lim_{x \to 1^-} \sum_{n=1}^{\infty} a_n x^n.$$

In order to use Abel's method, we need to make sure that the given numerical infinite series converges and then to consider the corresponding power series and find their limit as $x \to 1^-$ (find the limit of the sum when x goes to 1 from the left). The key condition for using Abel's method for finding sums of numerical series is convergence of numerical series, $S = \sum_{n=1}^{\infty} a_n$, otherwise the method cannot be applied. Indeed, consider alternating series $1 - 1 + 1 - 1 + 1 - 1 + \ldots = \sum_{n=1}^{\infty} (-1)^n$ that obviously diverges and hence the limit of the sequence of the partial sums does not exist. On the other hand, consider the power series $\sum_{n=0}^{\infty} (-1)^n x^n = 1 - x + x^2 - x^3 + \ldots$ that converges at $|x| < 1$ to the sum $S(x) = \frac{1}{1+x}$. At $x = 1$, $S(1) = \frac{1}{2}$.

Example. Consider the following known power series for a logarithmic function: $f(x) = \sum_{n=1}^{\infty} \frac{(-1)^{n+1} x^n}{n} = \ln(1 + x)$. Because the numerical series $\sum_{n=1}^{\infty} \frac{(-1)^{n+1}}{n}$ converges by the Leibniz Theorem (Theorem 3.9), then the sum of this numerical series can be found as $\sum_{n=1}^{\infty} \frac{(-1)^{n+1}}{n} = \lim_{x \to 1^-} f(x) = \lim_{x \to 1^-} \ln(1 + x) = \ln 2$.

Example. Let us consider another well-known function $g(x) = \arctan x = \sum_{n=0}^{\infty} \frac{(-1)^n x^{2n+1}}{2n+1}$. We know that the numerical series $\sum_{n=0}^{\infty} \frac{(-1)^n}{2n+1}$ converges by Leibniz.

How can we find this infinite sum? Using Abel's Method we have

$$\lim_{x \to 1^-} g(x) = \lim_{x \to 1^-} \arctan x = \frac{\pi}{4} = \sum_{n=1}^{\infty} \frac{(-1)^n}{2n+1}.$$

Finally, if we know power series for a function and know that the corresponding numerical series converges, then the sum of the numerical infinite series can be found easily. However, sometimes evaluation of a limit can be challenging task. Let us practice in Abel's method by solving the following problems.

Problem 148 Find the sum $1 - \frac{1}{2} + \frac{1\cdot3}{2\cdot4} - \frac{1\cdot3\cdot5}{2\cdot4\cdot6} + \dots \frac{(-1)^n(2n-1)!!}{(2n)!!} + \dots$

Solution. If $|x| < 1$, then the series is convergent. Consider the power series,

$(1 + x^2)^{-\frac{1}{2}} = 1 + \sum_{n=1}^{\infty} \frac{(-1)^n(2n-1)!!}{(2n)!!} x^{2n}$. By Abel's method the sum of the given alter-

nating numerical infinite series can be found by the limit of the power series above

as $x \to 1^-$ from the left so that $1 - \frac{1}{2} + \frac{1\cdot3}{2\cdot4} - \frac{1\cdot3\cdot5}{2\cdot4\cdot6} + \dots = \lim_{x\to1^-} (1 + x^2)^{-\frac{1}{2}}$

$= \frac{1}{\sqrt{1+1}} = \frac{1}{\sqrt{2}}$.

Answer. $\frac{1}{\sqrt{2}}$.

Problem 149 Evaluate $\sum_{n=0}^{\infty} \frac{(-1)^n}{3n+1}$.

Solution. This series converges by the Leibniz Theorem (Theorem 3.9). Consider the power series:

$\sum_{n=0}^{\infty} \frac{(-1)^n x^{3n+1}}{3n+1}$ with convergence interval $|x| < 1$. By Abel's method, the following

must be true

$$\sum_{n=0}^{\infty} \frac{(-1)^n}{3n+1} = \lim_{x\to1^-} \sum_{n=0}^{\infty} \frac{(-1)^n x^{3n+1}}{3n+1} = \sum_{n=0}^{\infty} \left(\int_0^x (-1)^n t^{3n} dt \right) = \int_0^x \frac{dt}{1+t^3}$$

In order to find this limit, we need take an improper integral. If we factor $1 + t^3 = (1 + t)(1 - t + t^2)$, we can further simplify the fraction under the integral as follows: $\frac{1}{1+t^3} = \frac{A}{1+t} + \frac{Bt+C}{1-t+t^2}$, where A, B, and C to be determined and can be obtained from the system,

$$\begin{cases} A + C = 1 \\ -A + B + C = 0 \\ A + B = 0 \end{cases} \Rightarrow A = \frac{1}{3}, \ B = -\frac{1}{3}, \ C = \frac{2}{3}$$

Next, substituting the values of A, B, and C into the improper integral, we have

$$\frac{1}{3}\int_0^x \frac{dt}{t+1} - \frac{1}{3}\int_0^x \frac{tdt}{1-t+t^2} + \frac{2}{3}\int_0^x \frac{dt}{t^2-t+1}$$

$$= \frac{1}{3}\int_0^x \frac{dt}{t+1} - \frac{1}{6}\int_0^x \frac{(2t-1)dt}{1-t+t^2} + \frac{1}{6}\int_0^x \frac{dt}{t^2-t+1} + \frac{2}{3}\int_0^x \frac{dt}{t^2-t+1}$$

$$= \frac{1}{3}\int_0^x \frac{dt}{t+1} - \frac{1}{6}\int_0^x \frac{(2t-1)dt}{1-t+t^2} + \frac{5}{6}\int_0^x \frac{dt}{t^2-t+1}$$

The first antiderivative is $\ln(1+t)$, the second integral, we rewrite as the sum of two integrals, one of which is easily taken by substitution, $u = 1 - t + t^2$, $du = (2t-1)dt$ and the second one can be combined with the third integral, that after completing the square can be rewritten as $\int_0^x \frac{dt}{\left(t-\frac{1}{2}\right)^2 + \left(\frac{\sqrt{3}}{2}\right)^2}$, the antiderivative of which can be recognized as the arctangent function. Finally, we obtain the following:

$$\int_0^x \frac{dt}{1+t^3} = \frac{1}{3}\ln(x+1) - \frac{1}{6}\ln(x^2-x+1) + \frac{1}{\sqrt{3}}\arctan\frac{2x-1}{\sqrt{3}} + \frac{\pi}{6\sqrt{3}}$$

$$= \frac{1}{6}\ln\frac{(x+1)^2}{x^2-x+1} + \frac{1}{\sqrt{3}}\arctan\frac{2x-1}{\sqrt{3}} + \frac{\pi}{6\sqrt{3}}\lim_{x\to 1^-}\int_0^x \frac{dt}{1+t^3}$$

$$= \frac{1}{6}\ln 2^2 + \frac{\pi}{6\sqrt{3}} + \frac{1}{\sqrt{3}}\arctan\frac{1}{\sqrt{3}}$$

$$= \frac{1}{3}\ln 2 + \frac{\pi}{3\sqrt{3}} = \frac{1}{3}\left(\ln 2 + \frac{\pi}{\sqrt{3}}\right)$$

Hence, $\sum_{n=0}^{\infty} \frac{(-1)^n}{3n+1} = \frac{1}{3}\ln 2 + \frac{\pi}{3\sqrt{3}}$.

Answer. $\sum_{n=0}^{\infty} \frac{(-1)^n}{3n+1} = \frac{1}{3}\ln 2 + \frac{\pi}{3\sqrt{3}}$.

3.4 Using Series for Approximation

In this section we show how infinite series can be used for approximation of irrational numbers, for finding approximate value of a function at a given point, for approximation of definite integrals and for solving differential equations.

3.4.1 An Approximation of an Irrational Number

Infinite series are very useful for approximation. For example, just fifty years ago, before great calculators were invented, if one wanted to know an irrational number such as e, π, $\ln 2$, etc up to 9 decimal points, he or she would use infinite series representation of known functions and get the number with the required accuracy following the rules of approximation.

Assume that a number is the sum infinite numeric series, $A = a_1 + a_2 + a_3 + \ldots + a_n + \ldots$, where $a_1, a_2, \ldots, a_n$ are "good" numbers, for example, rational numbers. Let us approximate number A by only n terms of the infinite series, $A \approx A_n = a_1 + a_2 + a_3 + \ldots + a_n$. The error of such estimation $\alpha_n = a_{n+1} + a_{n+2} + \ldots,$. If $n \to \infty \Rightarrow \alpha_n \to 0$, and A_n approximates A at any given accuracy.

If series is alternating with monotonically decreasing by absolute value terms (Leibniz type series), then as we have learned, the remainder of the approximation by the first n terms of the series, has the sign of the first dropped term (a_{n+1}) and by absolute value it is less than $|a_{n+1}|$, $|\alpha_n| < |a_{n+1}|$ (Theorem 3.9). Although alternating series have these properties, not all of them are suitable for an approximation of a given number A. For example, consider two well-known convergent infinite alternating series, $1 - \frac{1}{2} + \frac{1}{3} - \frac{1}{4} + \frac{1}{5} - \ldots = \ln 2$ or $1 - \frac{1}{3} + \frac{1}{5} - \frac{1}{7} + \frac{1}{9} - \ldots = \frac{\pi}{4}$.

However, both series converge very slowly and in order to use them for evaluating $\ln 2$ or $\frac{\pi}{4}$, we would have to take too many terms of the expansion. Let us approximate $\ln 2$ by only the first five terms, $\ln 2 \approx 1 - \frac{1}{2} + \frac{1}{3} - \frac{1}{4} + \frac{1}{5} = \frac{47}{60} = 0.7833...3...$ On the other hand, TI 84 calculator gives us $\ln 2 = 0.69314718$. We can try to use the first nine terms of the expansion as

$$\ln 2 = \underbrace{\left(1 - \frac{1}{2} + \frac{1}{3} - \frac{1}{4} + \frac{1}{5} - \frac{1}{6} + \frac{1}{7} - \frac{1}{8} + \frac{1}{9}\right)}_{0.74563} - \frac{1}{10} + \ldots. \quad \text{Our} \quad \text{approximation}$$

becomes a little closer to the actual number with an error $|\alpha_9| \leq 0.1 = \left|-\frac{1}{10}\right|$.

Is there any better method to approximate $\ln 2$? The answer is "Yes." Using results of Problem 132, let us rewrite the difference of two logarithms as infinite series:

$$\ln \frac{n+1}{n} = \ln(n+1) - \ln n$$

$$= \frac{2}{2n+1}\left(1 + \frac{1}{3} \cdot \frac{1}{(2n+1)^2} + \frac{1}{5} \cdot \frac{1}{(2n+1)^4} + \ldots\right)$$

If $n = 1$, we obtain on the left hand-side $\ln 2$ and its approximation on the right hand side, $\ln 2 = \frac{2}{3}\left(1 + \frac{1}{3} \cdot \frac{1}{9} + \frac{1}{5} \cdot \frac{1}{9^2} + \frac{1}{7} \cdot \frac{1}{9^3} + \frac{1}{9} \cdot \frac{1}{9^4} + \frac{1}{11} \cdot \frac{1}{9^5} + \ldots\right)$. Let us approximate $\ln 2$ by the first four terms, $\ln 2 \approx \frac{2}{3}\left(1 + \frac{1}{3 \cdot 9} + \frac{1}{5 \cdot 9^2} + \frac{1}{7 \cdot 9^3}\right) = 0.6928734731$. This approximation gives us correct the first two digits after the decimal point, which is certainly an improvement.

Next, let us find the error of such approximation.

$$\text{error} = \alpha_4 = \frac{2}{3}\left(\frac{1}{9}\cdot\frac{1}{9^4} + \frac{1}{11}\cdot\frac{1}{9^5} + \frac{1}{13}\cdot\frac{1}{9^6} + \cdots\right)$$

$$< \frac{2}{3}\left(\frac{1}{9^5} + \frac{1}{9^6} + \frac{1}{9^7} + \cdots\right)$$

$$= \frac{2}{3}\cdot\frac{1}{9^5}\left(\frac{1}{1-\frac{1}{9}}\right) = \frac{1}{78732} \ll \frac{1}{10000} = 10^{-4}.$$

The infinite series inside parentheses is compared with an infinite geometric progression with the same first term as the series and the common ratio of 1/9. Using terms of the infinite series $1 - \frac{1}{3} + \frac{1}{5} - \ldots$ to approximate π is also not efficient. Other faster convergent series are usually used.

Further, in order to find an approximate value of a function $f(x)$ at the point x_0 with a given accuracy proceed as follows: The function $f(x)$ is written as a powers series of $(x - x_1)$ on the convergence interval containing this point x_0. The point x_1 is the point at which the values of the function and its derivatives are calculated accurately. The variable x is given a value x_0. In the obtained series $\sum_{k=0}^{\infty} a_k(x_0 - x_1)^k$ we keep only the terms that guarantee the given accuracy of calculations. The minimal number n_0 of such terms is determined either from the corresponding estimate of the remainder $R_n(x_0)$ of the Taylor formula or from the remainder, $r_n(x_0)$ of the Taylor series since in the case of convergence of the power series of $f(x)$, the remainders are equal.

3.4.2 An Approximation of Integrals

Many indefinite integrals that are not expressed in terms of elementary functions can be calculated using series.

Problem 150 Evaluate with accuracy $\alpha = 0.00001$ the integral $\int\limits_{0}^{0.1} \frac{\sin x}{x}dx$.

Solution. Using Eq. 3.24 we obtain that $\sin x = x - \frac{x^3}{3!} + \ldots + (-1)^n \frac{x^{2k+1}}{(2k+1)!} + \ldots = \sum_{k=0}^{\infty}(-1)^k \frac{x^{2k+1}}{(2k+1)!}$. Now

$$\frac{\sin x}{x} = 1 - \frac{x^2}{3!} + \ldots + (-1)^n \frac{x^{2k}}{(2k+1)!} + \ldots = \sum_{k=0}^{\infty} (-1)^k \frac{x^{2k}}{(2k+1)!}, \quad x \in R$$

$$\int_0^{0.1} \frac{\sin x}{x} dx = \int_0^{0.1} 1 \cdot dx - \int_0^{0.1} \frac{x^2}{3!} dx + \ldots + (-1)^n \int_0^{0.1} \frac{x^{2k}}{(2k+1)!} dx + \ldots$$

$$= 0.1 - \frac{0.001}{3 \cdot 3!} + \frac{0.00001}{5 \cdot 5!} + \ldots + (-1)^k \frac{x^{2k+1} \cdot (0.1)^{2k+1}}{(2k+1)(2k+1)!} + \ldots$$

Because $0.1 > 0.00001$, $\frac{0.001}{3 \cdot 3!} = 0.000055 \ldots > 0.00001$, $\frac{0.00001}{5 \cdot 5!} = \frac{0.00001}{600} < 0.00001$

$$\int_0^{01} \frac{\sin x}{x} dx \approx 0.1 - 0.000055 = 0.09994.$$

Answer. 0.09994.

Problem 151 Evaluate integral $\int_0^{1/4} e^{-x^2} dx$ with accuracy of 10^{-4}.

Solution. Using the Maclaurin series, we substitute in Eq. 3.21 $-x^2$ instead of x,
$e^{-x^2} = 1 - \frac{x^2}{1!} + \frac{x^4}{2!} - \ldots + (-1)^n \frac{x^{2n}}{n!} + \ldots$ ($-\infty < x < +\infty$). This series can
be integrated between any finite limits. Thus,

$$\int_0^{1/4} e^{-x^2} dx = \int_0^{1/4} \sum_{n=0}^{\infty} (-1)^n \frac{x^{2n}}{n!} dx = \sum_{n=0}^{\infty} \frac{(-1)^n}{n!} \int_0^{1/4} x^{2n} dx = \sum_{n=0}^{\infty} \frac{(-1)^n}{n!} \cdot \left(\frac{x^{2n+1}}{2n+1} \Big|_0^{1/4} \right) = \sum_{n=0}^{\infty} \frac{(-1)^n}{n! \cdot (2n+1) \cdot 4^{2n+1}}.$$

The resulting series of numbers is an alternating series satisfying the Leibniz
Theorem (Theorem 3.9), so if we take for computing the first few terms of the
series, the error, which in this case will be made, will not surpass the absolute value
of the first discarded term.

Notice that the third term of the series $\frac{1}{2! \cdot 5 \cdot 4^5} = \frac{1}{10240} < 10^{-4}$. Therefore, in order
to calculate the integral with the required accuracy up 10^{-4}, it is enough to take only
two members of the series, $\int_0^{1/4} e^{-x^2} dx \approx \frac{1}{4} - \frac{1}{1! \cdot 3 \cdot 4^3} = \frac{1}{4} - \frac{1}{192} \approx 0.2448.$

Answer. $\frac{47}{192} \approx 0.2448.$

3.4.3 Integration of Differential Equations

The power series may also be used for solving differential equations, for example, if their solution cannot be written in terms of elementary functions.

Problem 152 Solve the equation $xy'' - y = (x-1)^2$, $y(1) = 1$, $y'(1) = 0$.

Solution. Assume that our solution is written as Taylor Series with center at point x_0,

$$y(x) = y(x_0) + \frac{y'(x_0)}{1!}(x - x_0) + \frac{y''(x_0)}{2!}(x - x_0)^2 + \ldots + \frac{y^{(k)}(x_0)}{k!}(x - x_0)^k + \ldots.$$

The first two coefficients we find from the initial value problem. Then, because $x_0 = 1$,

$$y(x) = y(1) + \frac{y'(1)}{1!}(x - 1) + \frac{y''(1)}{2!}(x - 1)^2 \ldots + \frac{y^{(4)}(1)}{4!}(x - 1)^4.$$
$$y(1) = 1, y'(1) = 0$$
$$1 \cdot y''(1) - 1 = 0 \Rightarrow y''(1) = 1$$

Differentiating the given differential equation, $y'' + xy''' - y' = 2(x - 1)$. Substituting in the equation $x = 1$ and using $y(1) = 1$, $y'(1) = 0$, $y''(1) = 1$, from the last equation, we obtain $y''(1) + 1 \cdot y'''(1) - y'(1) = 0 \Rightarrow y'''(1) = -1$. Differentiating the previous equation again, replacing $x = 1$, $2y'''(1) + 1 \cdot y^{(4)}(1) - y''(1) = 2 \Rightarrow y^{(4)}(1) = 5$. Therefore, $y(1) = 1, y'(1) = 0, y''(1) = 1, y'''(1) = -1, y^{(4)}(1) = 5$. Finally, we obtain the solution, $\tilde{y}(x) = 1 + \frac{(x-1)^2}{2!} - \frac{(x-1)^3}{3!} + \frac{5(x-1)^4}{4!}$.

Problem 153 Find the first five terms of series solution of $y' = x^2 + y^2$, satisfying the condition $y = \frac{1}{2}$ at $x = 0$.

Solution. Let us write the series solution to this equation as Maclaurin series,
$y(x) = y(0) + \frac{y'(0)}{1!}x + \frac{y''(0)}{2!}x^2 + \frac{y'''(0)}{3!}x^3 + \ldots + \frac{y^{(n)}(0)}{n!}x^n + \ldots$. The first three derivatives are found by differentiating the differential equation, $y'' = 2x + 2yy'$, $y''' = 2 + 2 \cdot (y')^2 + 2yy''$, $y^{(4)} = 6y'y'' + 2yy'''$. Evaluating the values of these derivatives at $x = 0$ and using the initial condition $y(0) = \frac{1}{2}$ and the given DE: $y' = x^2 + y^2$, we obtain $y'(0) = 0 + \left(\frac{1}{2}\right)^2 = \frac{1}{4}$; $y''(0) = 2 \cdot 0 + 2 \cdot \frac{1}{2} \cdot \frac{1}{4} = \frac{1}{4}$; $y'''(0) = 2 + 2 \cdot \frac{1}{4^2} + 2 \cdot \frac{1}{2} \cdot \frac{1}{4} = \frac{19}{8}$; $y^{(4)}(0) = \frac{1}{4} + \frac{1}{8} + \frac{19}{8} = \frac{11}{4}$. Substituting these values into the Maclaurin series, we obtain the following approximate solution: $y(x) = \frac{1}{2} + \frac{1}{4}x + \frac{1}{8}x^2 + \frac{19}{48}x^3 + \frac{11}{96}x^4 + \ldots$.

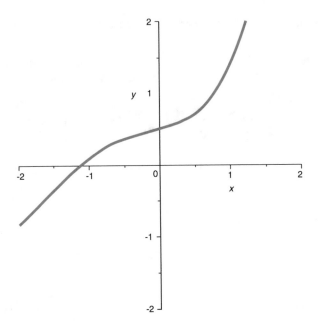

Figure 3.1 Approximate solution, $y(x) = \frac{1}{2} + \frac{1}{4}x + \frac{1}{8}x^2 + \frac{19}{48}x^3 + \frac{11}{96}x^4$

Figure 3.2 ODE plot by
MAPLE

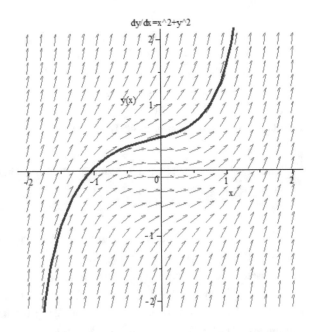

Let us take only the first five terms of this series and graph it (Figure 3.1).

MAPLE 15 solves this differential equation numerically with much better precision using more than five terms of the power expansion and plots the solution curve satisfying the IVP (Figure 3.2).

By comparing two graphs, we can see that both solutions looks almost identical, pass the point (0,1/2) and have the same visually X intercept. Finding the closed form solution in this case is not possible.

3.5   Generating Functions

Series where the nth term depends on some variable x are called functional series. In this section we focus on a particular kind of functional series called a power series. Consider a sum a geometric series again with the first term b_1 and common ratio r,

$$S_n = b_1 + b_1 r + b_1 r^2 + \ldots + b_1 r^n \tag{3.39}$$

If n is a specific number, say $n = 2$ or $n = 7$ or even $n = 1000$, then S_n is a polynomial of 2, 7, or 1000 terms, respectively. If n is an indefinite number of terms, the sum of the series can be written in the form

$$S_\infty = b_1 + b_1 r + b_1 r^2 + \ldots \tag{3.40}$$

In the case where the number of terms is indefinitely great (we say "as n approaches infinity") the expression on the right of Eq. 3.40 is called an infinite series.

If every coefficient of powers of r is different, and they are denoted by $a_0, a_1, a_2, \ldots$, then Eq. 3.40 becomes

$$S = a_0 + a_1 r + a_2 r^2 + a_3 r^3 + \ldots \tag{3.41}$$

Equation 3.41 is called a power series and the variable r is called the variable of the series. Also we can call Eq. 3.41 a power series in r, and we say that S is a sum to infinity. Let us do some analysis of Eq. 3.39 and Eq. 3.41. If an absolute value of r is greater than 1, then each term of Eq. 3.39 or Eq. 3.41 is greater than the preceding term and the sum S has no definite limiting value, but increases without bound as the number of terms increases.

Example. $1, 5, 25, 125, 625, \ldots$ is a geometric sequence with $b_1 = 1$, $r = 5 > 1$

$$S_3 = 1 + 5 + 25 = 31$$
$$S_5 = 1 + 5 + 25 + 125 + 625 = 781$$
$$S_{15} = \frac{b_1 \left(r^{15} - 1 \right)}{r - 1} = \frac{5^{15} - 1}{4} = 7,629,394,531.$$

If an absolute value of r is less than 1, then the successive terms are smaller and smaller and as the number of terms n becomes very big, "infinite," the last term approaches zero (n^{th} term is $b_1 r^{n-1}$, $|r| < 1$, $\lim\limits_{n \to \infty} r^{n-1} = 0$). In this case, the sum

S from Eq. 3.41 approaches a definite value as n becomes infinitely big and is said to be "the limit of the sum of the series as n approaches infinity." As we obtained earlier for a geometric series if $|r| < 1$, we have $S = \frac{b_1}{1-r} = b_1(1-r)^{-1}$ because if $n \to \infty$, $(1-r)^{-1} = 1 + r + r^2 + r^3 + \ldots$.

In the case when each coefficient is different, the way of finding a sum to infinity is different in each case and may be nontrivial. In this section, we look at the other approach to solving recursions using generating functions. Consider an infinite sequence of numbers:

$$a_0, a_1, a_2, \ldots \tag{3.42}$$

The corresponding power series is called a generating function for sequence of Eq. 3.42

$$G(x) = a_0 + a_1 x + a_2 x^2 + a_3 x^3 + \ldots \tag{3.43}$$

Thus, for the sequence, 1, 1, 1, 1, ..., the generating function $G(x)$ can be recognized as an infinite geometric series and can be written as

$$1 + x + x^2 + x^3 + \ldots = \frac{1}{1-x}. \tag{3.44}$$

For the infinite sequence of the natural numbers 1, 2, 3, 4, 5, ... a generating function can be found similarly to the series in Problem 135, i.e., $G(x) = 1 + 2x + 3x^2 + 4x^3 + \ldots + (n+1)x^n + \ldots$. Multiplying both sides by x, $x G(x) = x + 2x^2 + 3x^3 + 4x^4 + \ldots + nx^n + (n+1)x^{n+1} + \ldots$ and subtracting two equations, we obtain $(1-x)G(x) = 1 + x + x^2 + x^3 + \ldots + = \frac{1}{1-x}$. Solving,

$$G(x) = \frac{1}{(1-x)^2} \tag{3.45}$$

In order to find generating functions for general sequences, we would need to learn material that is beyond the scope of this book. However, we show how to use some generating functions in solving challenging problems on sequences.

Next, let us solve the following problem using generating functions.

Problem 154 A certain sequence of numbers $a_1, a_2, \ldots, a_n$ satisfies the condition: $a_0 = a_1 = 1$, $a_{n+2} = a_{n+1} + a_n$. Find a_n.

Solution. Let us generate some of the terms:

$$1, \ 1, \ 2, \ 3, \ 5, \ 8, \ 13, \ 21, \ 34, \ldots \tag{3.46}$$

Then consider the following infinite functional series in variable x:

$$F(x) = 1 + x + 2x^2 + 3x^3 + 5x^4 + 8x^5 + \ldots \tag{3.47}$$

Multiplying both sides by $(x + x^2)$ we obtain

$$(x + x^2)F(x) = x + x^2 + 2x^3 + 3x^4 + 5x^5 + 8x^6 + \ldots$$
$$x^2 + x^3 + 2x^4 + 3x^5 + 5x^6 + \ldots$$
$$(x + x^2)F(x) = x + 2x^2 + 3x^3 + 5x^4 + 8x^5 + \ldots \tag{3.48}$$

Subtracting Eq. 3.47 and Eq. 3.48 and solving it for $F(x)$ we have

$$F(x) = \frac{1}{1 - x - x^2} \tag{3.49}$$

Equation 3.49 is the sum of all terms of Eq. 3.47 or using new terminology, it is the generating function for sequence of Eq. 3.46.

Finding zeros of the denominator of Eq. 3.49, we obtain zeros of a quadratic equation,

$$1 - x - x^2 = 0$$
$$x_1 = \frac{-1 + \sqrt{5}}{2}, \quad x_2 = \frac{-1 - \sqrt{5}}{2} \tag{3.50}$$

Applying Vieta's Theorem to these roots more relationships can be obtained:

$$x_1 \cdot x_2 = -1$$
$$x_1 + x_2 = -1 \tag{3.51}$$
$$x_1 - x_2 = \sqrt{5}$$

Next, factoring the denominator as $1 - x - x^2 = -(x - x_1)(x - x_2)$. Using the relationships from Eq. 3.51, we can see that $x - x_1 - (x - x_2) = -\sqrt{5}$, then we can rewrite Eq. 3.49 as a difference of two fractional expressions:

$$\frac{1}{1-x-x^2} = -\frac{1}{(x-x_1)(x-x_2)} = -\frac{1}{\sqrt{5}}\left(\frac{1}{x-x_1} - \frac{1}{x-x_2}\right)$$

$$= \frac{1}{\sqrt{5}}\left(\frac{1}{x-x_2} - \frac{1}{x-x_1}\right) = \frac{1}{\sqrt{5}}\left(\frac{1}{x_1-x} - \frac{1}{x_2-x}\right)$$

$$= \frac{1}{\sqrt{5}}\left(\frac{1}{x_1\left(1-\frac{x}{x_1}\right)} - \frac{1}{x_2\left(1-\frac{x}{x_2}\right)}\right)$$

The last expression can be easily seen to be

$$F(x) = \frac{1}{\sqrt{5}x_1}\cdot\frac{1}{1-\frac{x}{x_1}} - \frac{1}{\sqrt{5}x_2}\cdot\frac{1}{1-\frac{x}{x_2}}. \tag{3.52}$$

Equation 3.52 gives us a different representation of the infinite sum $F(x)$. It is now the difference of two sums, each of which can be expanded as an infinite sum of a geometric series. Thus, Eq. 3.52 can be rewritten as

$$\begin{aligned}F(x) = {} & \frac{1}{\sqrt{5}x_1}\cdot\left(1+\frac{x}{x_1}+\left(\frac{x}{x_1}\right)^2+\left(\frac{x}{x_1}\right)^3+\dots\right) \\ & -\frac{1}{\sqrt{5}x_2}\cdot\left(1+\frac{x}{x_2}+\left(\frac{x}{x_2}\right)^2+\left(\frac{x}{x_2}\right)^3+\dots\right)\end{aligned} \tag{3.53}$$

Equations 3.47 and 3.53 represent the same function $F(x)$. Therefore, equating coefficients of $x, x^2, x^3, \dots$, we get the formula for any term of the Fibonacci sequence! Thus $a_0 = 1 = \frac{1}{\sqrt{5}x_1} - \frac{1}{\sqrt{5}x_2} = \frac{x_2-x_1}{\sqrt{5}x_1x_2} = \frac{-\sqrt{5}}{\sqrt{5}(-1)} = 1$ and $a_n = \frac{1}{\sqrt{5}}\left(\frac{1}{x_1^{n+1}} - \frac{1}{x_2^{n+1}}\right) = \frac{1}{\sqrt{5}}\left(\frac{x_2^{n+1}-x_1^{n+1}}{(x_1x_2)^{n+1}}\right)$. In order to make this formula more convenient for further usage, using Eq. 3.51, we put fractions over the common denominator and replace the product of the roots $x_1 \cdot x_2$ by (-1). Finally, we have an explicit formula for the nth term of the Fibonacci sequence,

$$a_n = \frac{(-1)^{n+1}}{\sqrt{5}}\left(\left(\frac{-1-\sqrt{5}}{2}\right)^{n+1} - \left(\frac{-1+\sqrt{5}}{2}\right)^{n+1}\right) \tag{3.54}$$

If we factor out $(-1)^{n+1}$ from each term inside parentheses, Eq. 3.54 becomes Eq. 1.19. You can see that the formula works for any member of the sequence. For example, if $n = 3$, then we find $a_3 = 3$, which can be obtained by the recursive relationship. On the other hand, substituting $n = 3$ into Eq. 3.54, we get

$$a_3 = \frac{(-1)^4}{\sqrt{5}} \left[\left(\frac{-1-\sqrt{5}}{2} \right)^4 - \left(\frac{-1+\sqrt{5}}{2} \right)^4 \right] = \frac{2 \cdot 2\sqrt{5} \cdot 12}{16 \cdot \sqrt{5}} = \frac{48}{16} = 3.$$

Maybe this approach is not as elegant as one presented in Chapter 1. Generating functions have a lot of applications, for example, in combinatorics and computer science. It never "hurts" to learn different ways of solving a problem.

Let us solve the following problem.

Problem 155 Given a sequence such that $a_{n+1} = 2a_n + 3n$, $a_0 = 1$, $n = 0$, $1, 2, 3, \ldots$. Find the nth term of the sequence.

Solution. Evaluating some of the first terms $1, 2, 7, 20, 49, \ldots$ we cannot find the formula right away, so we use the method of generating functions. Instead of looking for the sequence, we look for the function. Let us multiply the given formula by x^n and then using sigma notation write down an equivalent relationship,

$$\sum_{n=0}^{\infty} a_{n+1}x^n = 2\sum_{n=0}^{\infty} a_n x^n + 3\sum_{n=0}^{\infty} n x^n \tag{3.55}$$

The right side of it can be rewritten with the use of known generating functions. First, we know that $\sum_{n=0}^{\infty} a_n x^n = a_0 + a_1 x + a_2 x^2 \ldots = F(x)$ which can be used in the first term. Let us evaluate the second sum on the right side of Eq. 3.55,

$$\sum_{n=0}^{\infty} n x^n = x + 2x^2 + 3x^3 + 4x^4 + \ldots \tag{3.56}$$

Earlier we established that $\sum_{n=0}^{\infty} x^n = 1 + x + x^2 + x^3 + x^4 + \ldots = \frac{1}{1-x}$. Let us take the derivative of this formula with respect to x and then denote it by $P(x)$, i.e., $P(x) = \sum_{n=0}^{\infty} \frac{d}{dx}(x^n) = \frac{d}{dx}\left(\sum_{n=0}^{\infty} x^n\right) = \frac{d}{dx}\left(\frac{1}{1-x}\right) = \frac{1}{(1-x)^2}$. Then multiplying it by x will give   us   generating   function   for   Eq.   3.56,   i.e.,
$\sum_{n=0}^{\infty} n x^n = xP(x) = x + 2x^2 + 3x^3 + 4x^4 + \cdots + = \sum_{n=0}^{\infty} n x^n = \frac{x}{(1-x)^2}$

Next, we consider the left side of Eq. 3.55,

$$\sum_{n=0}^{\infty} a_{n+1}x^n = a_1 + a_2 x + a_3 x^2 + a_4 x^3 + \ldots \tag{3.57}$$

It is clear that Eq. 3.57 can be seen as $F(x)$ minus the very first term and then divided by x. Therefore, Eq. 3.55 can be written as $\frac{F(x)-a_0}{x} = 2F(x) + \frac{3x}{(1-x)^2}, a_0 = 1$. Solving for $F(x)$ we obtain the formula and decompose it into partial fractions,

$$F(x) = \frac{1 - 2x + 4x^2}{(1 - 2x)(1 - x)^2} = \frac{A}{1 - 2x} + \frac{B}{1 - x} + \frac{C}{(1 - x)^2} \tag{3.58}$$

Putting the fractions on the right over the common denominator and equating corresponding coefficients of powers of x,

$$A(1 - x)^2 + B(1 - x)(1 - 2x) + C(1 - 2x) = 1 - 2x + 4x^2$$
$$A(1 - 2x + x^2) + B(1 - 3x + 2x^2) + C(1 - 2x) = 1 - 2x + 4x^2$$
$$\begin{cases} A + B + C = 1 \\ -2A - 3B - 2C = -2 \Leftrightarrow A = 4, B = 0, C = -3 \\ A + 2B = 4 \end{cases}$$

Now Eq. 3.58 can be rewritten as

$$F(x) = \frac{4}{1 - 2x} - \frac{3}{(1 - x)^2}$$
$$= 4 \cdot \left[\left(1 + 2x + (2x)^2 + (2x)^3 + \dots \right) \right] \tag{3.59}$$
$$-3 \cdot [1 + 2x + 3x^2 + 4x^3 + 5x^4 + \dots]$$

Combining like terms in Eq. 3.59, we can rewrite it as $F(x) = \sum_{n=0}^{\infty} (4 \cdot 2^n - 3(n + 1))x^n$. From this, we immediately obtain the formula for the nth term,

$$a_n = 4 \cdot 2^n - 3(n + 1) = 4 \cdot 2^n - 3n - 3. \tag{3.60}$$

In order to make sure that Eq. 3.60 is correct, we can use it for finding several members of the sequence:

$$n = 0, \ a_0 = 4 \cdot 2^0 - 3(0 + 1) = 4 - 3 = 1$$
$$n = 1, \ a_1 = 4 \cdot 2^1 - 3(1 + 1) = 4 \cdot 2 - 3 \cdot 2 = 2$$
$$n = 2, \ a_2 = 4 \cdot 2^2 - 3(2 + 1) = 4 \cdot 2^2 - 3 \cdot 3 = 7, \text{ etc.}$$

Answer. $a_n = 4 \cdot 2^n - 3(n + 1)$.

Chapter 4
Real-Life Applications of Geometric and Arithmetic Sequences

Over the millenia, legends have developed around mathematical problems involving series and sequences. Here in this book we try to create an original fable about two children, Brian and Paul whose parents wanted to reward them for good grades in mathematics and saw how Paul by asking for 1 cent on day one and asking for doubling the amount each consecutive day was way ahead of his dull-witted brother Brian. Soon any parent would realize that there is not enough money on the planet Earth to supply Paul's request.

One of the most famous legends about series concerns the long inventor of chess whose name is lost the ages. According to the legend, an Indian king summoned the inventor and suggested that he chose the award for the creation of the interesting and wise game. The king was amazed by the "modest" request from the inventor who asked to give him for the first cell of the chessboard 1 grain of wheat, for the second—2 grains, for the third—4 grains, for the fourth—twice as much as in the previous cell, etc. As a result, the total number of grains per 64 cells of the chessboard would be number 18,446,744,073,709,551,615 (18 quintillion, 446 quadrillion, 744 trillion, 73 billion 709, and 551,000,615). If the king was able to have that much of wheat, he would have to plant it everywhere on the entire surface of the Earth including the territories of the seas and oceans, and mountains, and the desert, from the Arctic to the Antarctic in order obtain a satisfactory harvest, then, perhaps, he maybe could pay his amazing debt off to the chess inventor in over 5 years.

In our life, we deal with geometric progression in that or this form all the time without even thinking about it. For example, many of us like shopping and we all are attracted by the sales price of the items. Assume that the price of a dress of your dreams is reduced every week by five percent. If today the price is $300, what would be the price of the dress in 6 months? Obviously the price of the dress in each week will be multiplied by a factor of 0.95 that is equivalent to 5% reduction. The following will represent a few consecutive sales prices: $300, 0.95 \cdot 300$, $(0.95)^2 \cdot 300$, $(0.95)^3 \cdot 300$, $\ldots$. Therefore, the sales price after n reductions will

© Springer International Publishing Switzerland 2016
E. Grigorieva, *Methods of Solving Sequence and Series Problems*,
DOI 10.1007/978-3-319-45686-7_4

be $300(0.95)^n$, and if after a year the sale continues (52 weeks equals one year), you can buy the same dress for a just $20.83.

Notice that if we assume that the price of the dress is a continuous function of time, then it follows an exponential decay function where the discrete terms of the geometric progression are represented by points on a continuous curve. An exponential decay function with decay rate r and initial value P, can be written as

$$A(t) = P(1 - r)^t \tag{4.1}$$

Specifically, for radioactive decay, using the geometric progression approach, this formula can be written as

$$A(t) = P\left(\frac{1}{2}\right)^{\frac{t}{t_{1/2}}}. \tag{4.2}$$

Here $t_{1/2}$ is the half-life, the period specific for all radioactive substances over which the mass of the substance decays by half.

For example, for a decaying hypothetical radioactive substance, with $t_{1/2} = 1000$ years, the initial amount of 100 kg would become 50 kg in 1000 years, then 25 kg in the following 1000 years, 12.5 kg in the next 1000 years, etc. All these moments are discrete points on a continuous curve of the radioactive decay with represent the terms of infinitely decreasing geometric progression with common ratio ½ (Figure 4.1).

Have you ever thought of how archeologists in the movies, such as Indiana Jones can predict the age of different artifacts? Do not you know that the age of artifacts in real life can be established by the amount of the radioactive isotope of Carbon 14, $^{14}_{6}C^6 = {}^{14}C$ in the artifact found by a scientist? Carbon 14 has a very long half-lifetime which means that each half- life time of 5730 years or so, the amount of the isotope is reduced by half.

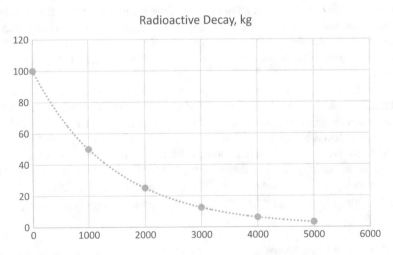

Figure 4.1 Radioactive decay

Equation 4.2 is very useful if the half-life time of an isotope is known. In a general exponential decay case given by Eq. 4.1, the half-lifetime can be easily found by setting the left side of the formula to $\frac{P}{2}$, then taking a logarithm of both sides and solving the formula for the corresponding time that will be the half-lifetime, $t_{1/2}$, $\frac{P}{2} = P(1 - r)^t \Rightarrow \frac{1}{2} = (1 - r)^t$.

$$t = t_{1/2} = \frac{\ln\left(\frac{1}{2}\right)}{\ln(1 - r)} \tag{4.3}$$

In this chapter, you will find some interesting problems on radioactive decay coming from theoretical part of the International Chemistry Olympiads.

If we again model the price reduction by a continuous decay exponential function, then the dress will have its own half-life time in the sense of the time when the price is exactly half of the original price. Thus, substituting 150 into Eq. 4.1 and solving it for time, we obtain that $t_{1/2} = \frac{\ln 0.5}{\ln 0.95} \approx 13.51$, which indicates that the dress will be 50% off approximately between 13th and 14th weeks of continuous reduction process.

On the other hand, the terms of an increasing geometric progression with common ratio greater than one will produce a curve of an exponential growth and terms of the progression will be represented by discrete points on the continuous curve. By analogy, a function of an exponential growth can be modeled as

$$A(t) = P(1 + r)^t. \tag{4.4}$$

Here P is the principal or original amount and r is the nominal growth rate.

Additionally, there are many examples of arithmetic progression around us in a real life. For example, the same Indiana Jones who wants to get a precious artifact from the deep 10 meters underground hires people to do digging for him. Assume that he promised to pay a $100 for the first meter and will pay $50 more for each following consecutive meter of dirt digging. How much would he pay for digging 10 meters deep underground? In this case we have an arithmetic sequence of the payments with the first term of $100 and common difference of $50: $100, $150, $200, $250, $300, $350, $400, $450, $500, $550. The total payment equals the sum of the terms of this arithmetic progression and it is $\frac{100+550}{2} \cdot 10 = \$3250$.

This chapter is for those who want to see applications of arithmetic and geometric progressions to real life. There are many applications for sciences, business, personal finance and even for health, but most people are unaware of these. We familiarize you with these by giving you five mini-projects and some related problems associated with the concepts afterwards. However, most of the problems of this chapter are not contest type problems and if you are not interested or already understand them, skip this chapter and go to the homework section.

4.1   Mini-Project 1: Radioactive Decay and its Applications

Natural radioactivity was discovered by Henry Becquerel and then studied in depth by French physicists Pierre and Marie Sklodowsca Curie. All three scientists received Nobel Prizes in physics in 1903. Willard Libby (American physicist) who worked on the Manhattan Project during World War II, invented the process of carbon dating in the early 1950 s at the University of Chicago. In 1960, he was awarded the Nobel Prize in Chemistry for this research. It was found that cosmic radiation generated neutrons that penetrate into our atmosphere, interact with nitrogen 14 in the air and produces radioactive isotope Carbon 14.

$$_{7}^{14}\text{N} + _{0}^{1}\text{n} \to _{6}^{14}\text{C} + _{1}^{1}\text{H}$$

$$_{6}^{14}\text{C} \xrightarrow{\beta^{-}} _{7}^{14}\text{N}$$

Since carbon C_{14} is radioactive, it can be detected using a Geiger counter. This radioactive carbon is mixed with carbon in a normal atmosphere, typically in the form of carbon dioxide (CO_2). Plants absorb carbon dioxide, and it becomes part of their tissue. Animals eat plants and CO_2 becomes a part of their body tissue. When a plant or animal dies, the arrival of the new CO_2 is terminated. In today's environment, there is very little radioactive carbon (only 0.0000765%) and its amount is further reduced after its decay into simpler components. About half of the CO_2 splits into simpler components in about 5,730 years. This is called the half-life of radioactive CO_2. After the two "half-life" (11.460 years) only one-fourth of the original amount of CO_2 will remain radioactive and so on.

As we mentioned above, radioactive isotope of Carbon, Carbon 14 dating by far the most important method of determination of age of artifacts in archeology. However, due to a very long half-lifetime (about 5730 years) and lower abundance of this isotope in the artifacts or natural samples, this method requires a very high sensitivity of measurements. If a sample is gaseous, it can be introduced directly into Geiger counter to increase sensitivity. Using the number of Geiger counter clicks as a reference point, scientists are trying to determine the age of the artifact. Since today's atmosphere produces about sixteen counter clicks per minute per one gram of Carbon 14, then Geiger counter has to click eight times per minute if the artifact is 5,730 years old and four times per minute if the artifact is 11,460 years of age.

When Carbon 14 decays, it forms Nitrogen 14 and loses one electron. This reaction is called "beta decay,"

$$_{6}^{14}\text{C}^{6} \to _{7}^{14}\text{N}^{7} + e^{-}.$$

Such a reaction, as any monomolecular chemical reaction can be modeled by a differential equation of the first order, $\frac{dx}{dt} = -k \cdot x$. Here C is the concentration of Carbon 14 or in general, concentration of any radioactive substance. The substance is decaying, proportionally to the current amount of the radioactive substance, which means that the rate of change equals minus constant of reaction, k, times the current concentration. It is not difficult to find the solution to this equation. We can separate the variables and after integration to obtain the following

$$x = x_0 e^{-kt} \tag{4.5}$$

Where x_0 is an initial concentration of a radioactive substance. If $x = \frac{x_0}{2} \Rightarrow t = t_{1/2}$ and we can evaluate the constant of the reaction, k,

$$k = -\frac{\ln\frac{1}{2}}{t_{1/2}} = \frac{\ln 2}{t_{1/2}}. \tag{4.6}$$

Using base change formula applied to Eq. 4.6, Eq. 4.5 can be easily rewritten equivalent to the form of Eq. 4.2 with base ½ and half-lifetime as a parameter,

$$x = x_0 e^{-kt} = x_0 \left(e^{\ln\frac{1}{2}} \right)^{\frac{t}{t_{1/2}}}$$
$$= x_0 \left(\frac{1}{2}\right)^{\frac{t}{t_{1/2}}}.$$

Let us solve the following problem. A similar problem was offered at preparation for International Chemistry Olympiad (IChO 1996) that took place in Moscow, Russia.

Problem 156 Indiana Jones wants to know the age of artifact that he was able to dig out. In one of his experiments, 0.011 moles of methane $^{14}CH_4$ was introduced into Geiger counter chamber. After 30 minutes the counter was switched on and the measurements continued for another five minutes to register 2000 decays. Write the equation describing β decay of carbon 14 nucleolus. Calculate the number of Carbon 14 atoms in the sample and the molar % of radioactive methane $^{14}CH_4$ taken to the experiment described.

Solution. Because the time of the experiment is 30 minutes is much less than the half lifetime of 5730 years, then given 2000 nuclei decayed within 5 minutes may indicate that approximately $400 = 2000/5$ will decay in one minute. Then about 12,000 would decay in 30 minutes. Assuming that $-\frac{dN}{dt} = kN = 400\text{min}^{-1} = $ const, then the initial number of nuclei, x_0, can be approximated by $x_0 \approx N = \frac{400\text{min}^{-1}}{k}$, where the constant of reaction can be found using Eq. 4.6, $k = \frac{\ln 2}{t_{1/2}} = \frac{0.693}{5730 \cdot 365 \cdot 24 \cdot 60} \approx 2.46 \cdot 10^{-10}\text{min}^{-1}$. Thus, $x_0 \approx N \approx \frac{400}{2.46 \cdot 10^{-10}} = 1.66 \cdot 10^{12}$.

However, let us find the answer in a more mathematical way. Though the Geiger counter did not work for the first 30 minutes, the radioactive Carbon 14 continued decaying. Let us see how we can use the information that when the Geiger counter switched on, that only within the last five minutes, the count was recorded as 2000. Using Eq. 4.5, we can subtract the actual number of nuclei decayed within 30 minutes and 35 minutes, respectively, and to obtain the number of atoms of Carbon 14 decayed between 30th and 35th minutes and equate it to 2000:

$$x_0 e^{-30k} - x_0 e^{-35k} = 2000$$
$$x_0 = \frac{2000}{e^{-30k} - e^{-35k}}$$

Because we know the half-lifetime, we substitute k from Eq. 4.6 and obtain that

$$x_0 = \frac{2000}{\left(\frac{1}{2}\right)^{\frac{30min}{5730years}} - \left(\frac{1}{2}\right)^{\frac{35min}{5730years}}} > \frac{2000}{\left(\frac{1}{2}\right)^{\frac{30}{3,011,688,000}} - \left(\frac{1}{2}\right)^{\frac{35}{3,011,688,000}}} = 1.67 \cdot 10^{12}.$$

It is interesting the both methods of calculation for the initial number of atoms of radioactive Carbon 14 gave practically the same answer.

Answer. $x_0 = 1.67 \cdot 10^{12}$ atoms of Carbon 14.

Remark. In order to find the molar percent of radioactive methane, we use the fact that one mole contains approximately $6.02 \cdot 10^{23}$ atoms of carbon (this is called Avogadro number). Hence, 0.011 moles of methane will have $6.6 \cdot 10^{21}$ atoms of Carbon. Because $x_0 = 1.67 \cdot 10^{12}$ of radioactive Carbon 14, then 0.0011 moles of methane contain the following % of radioactive Carbon 14.

$$\frac{1.67 \cdot 10^{12}}{6.6 \cdot 10^{21}} \cdot 100\% = \frac{253}{10000000000}\% = 2.53 \cdot 10^{-8}\%.$$

The following problem was offered at the 25th International Chemistry Olympiad, 1993, Italy.

Problem 157 A radioactive isotope of iodine, ^{131}I, is used in nuclear medicine for analytical procedures to determine thyroid disorders by scintigraphy. The decay rate constant, k, of ^{131}I is $9.93 \times 10^{-7} s^{-1}$. The decay equation is $^{131}I \rightarrow {}^{131}Xe + e^{-}$.

a. Calculate the half-lifetime of ^{131}I expressed in days.
b. Calculate the time necessary (expressed in days) for a sample of ^{131}I to reduce its activity to 30% of the original value.

Solution.
a. Using Eq.   4.6   we   can   find   the   half-lifetime   in   seconds   as
$t_{1/2} = \frac{\ln 2}{9.93 \cdot 10^{-7} s^{-1}} = 698033.41 s \approx 8.079$ days.

b. Using Eq. 4.5, replacing $x = 0.3x_0$ and after canceling the common factors, we obtain, $0.3 = e^{-kt}$, $k = 9.93 \cdot 10^{-7}$.

After taking natural logarithm of both sides we get

$$\ln 0.3 = -9.93 \cdot 10^{-7} \cdot t$$

$$t = \frac{\ln \dfrac{10}{3}}{9.93 \cdot 10^{-7}} \approx 1212460 \approx 14 \text{ days.}$$

Obviously, it takes longer to decay to 30% (14 days) than to 50 % of the original amount (8 days)

Answer. 8 days and 14 days.

Despite the fact that dating Carbon 14 sounds true, the process itself is based on several erroneous assumptions:

1. The amount of C_{14} in the atmosphere is in a state of equilibrium. This assumption is incorrect. According to estimates, the amount of C_{14} in the atmosphere will reach a state of equilibrium (when the rate of production is equal to the decay rate) during 30,000 years. Hence, the amount of C_{14} in the atmosphere is still increasing. This research points to the young age of the earth (probably less than ten thousand years).

    With the depletion of earth's magnetic field, more and more radiation penetrates into our atmosphere. Today, the Geiger counter clicks sixteen per minute per gram is usually denoted by living matter. The animals and plants that live on the Earth four thousand years ago, initially to be located far fewer C_{14}. With a small amount of C_{14}, they will look at a few thousand years older than they actually are. Several factors can alter the rate of C_{14}. One factor is the eleven-year sunspot cycle.

2. The decay rate is not changed. (Assume in Eq. 4.5 that $k = $ const.). Many times it has been shown that this assumption is uncertain. Since the rate of decomposition may vary, the age of the artifacts obtained by the C_{14} can be recognized only with the necessary safety precautions.

3. You can find the original amount of C_{14}. This assumption has been proven wrong many times. Different parts of a sample often produce different dates. A variety of live specimens produce quite unlike relations. The age of some items cannot be estimated by using carbon dating, even if the items contain carbon.

4. Checked samples sometimes were not radioactive (contaminated) that will change their original condition for thousands of years. This assumption is very difficult (if not impossible) to prove. The products of radioactive decay could be lost or gained by the sample. Laboratory tests confirmed a large number of what may occur.

    Some research demonstrates increase of radiocarbon in the atmosphere. It is contrary to the official data. The graph of the proportion of radiocarbon from time to time over the last century show that it declined to about the level of 1950 due to industrial emissions of fossil CO_2 which is no longer left radiocarbon, almost all fell apart. Then there was a sharp increase - then massively conducted due to nuclear tests. Now it is reduced again.

It is safe to use Carbon 14 for dating only inorganic artifacts because we know the exact proportion of isotopic CO_2 absorbed by plants from the atmosphere. Geiger counters are no longer used for determining the proportion of radioactive carbon because mass spectrometers give much better accuracy.

4.2 Mini-Project 2: Patients and Injections

Problem 158 After an ankle surgery, a patient is given an injection of 6 units of codeine. For each succeeding daily injection she is given 4 units of the same medication. The patient loses 80% of the medication between injections. How much of the medication will remain in the patient's system after the 3rd injection? How about after the 30th injection? What is the amount of the medication in the blood stream on the 31st and 35th days if the 30th injection was the last?

Solution. Let us think of what happens if we are given an injection of 6 units of a medication and will not get more. Will we have a portion of the medication in our blood on the second day?

For a specific medication different people have different elimination rates (say, a) that shows what % of a medication will be gone (eliminated) from our body. It can be 10%, 20%, 40%, even 90% etc. So some patients can lose 10%, 20% 40%, etc of the medication. But 90%, 80%, 60%,... of the medication respectively will remain in the blood stream on the next day. Let $x = 1 - \frac{a}{100}$, where a is an elimination rate in percent.

If we are given 6 mg of a medication on the 1st day and do not receive other injections we would have $6x$ mg of the medication on the second day, $6x \cdot x = 6x^2$ on the third day $6x^2 \cdot x = 6x^3$ on the fourth day and so on. However, if we continue injections the expression that gives the amount of the medication in a blood stream on each consecutive day will be some polynomial function in x:

During injections

Day 1	6
Day 2	$6x+4$
Day 3	$6x^2 + 4x + 4$
Day 4	$6x^3 + 4x^2 + 4x + 4$
Day 5	$6x^4 + 4x^3 + 4x^2 + 4x + 4$
...	
Day n ($1 \le n \le 30$)	$6x^{n-1} + 4x^{n-2} + 4x^{n-3} + \ldots + 4x + 4$
Day 30	$6x^{29} + 4x^{28} + 4x^{27} + \ldots + 4x + 4$

After stopping injections

Day 31	$6x^{30} + 4x^{29} + 4x^{28} + \ldots + 4x^2 + 4x$
Day 35	$6x^{34} + 4x^{33} + \ldots + 4x^5$
Day n ($n > 30$)	$6x^{n-1} + 4x^{n-2} + \ldots + 4x^{n-30}$

We notice that the amount of the medication, $A(n)$, on the nth day, $1 \le n \le 30$, can be simplified as $6x^{n-1}$ plus the sum of a geometric sequence with the first term of 4 and a common ratio of x.

$$A(n) = 6x^{n-1} + \frac{4(1 - x^{n-1})}{1 - x}, \quad 1 \le n \le 30 \tag{4.7}$$

Equation 4.7 gives us the amount of the medication in the blood system on any particular day n between the first and the last injections. ($1 \le n \le 30$). The amount of the medication in a blood stream on the nth day after stopping injections ($n > 30$) $A(n)$, ($n > 30$) again can be written as $A(n) = 6x^{n-1} + \frac{4x^{n-30}(1 - x^{29})}{1 - x}, n > 30$.

Table 4.1 demonstrates the influence of different x values on the amount of the medication in the system on some particular day between the first and the last injections. Also we can see that when $x \ll 1$ (see columns for $x = 0.05$, $x = 0.1$ or even for $x = 0.2$) the amount of the medication on the 10th and 30th day are almost the same. This fact can be explained from Eq. 4.7.

If $x < 1$ and $n \to \infty$, then $x^{n-1} \to 0$ and $A(n) \to \frac{4}{1-x}$ and is constant for specific x. If $x = 0.2$, $A(30) = 6x^{29} + \frac{4(1 - x^{29})}{1-x} \approx \frac{4}{1-x} = \frac{4}{1-0.2} = 5$, $A(10) = 6x^9 + \frac{4(1 - x^9)}{1-x} = 0.5 = A(30)$.

From this table we can see that for some people with a low elimination rate (x is bigger) there will be more residual medication in the system than for people with a higher elimination rate such as $a = 80\%$, for which $x = 0.2$. From the solution to this mini-project we learn two things:

- It is a good example of the application of geometric series.
- It illustrates something important about medications; even after we stop taking them we can still have the medication in our blood stream for many days (see the

Table 4.1 The amount of the medication remaining in a blood stream for different x values on certain days

Day	$x = 0.05$	$x = 0.1$	$x = 0.2$	$x = 0.4$	$x = 0.5$	$x = 0.6$	$x = 0.8$	$x = 0.9$
2	4.03	4.6	5.2	6.4	7	7.6	8.8	9.4
3	4.0215	4.46	5.04	6.56	7.5	8.56	11.04	12.46
10	4.02	4.44	5	6.67	8	9.96	18.121	26.83
30	4.02	4.44	5	6.66	8	10	19.978	38.4
31	0.02	0.44	1	2.66	4	6	15.98	34.5
35	E-11	4.4E-5	0.0016	0.068	0.25	0.77	6.54	22.67

table for days 31 and 35). This explains, for example, why even when a patient stops taking a drug, the doctor suggests not drinking any alcohol for a few days afterward.

At first glance, the next problem seems to be completely different from the patient problem. However, the same approach can be used.

Problem 159 Brian has decided to cut off 20% of the length of his hair today and will continue to cut off 20% every 2 months thereafter. Knowing that his hair is presently 30 centimeters long and that it will grow about 8 centimeters in 2 months time, how long will his hair be after his haircut one year from now?

Solution. Since Brian cuts 20% of his hair every 2 months, then mathematically it is like multiplying his recent length by 0.8. Let $x = 0.8$. Using the fact that his hair grows 8cm in two months, we obtain:

Now	30
Month 2	$x(30 + 8) = 30x + 8x$
Month 4	$x(30x + 8x + 8) = 30x^2 + 8x^2 + 8x$
Month 6	$30x^3 + 8x^3 + 8x^2 + 8x$
Month 8	$30x^4 + 8x^4 + 8x^3 + 8x^2 + 8x$
Month 10	$30x^5 + 8x^5 + 8x^4 + 8x^3 + 8x^2 + 8x$
Month 12	$30x^6 + 8x^6 + 8x^5 + 8x^4 + 8x^3 + 8x^2 + 8x$

The last expression can be written as $L(12) = 30x^6 + \frac{8x(1-x^6)}{1-x}$ with $x = 0.8$ so $L(12) = 31.5$ cm.

Answer. About 31.5 cm.

Remark. This can be used for any period of time. Noticing that the highest power of x is 1/2 of the number of the month we obtain,

Month $2k$:

$$L(2k) = 30x^k + 8x^k + 8x^{k-1} + \ldots\ldots + 8x = 30x^k + \frac{8x(1-x^k)}{1-x}$$

$$= 30 \cdot 0.8^k + \frac{8 \cdot 0.8(1-0.8^k)}{0.2}$$

Because $r = 0.8$ and the geometric sequence is decreasing as k increases, then the second term will get closer and closer to $\frac{8x}{1-x} = \frac{8 \cdot 0.8}{1-0.8} = 32$. The first terms depends on k and decreases very quickly as k increases. For example,

$$30 \cdot 0.8^{20} = .346,$$

$$30 \cdot 0.8^{30} = .037, \quad 30 \cdot 0.8^{40} = .004.$$

When $k > 40$, then the entire expression will approach 32 cm. In the long run, Brian's hair length would be 32 cm long.

4.3 Mini-Project 3: Investing Money

Sometimes nothing is more powerful and motivating than an example from real life. We all want to be wealthy and healthy. We work hard. We invest money. We want to know how much money we will have in our bank account if we stick with this or that rate, if our interest is compounded annually or continuously. We try to ask advice but sometimes do not know that all real life situations like investing money or planning retirement or buying a house can be reduced to math problems that we can solve ourselves!

4.3.1 Simple and Compound Interest

The best loan you can imagine is no loan at all or a loan from your grandfather or grandmother. Usually they would give you money just as a gift or by asking to pay them back when you can. Most people and businesses are not your grandparents and usually they lend you money at certain interest rate, which depends on many factors, including your credit history and market stability, etc.

First, there is a simple interest and compounded interest that you can expect. In the case of simple interest, each year, your payment back for a loan of $\$P$ will be increased by the same amount, rP, so after t years you will have to pay your principal (original loan amount) plus the interest times the number of the years, $P + t \cdot rP$, which represents the simple interest formula,

$$A = P + t \cdot rP = P(1 + tr) \tag{4.8}$$

If your interest is compound interest, then such a situation is more applicable for the growth of the balance of a bank account. For example, if you put $\$P$ on your bank account with annual percentage rate, r, then the amount of money to expect at the end of the first year will be $P + rP = P(1 + r)$. This principal will be increased by the interest rate and at the end of the second year will be $P(1 + r) + rP(1 + r) = P(1 + r)^2$.

If nothing changes and you do not take your money then this will be your new principal and at the end of the third year you would have $P(1 + r)^2 + rP(1 + r)^2 = P(1 + r)^3$, etc. By induction, after we continue the procedure t times we get

Eq. 4.4 for compound interest, $A = P(1 + r)^t$. Therefore, the amount of money one can take from his or her bank account form a geometric progression.

$$P, \ P(1+r), P(1+r)^2, \ P(1+r)^3, \ \dots \ , P(1+r)^n, \dots$$

However, regarding a personal bank account there is no such thing as the sum of this sequence, because each year the amount of money will be represented by only one term of the sequence and not by the sum of the terms.

The formula can be modified with knowledge of how the interest is compounded. For example, if the annual percentage rate is APR and the interest is compounded n times per year, then the formula can be rewritten as

$$A = P\left(1 + \frac{APR}{n}\right)^{nt}. \tag{4.9}$$

This formula will become a continuous interest formula if the number of compounded periods will increase without limit, $n \to \infty$. So your interest is not calculated monthly ($n = 12$), not even daily ($n = 365$) but continuously, more often that every second. It is as if the bank cares for you continuously.

Using the second formula of Eq. 3.6 from Chapter 3, we obtain that $\lim\limits_{n \to \infty} P$ $\left(1 + \frac{r}{n}\right)^n = Pe^{rn}$. Hence Eq. 4.9 will become

$$A = P \cdot e^{rt}, \tag{4.10}$$

where $r = APR$, $n = t$ are the annual percentage rates in decimal notation and time, respectively.

4.3.2 Saving Money by Periodic Deposits. Future Value of an Annuity

We have learned how to compute the future value and interest for a fixed sum of money deposited in an account that pays interest, compounded either periodically or continuously. But not many people are in a position to deposit a large sum of money at one time in an account. Most people save or invest money by depositing small amounts at different times. Consider the following problem:

Problem 160 Suppose $10,000 is deposited in a bank on January 1 of each year from 2011 through 2016, inclusive, where it earned an annual yield of 5%. What was the value of our account one year after the last deposit?

Table 4.2 Balance for Problem 155

Current year	Deposit	Number of years between current year and year 2017	Input from current year to year 2017
2011	10000	6	$10,000(1+0.05)^6$
2012	10000	5	$10,000(1+0.05)^5$
2013	10000	4	$10,000(1+0.05)^4$
2014	10000	3	$10,000(1+0.05)^3$
2015	10000	2	$10,000(1+0.05)^2$
2016	10000	1	$10,000(1+0.05)^1$

Solution. So we were checking our balance on January 1, 2017. Maybe we want to collect our money. A table can be helpful.

In this Table 4.2, the right column represents the future value of a particular deposit after accumulating the interest during certain period. Hence, the numbers in this column are all different, because each equal payment will be compounded for a different time!

As the table shows, if you put $10,000 in on January 1, 2011 and do no further deposits, on January 1, 2017 you would have $10,000(1+0.05)^6 = \$13,400$ (the last expression in the first row) which equals the accumulated return over 6 years. However, because you deposited $10,000 every year since 2011, we have to add all inputs in the last column. Look closely at this sum,

$$S = \$10,000(1+0.05)^6 + \$10,000(1+0.05)^5 + \$10,000(1+0.05)^4$$
$$+ \$10,000(1+0.05)^3 + \$10,000(1+0.05)^2 + \$10,000(1+0.05)$$

If you rewrite it backward, you will notice that we are looking for the sum of the first 6 terms of a geometric sequence with the first term, $b_1 = \$10,000(1+0.05) = \$10,500$, and the common ratio $r = 1 + 0.05 = 1.05$

Thus, $S_6 = \frac{\$10,500(1.05^6-1)}{1.05-1} = \$71,420$.

Answer. The value of the investment on January 1, 2017 will be equal to $71,420. Certainly, it is better than $60,000 that one can save in a pillow.

In business, the sum of all payments plus all interest earned is called the amount of the annuity or future value of the annuity. We consider payment intervals that coincide with the compounding period of the interest. P is the periodic payments (for example, annual deposits), j is the interest rate, and n the total number of payments. Let us create a similar table (Table 4.3).

At the end of the nth year we are accumulating money from each year so we have to add all inputs in the last column.

Table 4.3 An ordinary annuity

Current year	Deposit	Number of years between current year and the nth year	Input from Current year
1	P	$n-1$	$P(1+j)^{n-1}$
2	P	$n-2$	$P(1+j)^{n-2}$
3	P	$n-3$	$P(1+j)^{n-3}$
-----	-----	-----	-----
$n-1$	P	1	$P(1+j)$
n	P	0	P

$$S = P + P(1+j) + P(1+j)^2 + \ldots + P(1+j)^{n-2} + P(1+j)^{n-1}$$
$$= P\left\{1 + (1+j) + \ldots + (1+j)^{n-1}\right\}$$

Because the expression within parentheses is geometric series with the first term 1 and a common difference $(1+j)$, we can state that if periodic payments P are made for n periods at an interest rate j per period, the amount of the annuity will be given by $S = \frac{P\cdot((1+j)^n - 1)}{(1+j)-1} = P \cdot \left(\frac{(1+j)^n - 1}{j}\right)$. This formula can be rewritten in terms of the annual percentage rate (APR), number of payment periods per year (n), time of the investment in years (t), and the amount of the periodic payment (P),

$$S = \frac{P \cdot \left(\left(1 + \frac{APR}{n}\right)^{nt} - 1\right)}{\left(\frac{APR}{n}\right)}. \tag{4.11}$$

Equation 4.11 will be especially helpful if one saves P dollars periodically, consistently with a given APR compounded periodically, for example monthly. You may have seen such a formula in a financial context, but now you know how a geometric series can be used for its derivation!

Problem 161 Which is better for you: to deposit $10,000 at once, to be held for 10 years at 6% compounded annually, or to deposit $1,000 in each of the next 10 years, to be held in an account earning interest at the same rate?

Solution. If you already understand the idea of compounding, it may be obvious that the first choice would be better, but it is still interesting to see how much better! If you deposit $10,000 now, to be held for 10 years at 6% annual rate, at the end of the 10th year you will have $10000 \cdot (1+0.06)^{10} = 10000 \cdot 1.06^{10} \simeq \$17,908$. If you deposit $1000 each year for 10 years we can use Eq. 4.2 to obtain $1000 \cdot \left(\frac{(1.06)^{10} - 1}{0.06}\right) \simeq \$13,181$. We notice that in part b) our savings account will be $4727 less, but it is still better than keeping money at home (in which case you would have only $10 \cdot \$1000 = \$10,000$) and it will be safe, as well. So if you can put a large amount of money in your savings account it would be best, but if you cannot afford it, would it be better, to deposit a portion of money monthly or annually?

Problem 162   You have an opportunity to save $1200 each year. The interest rate is 6%, compounded annually. Which will give you more money after 10 years? a). You deposit $1200 annually for 10 years, or b). You deposit $100 monthly for 10 years?

Solution.   After 10 years you will have $\$1,200 \cdot \left(\frac{1.06^{10}-1}{0.06}\right) \simeq \$15,817$. If you deposit money monthly for 10 years you would do 120 deposits during the entire period. The interest rate in each of the periods would be $.06/12$ or $.005$. At the end of the 10th year using Eq. 3.53 you will have $\$100 \cdot \left(\frac{1.005^{120}-1}{0.005}\right) \simeq \$16,388$.

Answer.   Monthly deposits will work better for you. This is again due to the power of compounding.

Problem 163   To save money for college, Jim deposits most of his summer job earnings in a savings account at the end of each summer. He deposits $1,000 in the first year. In each of the next three years, he deposits $500 more than he did in the preceding year. Assume that he makes each deposit on the same day of the year, that the annual interest rate is x and remains constant, and that he makes no withdrawals or other deposits. Find a function in x that expresses Jim's total accumulated savings immediately after he makes his last deposit. Evaluate this expression if the interest rate is 5.8%

Solution.   Let us create a table again (Table 4.4).

Actual deposits form an arithmetic progression with the common difference of $500. The accumulated savings are $S(x) = 1000 \cdot (1+x)^3 + 1500 \cdot (1+x)^2 + 2000 \cdot (1+x) + 2500$. The total savings when the interest rate is 5.8% is $S = 500(2 \cdot 1.058^3 + 3 \cdot 1.058^2 + 4 \cdot 1.058 + 5) \approx \$7,479.33$.

Answer. $7,479.33.

Table 4.4   Balance for Problem 163

Current year	Deposit	Number of years between current year and year of withdrawal	Input to the day of withdrawal, annual rate x	Input to the day of withdrawal, annual rate 0.058
1st	1000	3	$1000(1+x)^3$	$1000(1.058)^3$
2nd	1500	2	$1500(1+x)^2$	$1500(1.058)^2$
3rd	2000	1	$2000(1+x)^1$	$2000(1.058)^1$
4th	2500	0	2500	2500

Problem 164 There are three friends with different investment habits. Mary deposits \$500 every month into her savings account at 6% APR compounded monthly for 5 years. Ann has a summer job as a dancer and has an opportunity to invest \$6000 annually at the same nominal rate of 6% compounded annually, and Colleen does not trust any banks and keeps all her available cash in her pillow by placing there each month \$500 for five years. Which of the three friends is better off at the end of the 5th year?

Solution. We will use Eq. 4.11 applied to the of cases of Mary and Ann.

Mary: $S_M = \dfrac{500\left(\left(1 + \frac{0.06}{12}\right)^{60} - 1\right)}{\left(\frac{0.06}{12}\right)} = \$34,885.02.$

Ann: $S_A = \frac{6000((1+0.06)^5 - 1)}{0.06} = \$33,822.56.$

Colleen will have $S_c = 500 \cdot 12 \cdot 5 = \$30,000.$

Answer. Mary will save more money.

4.4   Mini-Project 4: Thinking of Buying a House?

When we want to buy a house, many of us feel frustrated for many reasons. First, we usually do not have enough money to pay off any house right away. Second, we've heard a little bit about down payments from our experienced friends who have already purchased a house. Third, even if we can imagine all steps of this complicated procedure, it would be nice to know that arithmetic and geometric series can help us to select a mortgage company, an interest rate, the number of years for which we will finance, and a monthly payment.

Different financial advices from a mathematical point of view will be given here.

4.4.1   Present Value. Debt Payment Schedules

It is important to introduce now a so-called "present value." If your debt now is \$5000, then in two years, unless you have to pay the same amount to your nice grandpa, it will be increased by the law of compounding and will become $5000 \cdot (1 + r)^2$. If on the other hand, you have to pay someone \$A in t years under annual percentage rate r, then your debt can be recalculated to its present value, i.e., the amount you owe as off today. For this, we have to solve Eqs. 4.4, 4.9, or 4.10 for P:

$$P = \frac{A}{(1+r)^t} = A(1+r)^{-t} \tag{4.12}$$

This is the present value of the amount A and can be modified depending on the compounding process as

$$P = A\left(1 + \frac{APR}{n}\right)^{-nt}. \tag{4.13}$$

$$P = Ae^{-rt}. \tag{4.14}$$

Present value is used in business if we, based on the analysis of the cash flow and the knowledge of the initial investment, can decide whether or not this business is profitable or not.

Consider the following problem.

> **Problem 165** An initial investment of \$35,000 in a business guarantees the cash flows summarized by Table 4.5. Assume an interest rate of 5% compounded annually. Find the net present value of the cash flow. Is the investment profitable?

In order to decide on the profitability of this business, we find the net present value by recalculating all cash flow by its present value and then adding them together.

$$\text{NPV} = \$8000(1+0.05)^{-3} + \$10,000(1+0.05)^{-4} + \$14,000(1+0.05)^{-6}$$
$$\approx \$25,584.74.$$

Because this number is less than the initial investment of \$35,000, we can state that this business is not profitable. Moreover, it loses money.

Sometimes one needs to pay a debt now but does not have the money and asks for an opportunity to pay his or her debt either later, in certain number of years (months) or for example, to pay the debt partially now and then by the schedule offered by the lender. Despite the fact that many people believe that the following problem would never correspond to a real life scenario, it is not true. By solving this problem, you can make important financial decision in your life.

Table 4.5 Cash Flow for Problem 165

Year	Cash Flow
3	\$8,000
4	\$10,000
6	\$14,000

Problem 166 A debt of $5000 due four years from now and $3500 due 6 years from now is to be repaid by a payment of $2000 now and one an additional payment in five years from now. If the interest rate is 4.8% compounded annually, how much is that single payment?

Solution. The easiest way is to recalculate both payment scenarios for their present value at moment $t = 0$. However, we recalculate the balance at time $t = 5$ years. At which the final unknown payment must be made. When solving such problems, I always recommend sketching the problem first (Figures 4.2 and 4.3).

The first option will be called the upper schedule and the second scenario is lower schedule.

Upper schedule: $US = \$3500(1 + 0.048)^{-1} + \$5000(1 + 0.048)^{1}$

Lower Schedule: $LS = x + \$2000(1 + 0.048)^{5}$.

The balance obtained from the both schedules must be the same, from which we can find the final payment, x:

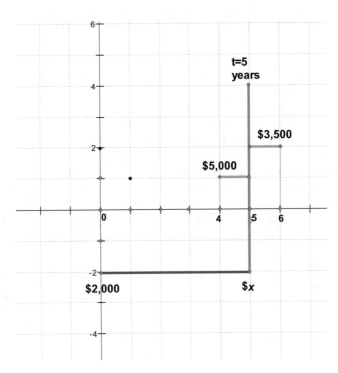

Figure 4.2 Balance for problem at year five

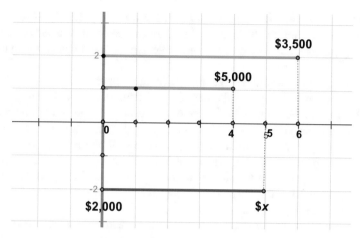

Figure 4.3 Balance at present time

$$x = 3500 \cdot 1.048^{-1} + 5000 \cdot 1.048 - 2000 \cdot 1.048^5$$
$$= \$6051.35.$$

We can see that the upper schedule would make one to pay \$8500. By the low schedule, one would pay \$8051.35, which is less by almost \$450. Obviously, if we recalculate either schedule to its present value (Figure 4.3).

$$PV = \$3500(1 + 0.048)^{-6} + \$5000(1 + 0.048)^{-4}$$
$$= \$6786.81.$$

Then this result can lead us to another consideration or a new problem. If one has the money (\$6786.81) then the debt could be paid immediately.

Answer. $x = \$6051.35.$

4.4.2   Present Value of an Annuity. Mortgage Payment

Let us take an example from real life. Say a couple finds a house in neighborhood that they like. The house is a two-story house. It is 3300 sq. ft., 5 bedrooms, 3 bathrooms, a bar, stained glass windows in the living room, a big balcony, and their dream- a nice swimming pool. The owner wants \$160,000 and wants it quickly. The couple has some savings in a bank account; their friend gave them the address and telephone of a good mortgage company that offers a rate of 7.2%. The couple has a stable income, but it is not very big and they do not yet know what kind of loan and payment to choose, or if they can even afford a house like this. He is 42 years old and she is 38. They hear that a 30-year loan would give them lower payments every month but they would be 72 and 68 respectively after 30 years. They really do not want to be in debt when they are senior citizens. So before going to a mortgage company they decide to estimate approximately what would be good for them.

The first question would be: What is the monthly payment?

How to estimate the size of the periodic payment? Let L be a loan for t years; $m = 12 \cdot t$ the number of months for repayment starting at the end of first month; r = interest rate per year in decimal notation; $j = r/12$ monthly interest rate; P = amount of repayment per month (starting at the end of first month), your monthly payment. The house would be yours as of today if you pay the entire mortgage or loan, L, now. If you have to make n periodic payments each month for the next t year (15, 20, or 30 years), then you will pay eventually $S = P \cdot t \cdot n$ dollars, which equals to your total payment.

Let us assume that P is our monthly payment. Each periodic payment P can be recalculated as its present value at moment 0. So if we pay P per month, then the total present value of all periodic payments after three months will be $P(1+j)^{-1} + P(1+j)^{-2} + P(1+j)^{-3}$ and so on, and the present value of your repayments after m months will be $P(1+j)^{-1} + P(1+j)^{-2} + P(1+j)^{-3} + \ldots + P(1+j)^{-m}$. The expression above is a geometric series with m terms, the first term equals $P(1+j)^{-1}$ and common ratio is $(1+j)^{-1}$, so the sum of m terms can be found as

$$P(1+j)^{-1}\left(\frac{1-(1+j)^{-m}}{1-(1+j)^{-1}}\right) = \frac{P}{(1+j)} \cdot \frac{(1-(1+j)^{-m})}{\left(1-\frac{1}{i+j}\right)} = \frac{P(1-(1+j)^{-m})}{j}$$

We must now equate the present value of all repayments above to the total value of the loan (mortgage) taken at moment zero, L. Replacing $j = \frac{APR}{n}$, $m = nt$, and L by A, we obtain the formula that is called the present value of an annuity:

$$L = A = \frac{P \cdot \left(1 - \left(1 + \frac{APR}{n}\right)^{-nt}\right)}{\left(\frac{APR}{n}\right)} \tag{4.15}$$

Solving this formula for P, we derive the expression for a periodic payment of P, n times a year during t years toward your original loan of A at the annual percentage rate APR compounded n times per year:

$$P = \frac{A \cdot \left(\frac{APR}{n}\right)}{1 - \left(1 + \frac{APR}{n}\right)^{-nt}} \tag{4.16}$$

For monthly payment, $n = 12$, and the formula becomes

$$P = A \cdot \left(\frac{\frac{APR}{12}}{1 - \left(1 + \frac{APR}{12}\right)^{-12t}} \right).$$

We will solve several problems related to mortgage, credit card, or car loan payment.

Problem 167 A couple buys a house for \$160,000. They make a \$16,000 down payment and agree to amortize the rest of the debt in monthly payments over the next 20 years. If the interest on the debt is 7.2%, compounded monthly, what will be the size of the monthly payment?

Solution. What will the couple's monthly payment be for different rates, down payments, and repayment periods? To estimate the monthly payment, we use Eq. 4.16 where L is the actual debt (the price of the house minus a down payment). Because $L = \$160,000 - \$16,000 = \$144,000$ $j = 0.072/12 = 0.006$ and $n = 12 \cdot 20 = 240$ then $P = \$144,000 \cdot \left(\frac{0.006}{1-1.006^{-240}}\right) = \$1133.78$ so the couple's monthly payments would be approximately \$1134.

"I think we could afford it." The wife said. For the same down payment, can they afford to pay off the loan in 10 years? Now only m will be changed, $m = 120$, $P = \$144,000 \cdot \left[\frac{0.006}{1-1.006^{-120}}\right] = \$1686.84$. It is almost \$1700; If we think about additional costs such as the utility bills and insurance, it could be difficult for them. What if they tried to pay off the loan in 15 years; then $m = 180$, $P = \$144,000 \cdot \left[\frac{0.006}{1-1.006^{-180}}\right] = \$1310.47$ that is approximately \$1311 per month. This seems more bearable.

What would happen if they put down 20% or \$32,000? Assuming that the annual percentage rate (an interest on the debt) is the same, 7.2%, let us estimate their monthly payments for 20 years, 10, and 15 respectively, $L = \$160,000 - \$32,000 = \$128,000$:

- Loan for 20 years, $n = 240$, $P = \$128,000 \cdot \left(\frac{0.006}{1-1.006^{-240}}\right) = \$1007.81$ or about \$1008 per month.
  You notice that there is not a big difference between \$1,134 and \$1008. Maybe it is not a good idea to put down 20%, they would lose \$16,000 from their saving account, but monthly payments would not be that much less.

- Loan for 10 years, $n = 120$, $P = \$128,000 \cdot \left(\frac{0.006}{1-1.006^{-120}}\right) = \$1499.42 \approx \$1500$.
  There is a difference of about \$200 every month, but it is still a pretty large monthly payment!

- Loan for 15 years, $n = 180$, $P = \$128,000 \cdot \left(\frac{0.006}{1-1.006^{-180}}\right) = \$1164.85 \approx \$1165$.

Again we do not see a big difference between monthly payments for a 10% down payment versus a 20% down payment. Let us look at the interest rate now. We would want to stick with a rate of 7.2% or less if it is possible. Let us try a rate of 6.5% and a down payment of 10% or $16,000. So $L = \$144,000$ and $i = 0.0054$:

- Loan for 10 years, $(n = 120)$; $P = \$144,000 \cdot \left(\frac{0.0054}{1-1.0054^{-120}}\right) = \$1633.63$

- Loan for 15 years $(n = 180)$; $P = \$144,000 \cdot \left(\frac{0.0054}{1-1.0054^{-180}}\right) = \$1252.81$.

- Loan for 20 years $(n = 240)$; $P = \$144,000 \cdot \left(\frac{0.0054}{1-1.0054^{-240}}\right) = \$1071.93$

What happens if a family cannot get a lower rate? Let us estimate how much they would pay monthly with a down payment of $16,000 (10%) and an agreement to amortize the rest of the debt in monthly payments over the next 20 years $(n = 240)$ with an interest rate on the debt of either 8.5% or 9.5%?

- Loan for 20 years, $L = \$144,000$, $j = 8.5\%$, $i = 0.071$, $P = \$144,000 \cdot \left(\frac{0.0071}{1-1.0071^{-240}}\right) = \$1251.5$.

We can look at other loan lengths:

$n = 120$ (10 years) $P = \$1787$
$n = 180$ (15 years) $P = \$1419$
$n = 300$ (25 years) $P = \$1162$
$n = 360$ (30 years) $P = \$1109$

- Loan for 20 years, $L = \$144,000$, $j = 9.5\%$, $i = 0.0079$; x payments, $P = \$144,000 \cdot \left(\frac{0.0079}{1-1.0079^{-x}}\right)$:

$x = 120$, $P(120) = \$1861.7$
$x = 180$, $P(180) = \$1501.9$
$x = 240$, $P(240) = \$1340.4$
$x = 300$, $P(300) = \$1256.1$
$x = 360$, $P(360) = \$1208.7$

Sometimes a home buyer wants to put down just 5%. For the same house this would result in a loan of $152,000. Let us say the loan is at 8% and estimate the monthly payment for different lengths of the loan. We use the formula,

$$P = \$152,000 \cdot \left(\frac{0.0067}{1 - 1.0067^{-x}}\right).$$

$x = 120$, $P(120) = \$1844.5$
$x = 180$, $P(180) = \$1452.9$
$x = 240$, $P(240) = \$1271.8$
$x = 360$, $P(360) = \$1115.7$.

Problem 168 A pre-owned 2010 Infinity G 37 is purchased for $3000 down and monthly payments of $450 for four years. If interest is at 5.2% compounded monthly, find the corresponding present cash price of the car.

Solution. First, we use Eq. 4.15 in order to calculate the present value of the loan

$$L = A = \frac{450 \cdot \left(1 - \left(1 + \frac{0.052}{12}\right)^{-12 \cdot 4}\right)}{\left(\frac{0.052}{12}\right)} \approx \$19,463.67.$$

Secondly we add a $3000 down payment to the number above in order to obtain today's cash price of the Infinity G37: $19,463.67 + $3000 = $22,463.67.

Answer. $22,463.67

4.5   Mini-Project 5: Loan Amortization

By doing this mini-project you will learn the best and most efficient way to pay off your debt of any kind, for example, a credit card or car loan. It would also be interesting to know how soon you will pay off your debt if you make extra payments every month or annually. Loan amortization will be briefly discussed. Additionally, we compare different options of paying for a new or used car and whether it is better for you to pay by credit card or take a loan from the dealership if both charge the same annual nominal interest rate.

4.5.1   Paying Off an Outstanding Credit Card Debt

Let us demonstrate how knowledge about arithmetic and geometric series helps us solving very important financial problems. Consider two different mathematically but similar real life scenarios. You want to buy a car of your dream but do not have all the money to pay the cash price of the car, so you need to borrow money.

1. Assume that you come to a private company that offers to lend you money (L) at $r\%$ with n payments per year (usually $n = 12$) and m total payments required (m would be the number of years for the loan times n). This might be a dealership that sells that dream car and they refer you to their financial office.
2. You decided to use your credit card that happens to charge the same nominal rate as the dealer. This can be rephrased as follows: You have an outstanding debt of L dollars on your credit card that charges APR of $r\%$ compounded monthly and you need to pay it off in n months, each month by an equal amount of L/n dollars plus an interest on unpaid balance.

Are these two scenarios mathematically equivalent? Are they financially equal to you? Which scenario is better?

The common thing is that you have an outstanding debt of L and you need to pay it off by making precisely n equal periodic payments. However, in case one you do not have the freedom to choose your monthly payment, usually it is the number of dollars that is calculated by the Loan company (for example, by the dealership office or by a bank).

Regarding the second scenario, one decided to pay off his or her credit card debt by making equal monthly payments of (L/n) dollars plus the corresponding monthly interests.

Let us start from the second scenario. Assume that L is the amount borrowed and (L/n) the periodic payment size, then $n(L/n) = L$ is our unpaid balance before the first payment, $(n-1)(L/n) = L - L/n$ is the unpaid balance before the second payment , and so on, with the balance before the last payment of (L/n).

Let us estimate the total interest (in dollars), I, charged by a credit card company. Because r is the rate, then (r/m) is the interest rate per payment period. We must multiply every unpaid balance by the coefficient (r/m) and then add them all together to get $I = n \cdot \left(\frac{L}{n}\right)\left(\frac{r}{m}\right) + (n-1) \cdot \left(\frac{L}{n}\right)\left(\frac{r}{m}\right) + \ldots + 2 \cdot \left(\frac{L}{n}\right)\left(\frac{r}{m}\right) + 1 \cdot \left(\frac{L}{n}\right)\left(\frac{r}{m}\right)$. Factoring out $\left(\frac{L}{n}\right)\left(\frac{r}{m}\right)$ we obtain, $I = \left(\frac{L}{n}\right)\left(\frac{r}{m}\right)(n + (n-1) + (n-2) + \ldots + 2 + 1)$. You notice that the expression within parentheses is the sum of an arithmetic sequence (a sum of all natural numbers from 1 to n) that equals $\frac{n(n+1)}{2}$. Thus,

$$I = \frac{L}{n} \cdot \frac{r}{m} \cdot \frac{n(n+1)}{2} = L \cdot \left(\frac{r}{m}\right) \cdot \frac{(n+1)}{2} \tag{4.17}$$

We can use Eq. 4.17 to solve for r and find the true annual percentage rate,

$$r = \frac{2I \cdot m}{L(n+1)} \tag{4.18}$$

In this formula $j = \frac{r}{m}$ and the total payment (TP) toward the outstanding principal after n payments is

$$TP = L + L \cdot j \cdot \frac{(n+1)}{2}. \tag{4.19}$$

Problem 169 Assume that you found a car in a local dealership and paid $12,000 for the car by a credit card at APR $= 12\%$. Now you want to pay it off in 4 years (48 months). You decided to pay equal monthly portions of $\frac{\$12,000}{48}$ $= \$250$ plus the interest on the outstanding principal. What will be your total payment to the credit card company after 48 months? Is your decision to use a credit card better than the first option you always have? (Scenario 1, sign for a 4 year contract with the dealership and pay your car loan by equal monthly payments).

Solution. Based on our Eq. 4.19 after 48 months your total payment to the credit card will be

$$\text{Loan} + \text{Interest} = TP = \$12,000 + \$12,000 \cdot 0.01 \cdot \frac{48 + 1}{2} = \$14,940.$$

If you sign a 4 year contract with the dealership, then first, we need the actual amount of money you must pay in order to liquidate your debt in 48 payments. If you are taking a car loan of $12,000 and agree to pay the loan plus an interest at APR $= 12\%$ by making equal payments over a 4-year period, your payment will be $316.01. (We used Eq. 4.16 for the present value of the annuity derived in the previous section to calculate this payment.) In four years, you would have paid 48· $316.01= $15,168.29.

When you decide to pay $250 each month toward your principal plus the corresponding interest, then each month your total payment will be different and will eventually decrease as time (in months) increases. Thus, for the first month you would pay $250 + 0.01 \cdot 12000 = \$370$, for the second month it would be $250 + 0.01 \cdot 11750 = \$367.50$, for the 3rd month it would be $250 + 0.01 \cdot 11500 = \$365$, for the 4th month it would be $250 + 0.01 \cdot 11250 = \$362.50$, for the 5th month you would have to pay $250 + 0.01 \cdot 11000 = \$360$, etc.

In general, the total monthly payment to a credit card for a month k, using your own schedule will include an equal payment of $\frac{L}{n} = \frac{12000}{48} = \$250$ and the interest charge on the unpaid balance for the kth month, which will be $(n - (k - 1)) \cdot \left(\frac{L}{n}\right) \cdot \frac{r}{m} = (49 - k) \cdot \frac{12000}{48} \cdot \frac{012}{12} = (49 - k) \cdot 2.5$. The total k^{th} payment can be calculated as

$$y_k = \frac{L}{n} + (n - (k - 1)) \cdot \left(\frac{L}{n}\right) \cdot \frac{r}{m}, \quad 1 \le k \le n. \qquad (4.20)$$

Using Eq. 4.20, the total k^{th} payment for $n = 48$, $m = 12$, $r = 0.12$, $L = \$12,000$ is described by

$$y_k = \$250 + (\$49 - \$k) \cdot 2.5. \qquad (4.21)$$

If we wish to know after what month, your payment will be less than $316, we can make the monthly payment given by Eq. 4.21 less than 316 and solve the inequality for k,

$$y_k = \$250 + (\$49 - \$k) \cdot 2.5 < \$316$$
$$\$49 - \$k < \frac{\$66}{2.5} = \$22.6$$
$$\$23 \le k \le \$48.$$

Starting from month 23, our payment will be less than $316. We can create a table of the total monthly payments given by Eq. 4.21 using Excel together with the second option of $316 each month (Table 4.6) and graph both cases for comparison (Figure 4.4). Curve Y1 represents your own schedule when the monthly payment is calculated by Eq. 4.21 and curve Y2 represent the same monthly payment of $316.

Table 4.6 Monthly
payments for Problem 169

k	Y1	Y2
1	370	316
2	367.5	316
3	365	316
4	362.5	316
5	360	316
6	357.5	316
7	355	316
8	352.5	316
9	350	316
10	347.5	316
11	345	316
12	342.5	316
13	340	316
14	337.5	316
15	335	316
16	332.5	316
17	330	316
18	327.5	316
19	325	316
20	322.5	316
21	320	316
22	317.5	316
23	315	316

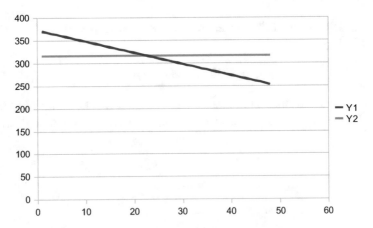

Figure 4.4 Credit card monthly payments schedules for Problem 169

Both, the table and the graph confirm our analytical result that the 23rd payment to be less than $316 and that any payment after that will be less than $316.

The total payment of the first method of repayment is $14,940 which is less than the total payment of $15,168 of the equal monthly payment amount of $316 for a total savings of $15,168.29 − $14,940 = $228.29 . Therefore, at the same nominal interest rate and under conditions such as time period, etc., the second scenario would be better for you and would save you a little money. This is probably why dealerships do not allow you to pay for new or old car using you credit card and generally prefer that you sign a financing contract. However, as you probably understand the second scenario requires more discipline and knowledge of mathematics. It is very easy to forget about your payment, especially if the amount is changing from one month to the other. Of course, credit cards are not always a better choice for you—only if you are a very disciplined person and would not miss any planned payment. What is best is to have control of your finances and understand your options.

4.5.2 Using a Computer to Build an Amortization Table

Suppose that you get a $1500 loan from a bank with the interest rate of 12% compounded monthly. The $1500 plus interest is to be repaid by equal payments of R dollars at the end of each month for three months. Usually such a schedule is initiated by yourself because you need to pay your debts as soon as possible. Using Eq. 4.16 we find that the monthly payment is $510.0332. We round the payment to $510.03, which will probably result is a slightly higher final payment. The bank can consider each payment as consisting of two parts: a. Interest on the outstanding loan and b. Repayment of part of the loan. This is called amortizing.

   A loan is amortized when a part of each payment is used to pay interest and the remaining part is used to reduce the outstanding principal. Since each payment reduces the outstanding principal, the interest portion of a payment decreases as times goes on. Let us analyze the loan just described. At the end of the first month, you pay $510.03. The interest on the outstanding principal is $0.01(1500) = \$15$. The balance of the payment, $510.03 − 15 = \$495.03$, is then applied to reduce the principal, hence, the outstanding principal is now $1500 − 495.03 = \$1004.97$ and interest is $10.05. At the end of the second month, $499.98 is applied toward the principal, and the outstanding balance is $1000.94 − 499.98 = \$504.99$, etc. We can make a table like Table 4.7.

Table 4.7 Credit card debt

Period	Principal outstanding at beginning of period	Interest for period	Monthly payment	Principal prepaid at the end of period
1	1500	15	510.03	495.03
2	1004.97	10.05	510.03	499.98
3	504.98	5.05	510.04	504.98
Total		30.10	1530.10	1500

Table 4.8 Credit card debt (your own schedule)

Period	Principal outstanding at beginning of period	Interest for period	Monthly payment	Principal prepaid at the end of period
1	1500	15	500	485
2	1015	10.15	500	489.85
3	525.15	5.25	500	494.75
4	30.4	0.3	30.7	30.4
Total		30.7	1530.7	1500

Table 4.9 Credit card debt (your own schedule, only 3 months)

Period	Principal outstanding at beginning of period	Interest for period	Monthly payment	Principal prepaid at the end of period
1	1500	15	500	485
2	1015	10.15	500	489.85
3	525.15	5.25	530.4 = 525.15 + 5.25	525.15
Total		30.4	1530.4	1500

The last row shows that during these three months you paid $30.10 of interest. The total payment is the sum of all numbers in the monthly payment column, and it differs from the principal by the interest amount.

Next, assume that instead of $510.03 you decided to pay just $500 each month, not equal payments of $500 plus corresponding interest, as it was done in Problem 169 about a $12,000 used car purchased by a credit card. Will you still be able to pay off your debt in three months? Let us make a new amortization table (Table 4.8). A simple guess would tell us that the answer is "No" because by paying $1500 you would not pay for any interest that will be your duty afterwards.

In order to pay off your debt in 3 months, not in four months as it is in Table 4.8, you had to pay more as the last payment. In Table 4.9, you must pay $530.4 = 525.15 + 5.25, and will owe no money after that.

You just learned how to create an amortization schedule for a loan. You can see that the interest paid depends on our payment amount. Different schedules are possible when you decide to pay off your credit card, and for each scenario, you can make similar amortization tables. However, if you have a long-term loan such as a car loan or a mortgage, using a computer and Excel spreadsheet would be helpful. Figure 4.5 shows an Excel spreadsheet for the loan amortization described in Table 4.7.

Using the same type of table as Tables 4.8 and 4.9, we can now solve any mortgage problem, including one with prepayments. Consider a $150,000 mortgage at APR = 4.8% compounded monthly for 15 years. We create a spreadsheet and rearrange the initial data on the left and change the formula for calculating a monthly payment cell B9 as (Figure 4.6)

Cell B9: Payment = $B$2*($B$3/$B$4)/(1−(1+$B$3/$B$4)^(−1*$B$4*$B$5))

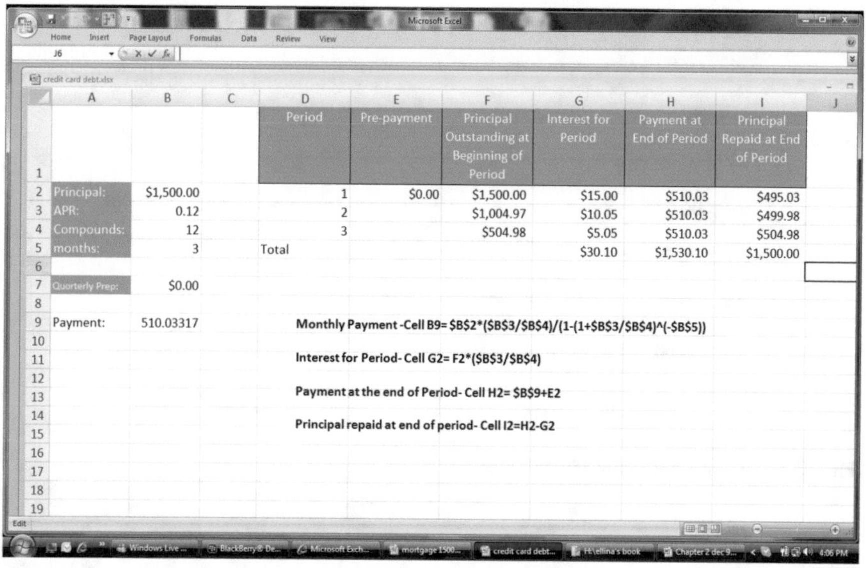

	A	B	C	D	E	F	G	H	I
1				Period	Pre-payment	Principal Outstanding at Beginning of Period	Interest for Period	Payment at End of Period	Principal Repaid at End of Period
2	Principal:	$1,500.00		1	$0.00	$1,500.00	$15.00	$510.03	$495.03
3	APR:	0.12		2		$1,004.97	$10.05	$510.03	$499.98
4	Compounds:	12		3		$504.98	$5.05	$510.03	$504.98
5	months:	3		Total			$30.10	$1,530.10	$1,500.00
6									
7	Quarterly Prep:	$0.00							
8									
9	Payment:	510.03317							

Monthly Payment -Cell B9= $B$2*($B$3/$B$4)/(1-(1+$B$3/$B$4)^(-$B$5))

Interest for Period- Cell G2= F2*($B$3/$B$4)

Payment at the end of Period- Cell H2= $B$9+E2

Principal repaid at end of period- Cell I2=H2-G2

Figure 4.5 Credit card debt amortization with Excel

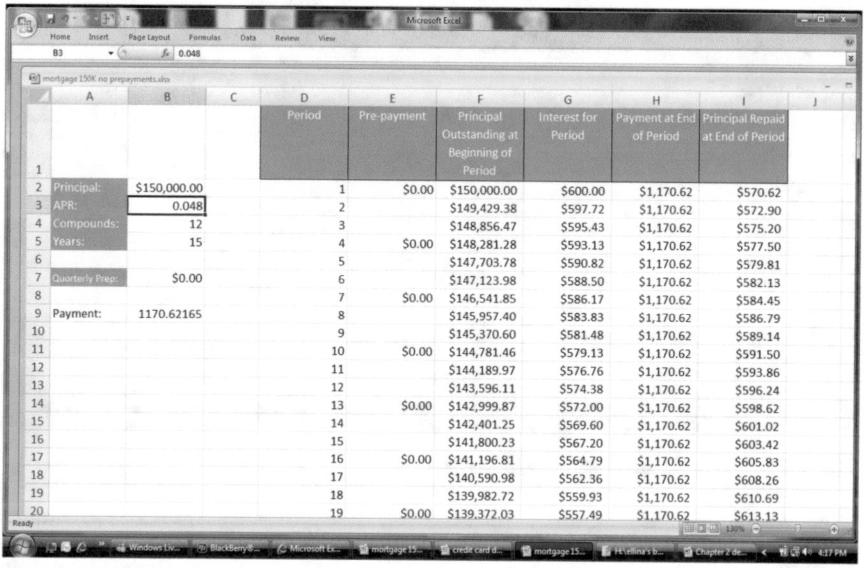

	A	B	C	D	E	F	G	H	I
1				Period	Pre-payment	Principal Outstanding at Beginning of Period	Interest for Period	Payment at End of Period	Principal Repaid at End of Period
2	Principal:	$150,000.00		1	$0.00	$150,000.00	$600.00	$1,170.62	$570.62
3	APR:	0.048		2		$149,429.38	$597.72	$1,170.62	$572.90
4	Compounds:	12		3		$148,856.47	$595.43	$1,170.62	$575.20
5	Years:	15		4	$0.00	$148,281.28	$593.13	$1,170.62	$577.50
6				5		$147,703.78	$590.82	$1,170.62	$579.81
7	Quarterly Prep:	$0.00		6		$147,123.98	$588.50	$1,170.62	$582.13
8				7	$0.00	$146,541.85	$586.17	$1,170.62	$584.45
9	Payment:	1170.62165		8		$145,957.40	$583.83	$1,170.62	$586.79
10				9		$145,370.60	$581.48	$1,170.62	$589.14
11				10	$0.00	$144,781.46	$579.13	$1,170.62	$591.50
12				11		$144,189.97	$576.76	$1,170.62	$593.86
13				12		$143,596.11	$574.38	$1,170.62	$596.24
14				13	$0.00	$142,999.87	$572.00	$1,170.62	$598.62
15				14		$142,401.25	$569.60	$1,170.62	$601.02
16				15		$141,800.23	$567.20	$1,170.62	$603.42
17				16	$0.00	$141,196.81	$564.79	$1,170.62	$605.83
18				17		$140,590.98	$562.36	$1,170.62	$608.26
19				18		$139,982.72	$559.93	$1,170.62	$610.69
20				19	$0.00	$139,372.03	$557.49	$1,170.62	$613.13

Figure 4.6 Mortgage amortization

We will obtain 181 rows in the Excel spreadsheet; you can view it as a spreadsheet (Figure 4.6) or as a graph (Figure 4.7) showing the declining balance of the loan.

With the spreadsheet, it is easy to determine that if we make additional payments of $3000 quarterly (Figure 4.8), we will pay off the mortgage in 80 months (Figure 4.9) and also will dramatically reduce the amount of interest paid.

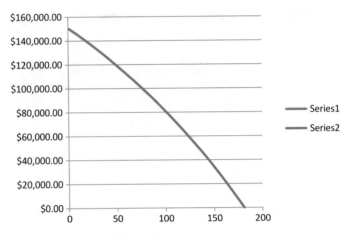

Figure 4.7 Loan amortization in 180 months

Figure 4.8 Mortgage amortization with prepayments

The advantages of using Excel for mortgage amortization are efficiency and flexibility. Many scenarios can be tested quickly once the spreadsheet is constructed. Though many useful websites exist, such http://www.dinkytown. com/java/MortgageLoan.html, the websites that give amortization tables and graphs usually are restricted to monthly or annual payments. For example, we can get an idea of how much our payment will be for a 30 year mortgage of $160,000 at APR = 5.2% and how soon it can be paid off if we pay an additional $400 a month towards the outstanding principal (Figure 4.10).

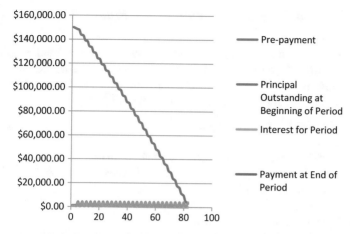

Figure 4.9 Loan amortization with prepayments

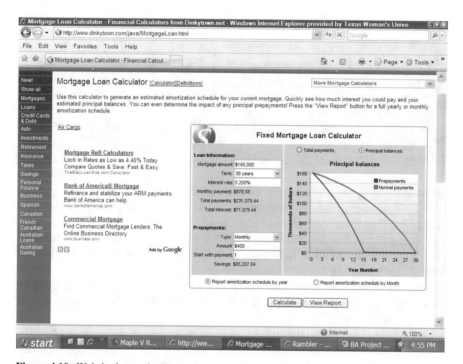

Figure 4.10 Website loan calculator

My favorite website www.dinkytown.com is very practical for getting a first calculation for your mortgage information and for checking our Excel results. However, sometimes in real life we do not follow any schedule and might want to make a payment this year all at once and next year maybe pay some every three months. Now using Excel you can calculate the effect on your mortgage yourself, since the spreadsheet can be so easily adjusted.

4.5.3   *Using a Graphing Calculator for Financial Estimates*

It is always nice to know that someone thought about us and created programs that simplify our life. If you have a TI 83/84 graphing calculator, then let us go to [APPS] applications, then to [Finance], and then to [TVM Solver] (Time Value Money). On the calculator screen you will see the following:

N = (TOTAL number of months you pay)
I% = (interest rate)
PV = (the present value of an annuity, mortgage or loan)
PMT = (monthly payments, must be negative because the present value will be decreased by this amount)
FV = 0 (Future value. Must be 0 someday)
P/Y = 12 (payments per year usually 12)
C/Y = 12 (compounded periods per year and c/y must be the same as p/y)
PMT: END BEGIN

Let us solve the following problem now.

Problem 170 A couple took a loan of $152,000 at a fixed interest rate of 8%, for 30 years. Property taxes are $3600 for the first year and insurance is $75 per month. What will the monthly payment be for the first year? Ignoring property tax and insurance, how soon would the mortgage be paid if the couple makes $1500 monthly payment instead of the scheduled periodic monthly payment?

Solution. Using the TVM Solver we enter all data and we leave an empty space after PMT, as in Figure 4.11. Then we put a cursor at PMT and press [ALPHA] then [ENTER]. That gives a payment of PMT = −1115.32 (Figure 4.12).

Adding 1115.32 + 75 + 3600/12 = $1490.32 we obtain the monthly payment including property tax and insurance. How fast would you pay off the loan of $152,000 if instead of $1115.32 per month you pay $1500?

Now we change our data. Enter everything except for N, then clear the field for N, and with the cursor there, press [ALPHA] then [ENTER]. The calculator screen will look like Figure 4.13.

Regarding Figure 4.13, we found that if we make monthly payments of $1500 we would pay our 30 year mortgage in N = 169.4 months which is approximately 14 years. If you will completely pay off your loan in 30 years but wish to know how much of the loan will be paid off after 1 year, 5 years, and so on. The BAL command (balance) gives the answer

$$[APPS] \ [Finance] \ [Bal] \ [ENTER].$$

Bal(n) is a function whose variable n is the number of payments. After five years the balance is $Bal(5*12) = Bal(60) = \$116,241.26$ if you are making monthly

Figure 4.11 Enter the mortgage ($152,000), interest rate (APR = 8%), the number of total payments over 30 years (360), and the number of compounding periods per year (12)

```
N=360
I%=8
PV=152000
PMT=
FV=0
P/Y=12
C/Y=12
PMT:BEGIN BEGIN
```

Figure 4.12 Calculation of PMT = (we put here a cursor and press [ALPHA] then [ENTER]). That gives a payment of PMT= − 1115.32

```
N=360
I%=8
PV=152000
■PMT=-1115.3221...
FV=0
P/Y=12
C/Y=12
PMT:BEGIN BEGIN
```

Figure 4.13 Monthly payment of $1500 N = (after entering all information we press [ALPHA] then [ENTER]) PV = 15200 PMT = −1500 FV = 0 P/Y = 12 C/Y = 12

```
■N=■69.4083414
I%=8
PV=152000
PMT=-1500
FV=0
P/Y=12
C/Y=12
PMT:BEGIN BEGIN
```

payments of $1500 (you can change PMT back to −1115.32 if you wish to make only the required payment). The Command $\sum \text{Prn}(1, 60)$ gives the principal paid off over 5 years ($35,758.74). Option $\sum \text{Int}(1, 60)$ gives the amount of interest paid over 5 years ($54,241.26). If we graph Bal($12T$) versus T as parametric function we can observe our remaining balance at any period of time T. We select [MODE]-parametric.

Figure 4.14 shows how to enter a balance function in parametric mode and Figure 4.15 helps with the appropriate window setting. Here $T_{\min} = 0$. If you used the Bal(n) function within the TVM solver, you need to restore the values we had before.

Let us look at the graph of Bal($12T$) versus T. Tracing T (pressing [TRACE]) we can observe our balance for any particular year. Thus, for $T = 0$ our remaining (original mortgage) balance is $152,000 (Figure 4.16). After 10 years your balance

Figure 4.14 Entering
balance function in
parametric mode

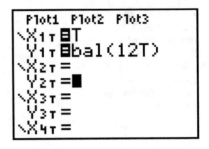

Figure 4.15 Window for a
balance function

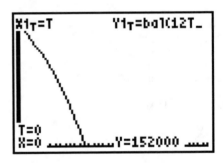

Figure 4.16 Original
balance (mortgage) ($T = 0$)

Figure 4.17 $62, 966.26
balance after 10 years

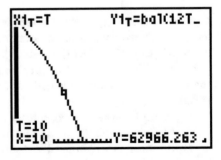

is about $63,000 (Figure 4.17), after 13 years it will be $19,178 (Figure 4.18), and
after 15 years less than 0—we would pay off the loan!

By pressing [2nd] and [Graph] we can look at the table that shows the remaining
balance at a particular time (Figures 4.19 and 4.20).

Figure 4.18 $19,178.74
balance after 13 years

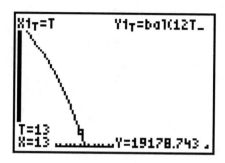

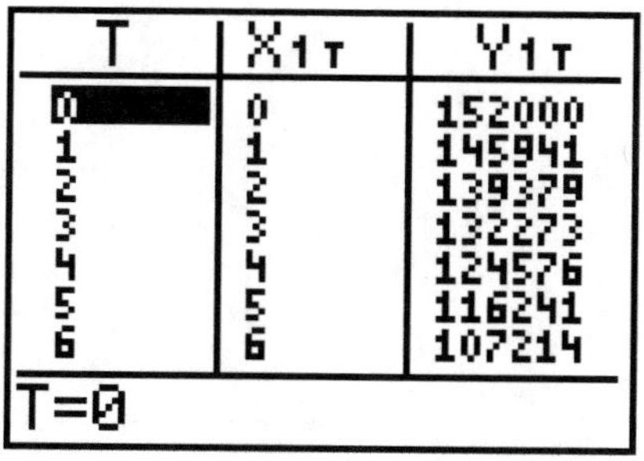

Figure 4.19 Balance table
for $T = 1, 2, 3, 4, 5, 6$ years

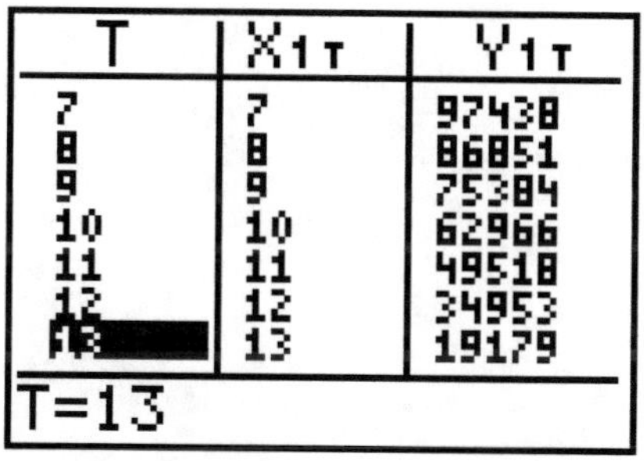

T	X1ᴛ	Y1ᴛ
0	0	152000
1	1	145941
2	2	139379
3	3	132273
4	4	124576
5	5	116241
6	6	107214

T=0

Figure 4.20 Balance table
for $T = 7$-13 years

T	X1ᴛ	Y1ᴛ
7	7	97438
8	8	86851
9	9	75384
10	10	62966
11	11	49518
12	12	34953
13	13	19179

T=13

The calculator or PC are great tools to help when we are making some of these financial decisions, but sometimes it is more fun to solve a problem analytically. Math olympiad problems demand the exact answer which can be obtained only analytically so I encourage you to work in this direction and develop your problem solving skills.

Chapter 5
Homework

1. Find the sum, $\frac{1}{2 \cdot 7} + \frac{1}{7 \cdot 12} + \ldots + \frac{1}{122 \cdot 127}$.

Answer. 25/254

2. Find the sum of the first four terms of a geometric sequence such that its first three terms are simultaneously the 1st, 4th, and the 8th terms of some arithmetic sequence and their sum is $16\frac{4}{9}$.

Answer. $S_4 = 25\frac{25}{27} = \frac{700}{27}$

3. Suppose 60 mg of a medication is taken daily for three days. Then the dose is reduced to 40 mg on the fourth day and 25 mg on the fifth day. Allowing for different elimination rates ($a\%$) for different people, let $x = 1 - \frac{a}{100}$. What is the expression that gives the amount of medication in the system on the first day, the second, the fifth day, and so on to the nth day? Assume that $a = 50\%$ and find the actual amounts in the system on days 1, 2, 5, and 10.

Solution.

Day 1	60
Day 2	$60 + 60x$
Day 3	$60 + 60x + 60x^2$
Day 4	$40 + 60x + 60x^2 + 60x^3$
Day 5	$25 + 40x + 60x^2 + 60x^3 + 60x^4$
Day n	$25x^{n-5} + 40x^{n-4} + 60x^{n-3} + 60x^{n-2} + 60x^{n-1}$
Day 10	$25x^5 + 40x^6 + 60x^7 + 60x^8 + 60x^9$

Answer. If $x = 0.5$, then $F(1) = 60$ mg, $F(2) = 90$ mg, $F(5) = 71/25$ mg, $F(10) = 2.23$ mg.

Day 1: 60 mg, Day 2: 90 mg, Day 5: 71.25 mg, Day 10: 2.23 mg

© Springer International Publishing Switzerland 2016
E. Grigorieva, *Methods of Solving Sequence and Series Problems*,
DOI 10.1007/978-3-319-45686-7_5

4. Dennis lives near a bus stop A. Bus stops A, B, C, and D are on the same street. Dennis starts at A with a speed of 3 km per hour and goes to D. Reaching D he turns back and goes to C. Walking this rout (A-D-C) requires 8 h and 20 min. It is known that he can cover the distance between B and D in 7 h. The distances between A and B, C and D and half of the distance between B and C form a geometric sequence in the given order. Find the distance between A and C.

Hint. See Problem 78.

Answer. 19 km.

5. A coroner arrives at midnight and finds that the victim's body temperature has dropped from 98.6 to 68.3 °F. Under these conditions the coroner knows that the body temperature has dropped 4 % every hour. Inspector Horace Pluckett found that there were very few visitors to the shop that day. Ron D'Bear was there at noon, Connie Wonka came about 1 p.m., Tiger Papier showed up about 2 p.m., LeRoy Goldberg arrived at 3 p.m., Allie Gator and Morris McMuffin came in together around 4 p.m., and Rusty Woods came at closing time, 5 p.m. Who should Horace arrest?

Answer. He should arrest LeRoy Goldberg.

Solution. Because body temperature drops 4 % every hour, the consecutive temperatures are terms of a geometric progression with the common ratio $r = 0.96$ and the first term $b_1 = 98.6°F$. Using the formula for the nth term of a geometric sequence we can figure out how many hours it would take for the temperature to drop from 98.6 to 68.3 °F:

$$98.6r^t = 68.3$$
$$0.96^t = \frac{68.3}{98.6}$$
$$t = \frac{\ln\frac{68.3}{98.6}}{\ln 0.96}$$
$$t = 8.99 \approx 9$$

This means that the murder happened at 3 p.m., approximately 9 h before 12 p.m. Therefore, Goldberg must be arrested.

6. Given the sequence $\{x_n\}$ such that $x_0 = 2$, $x_1 = \frac{3}{2}$, $x_{n+1} = \frac{3}{2}x_n - \frac{1}{2}x_{n-1}$, find x_n.

Hint. Find the exact formula for x_n using the approach of Section 1.5.

Solution. Solving a quadratic equation, $r^2 - \frac{3}{2}r + \frac{1}{2} = 0$ we obtain the roots: $r = \frac{1}{2}; r = 1$, then $x_n = A\left(\frac{1}{2}\right)^n + B$. Using that fact that $x_0 = 2, x_1 = 3/2$ we obtain the following system to solve for A and B:

$$\begin{cases} A + B = 2 \\ A \cdot \dfrac{1}{2} + B = \dfrac{3}{2} \end{cases}.$$

$A = 1, B = 1$. Therefore, $x_n = \left(\frac{1}{2}\right)^n + 1 = 1 + 2^{-n}$

7. Is there any increasing geometric sequence such that the first 10 terms are integers and the remaining terms are not integers?

Hint. Consider the 10th term of the sequence.

Answer. Yes, for example, $b_1 = 2^9$ $r = \frac{3}{2}$. In general, $b_1 = a^9$, $r = \frac{b}{a}$, $(a, b) = 1$. If a and b are relatively prime then $b_{11} = a^9 \cdot \left(\frac{b}{a}\right)^{10} = \frac{b^{10}}{a}$ is not an integer.

8. (MGU Entrance exam Chemistry department 1989) Numbers $a_1, a_2, \ldots, a_n$ are consecutive terms of an arithmetic sequence. It is known that $a_1 + a_5 + a_{15} = 3$. Find $a_5 + a_9$

Answer. 2

9. (MGU VMK 1988) Find the sum of the first 20 terms of an arithmetic sequence if the sum of its third, seventh, fourteenth, and eighteenth terms is 10.

Answer. 50

10. (MGU Biology department 1991) Victor is riding a bicycle. Because he gets tired, the time that it takes him to cycle every mile of the track is longer than the time it has taken for the previous mile by the same amount each time (the second mile took s seconds more than the first, the third mile took s seconds more than the second, etc.). We know that the time it takes him to cycle the second and the fourth miles can be added together to get 3 minutes and 20 s. How long will it take for him to ride the first 5 miles?

Solution. The time intervals for the mile segments form an arithmetic progression $\{t_n\}$, so $t_1 + t_5 = t_2 + t_4 = 2t_3$. Thus, $t_1 + t_2 + t_3 + t_4 + t_5 = 2(t_2 + t_4) + \frac{t_2 + t_4}{2} = \frac{5}{2}(t_2 + t_4) = \frac{5}{2} \cdot 200 \sec = 500 \sec = 8\min20\sec$.

Answer. 8 min 20 s

11. A geometric sequence $b_1, b_2, b_3, \ldots\ldots$ has the properties: $b_2 \cdot b_4 = 25$ and $b_3 + b_5 = 15$. Find b_1.

Answer. 5/2

12. (MGU Entrance Exam Geography department 1991) Numbers a_1, a_2, a_3 form an arithmetic sequence, but their squares in the same order form a geometric sequence. Find these numbers, if $a_1 + a_2 + a_3 = 21$.

Answer.

$$a_1 = a_2 = a_3 = 7$$
$$a_1 = 7(1 - \sqrt{2}), a_2 = 7, a_3 = 7(1 + \sqrt{2})$$
$$a_1 = 7(1 + \sqrt{2}), a_2 = 7, a_3 = 7(1 - \sqrt{2})$$

13. Find the sum of the first 20 terms of an arithmetic progression, if it is known that the sum of the third, seventh, fourteenth, and eighteenth terms is 10.

Answer. 50

14. (MGU Entrance exam, pure math, 1993) The sum of the first five terms of a geometric progression equals its first term multiplied by 5, but the sum of its first fifteen terms is 100. Find the sum of the first, sixth, and the eleventh terms of the progression.

Answer. 20

15. (Rivkin) Prove that $\sqrt{\underbrace{11...1}_{2n} - \underbrace{22...2}_{n}} = \underbrace{33...3}_{n}$

Hint. See Problem 49.

Proof. Consider the expression under the radical:

$$\underbrace{11...1}_{2n} - \underbrace{22...2}_{n} = \frac{1}{9}\left(\underbrace{99...9}_{2n} - \underbrace{99...9}_{n} \right) =$$

$$\frac{1}{9}[10^{2n} - 1 - 2(10^n - 1)] = \frac{1}{9}[10^{2n} - 2 \cdot 10^n + 1)] = \frac{1}{9}(10^n - 1)^2$$

Taking the square root of this we obtain $D = \frac{1}{3}[10^n - 1] = \frac{1}{3}\left(\underbrace{99...9}_{n \text{ digits}} \right) = \underbrace{33...3}_{n \text{ digits}}.$

16. (MGU 2007 Entrance Exam. 5.1) A farmer got a strategic loan from the bank in order to expand his production. The loan has to be paid off in two years with no monthly payment requirements. After the first year the farmer returned to the bank 1/6 of the total debt he owed to the bank at that time, and at the end of the 2nd year he paid off his debt by giving to the bank some amount that was 20 % more than the original loan. What was the annual percentage rate (APR)?

Solution. Let S be the total amount of the loan and $p\%$ is the APR. Then at the end of the year our farmer owes to the bank $S(1 + \frac{p}{100})$ and after paying 1/6 of it his debt the outstanding amount becomes $\frac{5}{6}S(1 + \frac{p}{100})$. At the end of the 2nd year his debt is $\frac{5}{6}S(1 + \frac{p}{100})^2$. Since the farmer paid off his debt by putting 120 % of the original loan in the bank, we have $\frac{5}{6}S(1 + \frac{p}{100})^2 = S + \frac{1}{5}S = \frac{6}{5}S$, so $1 + p/100 = 6/5$, and $p = 20\%$.

Answer. 20 %

17. Integers x, y, z form a geometric progression but $(5x - 4)$, y^2, and $(3z + 2)$ are members of an arithmetic progression. Find x, y, z.

Hint. See Chapter 1 of the book.

Answer. $(x, y, z) = \{(2,4,8), (2,-4,8)\}$

18. (2008 MGU Entrance Exam) The product of the first 11 terms of a geometric progression is $243\sqrt{3}$. What terms of the progression can be found exactly from this information and what are they?

Solution. Multiplying all 11 terms we obtain,

$$(b_1)^{11} r^{55} = (b_1 r^5)^{11} = 243\sqrt{3} = (\sqrt{3})^{11}$$
$$b_6 = \sqrt{3}.$$

Answer. $b_6 = \sqrt{3}$.

19. Peter loaned $10,000 to Paul at a 6% annual rate under the agreement that Paul would pay the debt plus interest in five years. Paul decided to save money each month by placing a certain amount in a saving account that pays 3.6% interest rate compounded monthly in order to pay Peter $10,000 plus interest. How much monthly should Paul put on his saving account? Would it be better for Paul to take a loan from a bank instead of getting money from Peter?

Hint. See Chapter 4.

Answer. After 5 years Paul will have to pay back $10,000 \cdot 1.06^5 = \$13,382.26$. Using formula for this future value of an annuity but a different annual percentage rate, APR $= 3.6\%$ with interest compounded monthly, we obtain his monthly investment as

$$R = \$13,382.26 \cdot \frac{\left(\frac{0.036}{12}\right)}{\left(\left(1 + \frac{0.036}{12}\right)^{60} - 1\right)} = \$203.90.$$

With total investments of $TP = 60 \cdot \$203.90 = \$12,233.98$. If he had a loan of $10,000 from a bank at 6% APR and interest compounded monthly, then his monthly payment would be

$$\$10,000 \cdot \frac{\frac{0.06}{12}}{\left(1 - \left(1 + \frac{0.06}{12}\right)^{-60}\right)} = \$193.33.$$

for a total payment of $11,599.80. Therefore, Peter's loan is not the best.

20. Evaluate $\displaystyle\sum_{n=1}^{100} \frac{3n^2 - 200n}{101}$.

Solution.

$$\sum_{n=1}^{100} \frac{3n^2 - 200n}{101} = 3\sum_{n=1}^{100} \frac{n^2}{101} - 200\sum_{n=1}^{100} \frac{n}{101}$$

$$= 3 \cdot \frac{100 \cdot 101 \cdot 201}{101 \cdot 6} - 200 \cdot \frac{100 \cdot 101}{101 \cdot 2} = 50$$

Answer. 50

21. Prove using mathematical induction that $\displaystyle\sum_{n=1}^{n} n^2 = \frac{n(n+1)(2n+1)}{6}$.

Hint. See Section 2.3.

22. (Lidsky) Evaluate the sum, $S_n = \sin x + \sin 2x + \sin 3x + \ldots + \sin nx$

Hint. Use de Moivre's Formula or see Problem 71.

Solution. Consider $S = (\cos x + i \sin x) + (\cos 2x + i \sin 2x) + (\cos 3x + i \sin 3x) + \ldots + (\cos nx + i \sin nx)$

Using de Moivre's Formula, $(\cos x + i \sin x)^n = \cos nx + i \sin nx$ we can evaluate S as the sum of a geometric series and obtain $S = \frac{(\cos x + i \sin x)^{n+1} - (\cos x + i \sin x)}{\cos x + i \sin x - 1}$ then the sum $S_n = \sin x + \sin 2x + \sin 3x + \ldots + \sin nx$ is the imaginary part of the expression, or $S_n = 2 \sin \frac{nx}{2} \sin \frac{(n+1)x}{2}$.

Answer. $S_n = 2 \sin \frac{nx}{2} \sin \frac{(n+1)x}{2}$

23. (Lidsky) Find the sum,

$$S_n = nx + (n-1)x^2 + (n-2)x^3 + \ldots + 2x^{n-1} + x^n \tag{5.1}$$

Solution. The sum can be written as $S_n = \displaystyle\sum_{k=1}^{n} (n-k+1)x^k$. Based on ideas given in this book, we multiply the original sum by x,

$$xS_n = nx^2 + (n-1)x^3 + (n-2)x^4 + \ldots + 2x^n + x^{n+1} \tag{5.2}$$

Subtracting Eq. 5.1 from Eq. 5.2 we have

$$(x-1)S_n = (n-(n-1))x^2 + \ldots + x^n + x^{n+1} - nx$$

$$(x-1)S_n = \frac{x^2(x^n - 1)}{x - 1} - nx$$

$$S_n = \frac{x^{n+2} - x^2 - nx^2 + nx}{(x-1)^2} = \frac{x^{n+2} - x^2(n+1) + nx}{(x-1)^2}$$

In 24–25, investigate whether or not the given infinite series are convergent or divergent. When you are making your statement, provide the corresponding theorem that you used.

24. $\displaystyle\sum_{n=1}^{\infty} \frac{(n!)^2}{(2n)!}$

Solution. By the sufficient D'Alembert Ratio Test (Corollary 3.4) we have

$$\frac{u_{n+1}}{u_n} = \frac{[(n+1)!]^2(2n)!}{(2(n+1))!(n!)^2} = \frac{(n+1)^2}{(2n+2)!} \cdot \frac{(2n)!}{n! \cdot n!} = \frac{(n+1)^2(2n)!}{(2n)!(2n+1)(2n+2)}$$
$$= \frac{(n+1)^2}{(2n+1)(2n+2)} \rightarrow \frac{1}{4} < 1$$

The series is convergent.

25. $\displaystyle\sum_{n=1}^{\infty} \frac{1}{n^4} \frac{3^n}{2^n}$

Solution. By the sufficient D'Alembert ratio test we have $\dfrac{u_{n+1}}{u_n} = \dfrac{3^{n+1}}{(n+1)^4 \cdot 2^{n+1}} \div \dfrac{3^n}{n^4 \cdot 2^n} = \dfrac{3}{2} \cdot \dfrac{n^4}{(n+1)^4} \rightarrow \dfrac{3}{2} > 1$. The series diverges.

26. Evaluate the finite sum, $\frac{1}{2 \cdot 5} + \frac{1}{3 \cdot 6} + \ldots + \frac{1}{50 \cdot 53}$.

Hint. See the similar Problem 50.

Solution.

$$\frac{1}{2 \cdot 5} + \frac{1}{3 \cdot 6} + \ldots + \frac{1}{50 \cdot 53} = \frac{1}{3}\left\{\frac{1}{2} + \frac{1}{3} + \frac{1}{4} + \frac{1}{5} + \frac{1}{6} + \ldots \frac{1}{50} - \frac{1}{5} - \frac{1}{6} - \ldots\right.$$
$$\left. - \frac{1}{50} - \frac{1}{51} - \frac{1}{52} - \frac{1}{53}\right\}$$
$$= \frac{1}{3}\left\{\frac{1}{2} + \frac{1}{3} + \frac{1}{4} - \frac{1}{51} - \frac{1}{52} - \frac{1}{53}\right\}$$
$$= \frac{72,079}{210,834} \approx 0.341876.$$

27. What do you think about the convergence of the series: $\displaystyle\sum_{n=1}^{\infty} \frac{1}{(n+1)(n+4)}$? Can you evaluate its partial sums directly using the problem above?

Hint. For any real c and d such that $c \neq d$ $\frac{1}{c \cdot d} = \frac{1}{d-c} \cdot \left[\frac{1}{c} - \frac{1}{d}\right]$ (Eq. 2.9)

Solution.

$$\sum_{n=1}^{\infty}\frac{1}{(n+1)(n+4)}=\frac{1}{3}\left\{\frac{1}{2}+\frac{1}{3}+\frac{1}{4}+\frac{1}{5}+\frac{1}{6}+\cdots\frac{1}{n+1}-\frac{1}{5}-\frac{1}{6}-\cdots\right.$$

$$\left.-\frac{1}{n+1}-\frac{1}{n+2}-\frac{1}{n+3}-\frac{1}{n+4}\right\}$$

Therefore, the partial sum equals

$$S_n=\frac{1}{3}\left\{\frac{1}{2}+\frac{1}{3}+\frac{1}{4}-\frac{1}{n+2}-\frac{1}{n+3}-\frac{1}{n+4}\right\}$$

We can see that its limit exists and can be evaluated as

$$\lim_{n\to\infty}S_n=\frac{1}{3}\left\{\frac{1}{2}+\frac{1}{3}+\frac{1}{4}\right\}=\frac{13}{36}.$$

28. Evaluate the following sum $S_n=1^4+2^4+3^4+\ldots+n^4$.

Hint. See Probs. 47 and 48 and consider the differences of the fifth powers of n and $(n-1)$: $n^5-(n-1)^5=5n^4-10n^3+10n^2-5n+1$.

Solution.

$$1^5-0^5=5\cdot1^4-10\cdot1^3+10\cdot1^2-5\cdot1+1$$
$$2^5-1^5=5\cdot2^4-10\cdot2^3+10\cdot2^2-5\cdot2+1$$
$$3^5-2^5=5\cdot3^4-10\cdot3^3+10\cdot3^2-5\cdot3+1$$

$$\cdots$$

$$n^5-(n-1)^5=5n^4-10n^3+10n^2-5n+1$$

Add   the   left   and   right   sides   and   use   sigma   notation   to   get
$n^5=5\sum_{n=1}^{n}n^4-10\sum_{n=1}^{n}n^3+10\sum_{n=1}^{n}n^2-5\sum_{n=1}^{n}n+n$. Solve for $5\sum_{n=1}^{n}n^4$ and replace
known sums by their equivalent expressions to obtain

$$5\sum_{n=1}^{n}n^4=n^5+10S_n^3-10S_n^2+5S_n^1-n=n(n^4-1)+10S_n^3-10S_n^2+5S_n^1$$

$$=n(n+1)(n-1)(n^2+1)+10S_n^3-10S_n^2+5S_n^1.$$

Here $S_n^1 = \sum_{n=1}^{n} n = \frac{n(n+1)}{2}$, $S_n^2 = \sum_{n=1}^{n} n^2 = \frac{n(n+1)(2n+1)}{6}$, $S_n^3 = \sum_{n=1}^{n} n^3 = \frac{n^2(n+1)^2}{4}$, and

$5S_n^4 = \frac{n(n+1)}{6} \cdot (6n^3 + 9n^2 + n - 1) = \frac{n(n+1)(2n+1)(3n^2+3n-1)}{6}$.

Answer. $S_n^4 = \frac{n(n+1)(2n+1)(3n^2+3n-1)}{30}$.

29. The sequence, $\{a_n\}$, is an arithmetic progression in which $a_3 = -13$ and $a_7 = 3$. Find for what number of terms the sum of the series will be smallest. Find this sum.

Hint. Find the first term and common difference then evaluate the sum of the first n terms and find its minimum.

Solution.

$$a_3 = a_1 + 2d = -13$$
$$a_7 = a_1 + 6d = 3$$
$$d = 4, a_1 = -21$$
$$S_n = \frac{2a_1 + (n-1)d}{2} \cdot n = (2n - 23)n = 2n^2 - 23n$$

Since this parabola opens upward, then its minimum is at its vertex, $n = 23/4 = 5.75$. Rounding up we obtain $n = 6$ and $\min(S_n) = -66$.

Answer. -66

30. Decide whether $\sum_{n=1}^{\infty} \frac{n!}{n+4}$ converges or diverges.

Hint. See Section 3.1 and check Theorem 3.1.

Answer. Diverges

31. Using mathematical induction prove that $\sum_{n=1}^{n} n = \frac{n(n+1)}{2}$.

32. There are two vessels containing a mixture of water and sand. In the first vessel there is 1000 kg of the mixture and in the second 1960 kg of the mixture. Water was added in both vessels. After that the percent content of sand in the first vessel was reduced k times and in the 2nd l times. It is known that $kl = 9 - k$. Find the minimum amount of water that could be added to both vessels together.

Hint. Use the inequality between an arithmetic and geometric means.

Answer. 3480 kg.

33. Given the series, $\frac{1}{2} + \frac{2}{2^2} + \frac{3}{2^3} + \ldots + \frac{n}{2^n}$

a. Evaluate the partial sum
b. Can you evaluate the infinite series sum?

Hint. See Problem 59.

Solution. We follow the same method we used for Problem 59. Denote $S_n = \frac{1}{2} + \frac{2}{2^2} + \frac{3}{2^3} + \ldots + \frac{n}{2^n}$. Multiplying this by two and regrouping terms we obtain $2S_n = 1 + \frac{2}{2} + \frac{3}{2^2} + \ldots + \frac{n}{2^{n-1}}$. Within this sum we recognize a geometric series and the original sum minus its last term,

$$1 + \left(\frac{1}{2} + \frac{1}{2}\right) + \left(\frac{1}{2^2} + \frac{2}{2^2}\right) + \left(\frac{1}{2^3} + \frac{3}{2^3}\right) + \ldots + \left(\frac{1}{2^{n-1}} + \frac{n-1}{2^{n-1}}\right)$$

$$= \frac{1 \cdot \left(1 - \dfrac{1}{2^n}\right)}{1 - \dfrac{1}{2}} + S_n - \frac{n}{2^n} = 2S_n.$$

Solving for the partial sum, $S_n = 2 - \frac{n+2}{2^n}$. This series is convergent because if n increases the second term will approach zero and the limit of partial sums will approach 2, i.e., $\lim\limits_{n \to \infty} S_n = 2$.

Answer. $S_n = 2 - \frac{2+n}{2^n}$.

34. (MGU VMK June 2009 Entrance Exam) Positive numbers a and b are such that the numbers a, x, and b form an arithmetic progression and the numbers a, y, b form a geometric progression. Can the difference $(x - y)$ take the following values: (a) -2009, (b) 0, (c) 2009?

Hint. Use arithmetic and geometric mean properties.

Solution. From the condition of the problem we obtain the following relationships for x and y:

$$\begin{cases} x = \dfrac{a+b}{2} > 0 \\ y = \sqrt{ab} > 0 \end{cases} \tag{5.3}$$

Therefore, $x - y > 0$. Considering different cases we see that

a) $x - y = -2009$ is false;
b) $x - y = 0$ can be true if and only if $a = b$, but if for any arithmetic sequence we assume that the common difference is nonzero, then (b) is false.
c) $x - y = 2009 \cdot \frac{a+b}{2} - 2009 = \sqrt{ab}$

From the system of Eq. 5.3 we obtain that

$$\begin{cases} a + b = 2x \\ ab = y^2 \end{cases}$$

and this can be rewritten as

$$x - y = \frac{a+b}{2} - \sqrt{ab} = \frac{1}{2}\left(\sqrt{a} - \sqrt{b}\right)^2 \geq 0, \text{ if } y > 0$$

$$x - y = \frac{a+b}{2} + \sqrt{ab} = \frac{1}{2}\left(\sqrt{a} + \sqrt{b}\right)^2 > 0, \text{ if } y < 0$$

This is another proof that cases (a) and (b) are impossible. On the other hand, case (c) is possible. Let $a > b$, then $\left(\sqrt{a} - \sqrt{b}\right)^2 = 2 \cdot 2009 \Leftrightarrow \sqrt{a} - \sqrt{b} = \sqrt{2 \cdot 2009}$ or in the equivalent form $a = \left(\sqrt{b} + \sqrt{2 \cdot 2009}\right)^2$. This relationship means that for any number $b > 0$, there can be found such a number $a > 0$ that $x - y = 2009$.

35. The following sets of numbers are given

$$1$$
$$1 + 2$$
$$1 + 2 + 3$$
$$1 + 2 + 3 + 4$$
$$\cdots$$
$$1 + 2 + 3 + \ldots + n$$

Find the value of n if the sum of all numbers equals 286.

Hint. Notice that each row is the sum of an arithmetic series

Solution.

Method 1. Denote S as the total sum and use the formula for the sum of an arithmetic series, $S = 1 + \frac{1+2}{2} \cdot 2 + \frac{1+3}{2} \cdot 3 + \frac{1+4}{2} \cdot 4 + \ldots + \frac{1+n-1}{2} \cdot (n-1) + \frac{1+n}{2} \cdot n$. Factoring out $\frac{1}{2}$ and replacing $k(k+1)$ by $k^2 + k$, we have

$$S = \frac{1}{2} \cdot \left\{ (1^2 + 1) + (2^2 + 2) + (3^2 + 3) + \ldots + (n^2 + n) \right\}$$

$$= \frac{1}{2} \left\{ \sum_{n=1}^{n} n^2 + \sum_{n=1}^{n} n \right\} = \frac{1}{2} \left\{ \frac{n(n+1)(2n+1)}{6} + \frac{n(n+1)}{2} \right\}$$

$$= \frac{n(n+1)(n+2)}{6} = 286$$

$n(n+1)(n+2) = 1716 = 11 \cdot 12 \cdot 13$, so $n = 11$.

Method 2. The sum is the sum of n triangular numbers,

$$S = \sum_{k=1}^{n} T_k = \sum_{k=1}^{n} \frac{k(k+1)}{2} = \frac{1}{2}\left(\sum_{k=1}^{n} k^2 + \sum_{k=1}^{n} k\right) = \frac{n(n+1)(n+2)}{6} = 286.$$

Therefore $n = 11$.

Answer. $n = 11$

36. Let $x^2 + y^2 = a^2$, $z^2 + u^2 = b^2$ and prove that $-ab \le xz + yu \le ab$

Hint. Multiply the left and the right sides of both equations, then use the relationship between the arithmetic and geometric means.

Proof.

$$(x^2 + y^2)(z^2 + u^2) = a^2 b^2$$
$$(x^2 z^2 + y^2 u^2) + (x^2 u^2 + y^2 z^2) = a^2 b^2$$

Rewriting the last expression in a different form:

$$\left[x^2 z^2 + y^2 u^2 + 2xuyz\right] + \left\{x^2 u^2 + y^2 z^2 - 2xyzu\right\} = a^2 b^2 \tag{5.4}$$

Since the expression inside the braces is $x^2 u^2 + y^2 z^2 - 2xyzu = (xu - yz)^2 \ge 0$, for Eq. 5.4 to hold the expression inside brackets must be less than or equal to $a^2 b^2$ $x^2 z^2 + y^2 u^2 + 2xuyz = (xz + yu)^2 \le a^2 b^2$. This   can   be   also   written   as $-ab \le xz + yu \le ab$, which completes the proof.

37. (Kaganov) Two progressions, one arithmetic and one geometric, have three terms each. The first and the third terms of the progressions are equal. For which of the progressions is the sum of the three terms bigger?

Solution. Consider two progressions:

Arithmetic $a_1, a_1 + d, a_1 + 2d$ and denote its sum as S_a
Geometric $a_1, a_1 r, a_1 r^2$ and denote its sum as S_g

By the condition $a_1 r^2 = a_1 + 2d$ we can find $d = \frac{a_1(r^2-1)}{2}$. Consider the ratio of the second terms,

$$\frac{a_1 + d}{a_1 r} = \frac{a_1 + \frac{a_1(r^2-1)}{2}}{a_1 r} = \frac{r^2 + 1}{2r} \ge 1$$

because $(r-1)^2 \ge 0$, $r > 0$. We have the cases:

1. if $r = 1$, then $S_a = S_g$
2. if $r \neq 1$, then $S_a > S_g (r > 0)$ or $S_a < S_g (r < 0)$.

38. (Kaganov) Let the sum of the first m terms of an arithmetic progression equal S_m. Prove that $S_{3m} = 3(S_{2m} - S_m)$.

Hint. Consider the formula for the sums of the first m, $2m$, and $3m$ terms of an arithmetic series.

Solution.

$$S_m = \frac{2a_1 + (m-1)d}{2} \cdot m$$

$$S_{2m} = \frac{2a_1 + (2m-1)d}{2} \cdot 2m$$

$$S_{3m} = \frac{2a_1 + (3m-1)d}{2} \cdot 3m$$

$$S_{2m} - S_m = \frac{2a_1 + (3m-1)d}{2} \cdot m = \frac{S_{3m}}{3}$$

39. For a geometric progression $b_{m+n} = A$, $b_{m-n} = B$ evaluate b_m, b_n.

Answer. $b_m = \sqrt{AB}$, $b_n = \sqrt[2n]{\left(\frac{B}{A}\right)^m}$.

40. The sum of consecutive natural numbers from one to some number n is a three digit number N with all digits the same. How many numbers did we add? Find N.

Solution. Any three-digit number with the same digits can be written as $N = xxx = 111 \cdot x = 100x + 10x + 1 \cdot x$. Since N is a result of adding n natural numbers, then $\frac{1+n}{2} \cdot n = 111x$, $x = \frac{n(n+1)}{2 \cdot 3 \cdot 37}$. Because 37 is prime then either n or $n+1$ is multiple of 37. Therefore, $n = 36$, $x = 6$, and $N = 666$.

Answer. $n = 36$; $N = 666$.

41. Find an arithmetic progression such that the sum of the first four terms is 26 the sum of the last four terms is 110, and the sum of all terms is 187.

Answer. 2, 5, 8, 11, ..., 23, 26, 29, 32

42. If x, y, z are terms of a geometric progression, prove that $(x+y+z)(x-y+z) = x^2 + y^2 + z^2$

Hint. Use a direct proof with terms: x, $y = kx$, $z = k^2x$.

Solution.

$$\left(x + kx + k^2x\right)\left(x - kx + k^2x\right)$$
$$= x^2 + (kx)^2 + \left(k^2x\right)^2$$
$$= x^2 + y^2 + z^2$$

43. In an arithmetic progression the sum of the first three terms equals 15. If this sum is divided by the product of the terms the result is 1/7. Find the progression.

Answer. 3, 5, and 7.

44. Prove that if numbers a^2, b^2, and c^2 are terms of an arithmetic progression then $\frac{1}{b+c}, \frac{1}{a+c}, \frac{1}{b+a}$ are also terms of an arithmetic progression.

Proof. By the condition $c^2 - b^2 = b^2 - a^2$. Consider the differences:

$$\frac{1}{a+b} - \frac{1}{a+c} = \frac{c-b}{(b+a)(a+c)}$$

$$\frac{1}{a+c} - \frac{1}{b+c} = \frac{b-a}{(c+a)(b+c)}$$

Equating these differences we obtain

$$\frac{b-a}{(c+b)(a+c)} = \frac{c-b}{(b+a)(a+c)}$$
$$(b-a)(b+a) = (c+b)(c-b)$$
$$c^2 - b^2 = b^2 - a^2.$$

This completes the proof.

45. Prove that the sum of squares of the first n natural numbers cannot be equal to the square of the sum of these natural numbers.

Hint. Prove it by contradiction

Proof. Assume contradiction $\sum_{n=1}^{n} n^2 = \left(\sum_{n=1}^{n} n\right)^2, n > 1, n \in N$. Then expanding these sums we obtain

$$\frac{n(n+1)(2n+1)}{6} = \frac{n^2(n+1)^2}{2^2}$$
$$\frac{2n+1}{3} = \frac{n^2+n}{2}$$
$$3n^2 - n - 2 = 0$$
$$n = \frac{1 + \sqrt{1+24}}{6} = 1.$$

Therefore such a case does not exist.

46. The following sequence is given: $1, \frac{1}{2}, \frac{1}{4}, \frac{1}{8}, \ldots$ Can some terms of this sequence form an infinite geometric series with the sum of 1/5 or 1/7?

Solution. Assume that such a sequence exists $b_1 = \frac{1}{2^k}$; $r = \frac{1}{2^m}$, $k, m \in \mathbb{Z}$. Then the sum of the infinite series is $S = \frac{b_1}{1-r} = \frac{1}{2^k} \div \left(1 - \frac{1}{2^m}\right) = \frac{1}{2^k - 2^{k-m}}$.

If $k > m$, then expression $2^k - 2^{k-m}$ is an even number and we do not consider this case.

If $k < m$, then $2^k - 2^{k-m}$ is a fraction and again will not be considered.

If $k = m$, then $k - m = 0$, then in order to satisfy the condition the expression $2^k - 1$ must be either 5 or 7.

If $2^k - 1 = 5$, then $2^k = 6$ there are no solutions.

If $2^k - 1 = 7$, then $2^k = 8$ and $k = 3$.

Therefore, such an infinite series can be selected from the given sequence and the terms are

$$\frac{1}{8}, \frac{1}{64}, \frac{1}{512}, \ldots$$

$$S_\infty = \frac{1}{7}.$$

Answer. Yes, it can. For example, 1/8, 1/64, 1/512... converges to 1/7. No subset converges to 1/5.

47. Insert inside the number 49 another number 48 to obtain 4489. Next, we insert the number 48 inside that new number, 4489, obtaining 444,889, and continue inserting 48 inside that number, etc. Prove that all these numbers are perfect squares.

Proof.

$$44...488...89 = 9 + 8 \cdot 10 + \ldots + 8 \cdot 10^k + 4 \cdot 10^{k+1} + \ldots + 4 \cdot 10^{2k+1} =$$

$$9 + 8 \cdot \frac{10(10^k - 1)}{10 - 1} + 4 \cdot \frac{10^{k+1}(10^{k+1} - 1)}{10 - 1} =$$

$$\frac{81 + 8 \cdot 10^{k+1} - 80 + 4 \cdot 10^{2k+2} - 4 \cdot 10^{k+1}}{9} =$$

$$\frac{1 + 4 \cdot 10^{k+1} + 4 \cdot 10^{2k+2}}{9} = \left(\frac{1 + 2 \cdot 10^{k+1}}{3}\right)^2 = \left(\frac{200...0 + 1}{3}\right)^2 = 66...67^2.$$

48. Find the sum: $7 + 77 + 777 + 7777 + \ldots + 77...777$, where the last number consists of n repetitions of digit 7.

Hint. See Problem 49. The number can be written as $7 \cdot (1 + 11 + 111 + 1111 + \ldots + 111...111)$.

Solution. We can notice that $9 = 10 - 1$, $99 = 100 - 1$, $999 = 1000 - 1$, etc. If we multiply and divide the given sum by 9 we can easily evaluate it using the formula for geometric series

$$S = \frac{7}{9}(10 - 1 + 100 - 1 + 1000 - 1 + 10000 - 1 + \ldots + 100\ldots0 - 1)$$

$$= \frac{7(10 + 10^2 + \ldots + 10^n - n)}{9}$$

$$S = \frac{7}{9}\left[\frac{10(10^n - 1)}{9} - n\right] = \frac{7}{9}\left(\underbrace{11\ldots1}_{n \text{ digits}}0\right)$$

49. For the sequence $3 + 8 + 15 + 24 + 35 + \ldots$

a. Evaluate the sum of the first 200 terms of the series.
b. Find the sum of all members between 25th and 50th.

Hint. Notice that the nth term can be written as a product of two integers that differ by 2, i.e., $n(n + 2)$.

Solution.
Part a.

$$1 \cdot 3 + 2 \cdot 4 + 3 \cdot 5 + 4 \cdot 6 + 5 \cdot 7 + \ldots + \sum_{n=1}^{200} n(n + 2) = \sum_{n=1}^{200} n^2 + 2\sum_{n=1}^{200} n$$

$$= \frac{200 \cdot 201 \cdot 401}{6} + 2 \cdot \frac{200 \cdot 201}{2} = 2,726,900.$$

Part b.

$$\sum_{n=25}^{50} n(n + 2) = \sum_{n=1}^{50} n^2 + 2\sum_{n=1}^{50} n - \sum_{n=1}^{24} n^2 - 2\sum_{n=1}^{24} n$$

$$= \frac{50 \cdot 51 \cdot 101}{6} + 2 \cdot \frac{50 \cdot 51}{2} - \frac{24 \cdot 25 \cdot 49}{6} - 2 \cdot \frac{24 \cdot 25}{2} = 39,975.$$

50. Find the formula for the nth term of the sequence given recursively by $a_{n+1} = 2a_n + 1$, $a_0 = 0$, $n \geq 0$.

Hint. Write down some of the members and recognize the pattern and then prove it using mathematical induction or by using a generating function (longer way).

Answer. For the sequences $0, 1, 3, 7, 15, 31, \ldots$, $a_n = 2^n - 1$, $a_0 = 0$, $n \geq 0$.

51. Find the formula for the nth term of the sequence given recursively by $a_{n+1} = 2a_n + n$, $a_0 = 1$, $n \geq 0$

Hint. Using a generating function you can solve it similarly to the way we solved Problem 155.

Solution. Using ideas similar to those in Problem 155 we consider $\sum a_{n+1}x^n = 2\sum a_n x^n + \sum n x^n$ $a_0 = 1,\ n \geq 0$ and denote $F(x) = \sum a_n x^n$. Equating the generating functions of the left and right sides we obtain

$$\frac{F(x) - 1}{x} = 2F(x) + \frac{x}{(1-x)^2}$$

$$F(x) = \frac{1 - 2x + 2x^2}{(1-x)^2(1-2x)} = \frac{-1}{(1-x)^2} + \frac{2}{(1-2x)}$$

Answer. For $1, 2, 5, 12, 27, 58, 121, \ldots$ $a_n = 2^{n+1} - n - 1,\ a_0 = 0,\ n \geq 0$

52. Find the nth term of the sequence given by the recursion, $u_{n+1} = 2u_n + u_{n-1}$, $u_1 = 1,\ u_2 = 2$.

Solution. Consider the quadratic equation, $r^2 - 2r - 1 = 0$ with solutions: $r_1 = 1 - \sqrt{2},\ r_2 = 1 + \sqrt{2}$. For the nth term, $u_n = A(1 - \sqrt{2})^n + B(1 + \sqrt{2})^n$. Using the condition of the problem and substituting $u_1 = 1,\ u_2 = 2$ into the expression above, we obtain the system for A and B:

$$\begin{cases} 1 = A(1 - \sqrt{2}) + B(1 + \sqrt{2}) \\ 2 = A(1 - \sqrt{2})^2 + B(1 + \sqrt{2})^2 \end{cases} \Leftrightarrow \begin{cases} (A + B) - \sqrt{2}(A - B) = 1 \\ 3(A + B) - 2\sqrt{2}(A - B) = 2 \end{cases}$$

$$\Leftrightarrow \begin{cases} A + B = 0 \\ A - B = -\dfrac{1}{\sqrt{2}} \end{cases} \begin{bmatrix} A = \dfrac{-1}{2\sqrt{2}} \\ B = \dfrac{1}{2\sqrt{2}} \end{bmatrix}$$

$$u_n = -\frac{1}{2\sqrt{2}}(1 - \sqrt{2})^n + \frac{1}{2\sqrt{2}}(1 + \sqrt{2})^n$$

$$u_n = \frac{(1 + \sqrt{2})^n - (1 - \sqrt{2})^n}{2\sqrt{2}}.$$

We can see that the formula is correct because

$$u_1 = \frac{(1 + \sqrt{2})^1 - (1 - \sqrt{2})^1}{2\sqrt{2}} = 1$$

$$u_2 = \frac{(1 + \sqrt{2})^2 - (1 - \sqrt{2})^2}{2\sqrt{2}} = \frac{3 + 2\sqrt{2}}{2\sqrt{2}} - \frac{3 - 2\sqrt{2}}{2\sqrt{2}} = 2,\ \text{etc.}$$

53. Evaluate the infinite series, $S = 1 + \ln 2 + \frac{(\ln 2)^2}{2!} + \frac{(\ln 2)^3}{3!} + \frac{(\ln 2)^2}{4!} + \cdots$

Solution. Comparing this series with the Maclaurin series (Section 3.2.2, Eq. 3.21) for $e^x, x = \ln 2$, we can see that $S = e^{\ln 2} = 2$.

Answer. $S = 2$

54. (Kaganov) Let a_i form an arithmetic progression such that

$$a_1 + a_2 + \ldots + a_n = a$$
$$a_1^2 + a_2^2 + \ldots + a_n^2 = b^2$$

Find the progression.

Solution. Using properties of the arithmetic progression, we have the following true equations:

$$a_2 = a_1 + d$$
$$a_3 = a_1 + 2d$$
$$a_n = a_1 + (n-1)d$$

Substituting these into the second equation of the problem, we have

$$b^2 = a_1^2 + \left(a_1^2 + 2a_1 d + d^2\right) + \left(a_1^2 + 4a_1 d + 4d^2\right) + \ldots$$
$$+ \left(a_1^2 + 2(n-1)a_1 d + (n-1)^2 d^2\right)$$
$$= na_1^2 + 2a_1 d(1 + 2 + 3 + \ldots + (n-1)) + d^2\left(1 + 4 + 9 + \ldots (n-1)^2\right)$$

Using Eqs. 1.29 and 1.30 to simplify the expressions inside the parentheses, we obtain $na_1^2 + a_1 d(n-1)n + d^2 n(n-1)(2n-1)$ or

$$\frac{6b^2}{n} = 6a_1^2 + 6a_1 d(n-1) + d^2(n-1)(2n-1) \tag{5.5}$$

Next, we can rewrite the first equation of the problem as

$$2a_1 + d(n-1) = \frac{2a}{n} \tag{5.6}$$

Squaring both sides of Eq. 5.6, multiplying by 1.5, and with the use of Eq. 5.5, we obtain the system:

$$\begin{cases} 6a_1^2 + 6a_1 d(n-1) + 1.5d^2(n-1)^2 = \dfrac{6a^2}{n^2} \\[2mm] 6a_1^2 + 6a_1 d(n-1) + d^2(n-1)(2n-1) = \dfrac{6b^2}{n} \end{cases}$$

Solving the system, we have $d^2(n-1)(1.5n-1.5-2n+1) = \frac{6}{n}\left(\frac{a^2}{n} - b^2\right)$. From this we can find d and the first term of the progression:

$$d^2 = \frac{12(b^2n - a^2)}{n^2(n-1)(n+1)}, \quad d = \pm\frac{2}{n}\sqrt{\frac{3(b^2n - a^2)}{n^2 - 1}}$$

$$a_1 = \frac{a}{n} - \frac{n-1}{2}\cdot d = \frac{a}{n} \mp \frac{(n-1)}{n}\sqrt{\frac{3(b^2n - a^2)}{n^2 - 1}}.$$

55. Suppose that a, b, and c are different primes greater than 3. Prove that they cannot be consecutive terms of some arithmetic progression.

Hint. See Problem 94.

Solution. The statement is false because 47, 53, 59 are primes and consecutive members of arithmetic progression with the difference $d=6$. It has been conjectured that there exist arithmetic progressions of finite (but otherwise arbitrary) length, composed of consecutive prime numbers. Examples of such progressions consisting of three and four primes, respectively, are 5, 101, 197 ($d=96$) and 251, 257,263, 269 ($d=6$).

56. Evaluate the following sum. $S = \frac{1}{1\cdot2\cdot3\cdot4} + \frac{1}{2\cdot3\cdot4\cdot5} + \frac{1}{3\cdot4\cdot5\cdot6} + \cdots + \frac{1}{(n-3)(n-2)(n-1)n}$

Hint. See Problem 56.

Solution. It can be shown that $\frac{1}{k(k+1)(k+2)(k+3)} = \frac{1}{3}\left[\frac{1}{k(k+1)(k+2)} - \frac{1}{(k+1)(k+2)(k+3)}\right]$ which yields

$$\frac{1}{1\cdot2\cdot3\cdot4} = \frac{1}{3}\left[\frac{1}{1\cdot2\cdot3} - \frac{1}{2\cdot3\cdot4}\right]$$

$$\frac{1}{2\cdot3\cdot4\cdot5} = \frac{1}{3}\left[\frac{1}{2\cdot3\cdot4} - \frac{1}{3\cdot4\cdot5}\right]$$

$$\frac{1}{3\cdot4\cdot5\cdot6} = \frac{1}{3}\left[\frac{1}{3\cdot4\cdot5} - \frac{1}{4\cdot5\cdot6}\right]$$

$$\cdots$$

$$\frac{1}{(n-3)(n-2)(n-1)n} = \frac{1}{3}\left[\frac{1}{(n-3)(n-2)(n-1)} - \frac{1}{(n-2)(n-1)n}\right]$$

Adding together the left and right sides of all equations we obtain

$$S = \frac{1}{3}\left[\frac{1}{1\cdot2\cdot3} - \frac{1}{(n-2)(n-1)n}\right] = \frac{1}{3}\left[\frac{1}{6} - \frac{1}{(n-2)(n-1)n}\right]$$

We can see that if the number of terms will increase without bound then the sum will go to 1/18.

57. Investigate if the series $\sum\limits_{n=1}^{\infty} \frac{\sqrt{n^3+2}}{n^2 \sin^2 n}$ convergent or divergent.

Hint. Using inequalities, try to find the boundary for the common term.

Solution. $\frac{\sqrt{n^3+2}}{n^2 \sin^2 n} \geq \frac{\sqrt{n^3+2}}{n^2} \geq \frac{n^{3/2}}{n^2} = \frac{1}{n^{1/2}}$. By Comparison Criterion (Theorem 3.3), the given series diverges as $\sum\limits_{n=1}^{\infty} \frac{1}{n^{1/2}}$.

Answer. Diverges.

58. Investigate if the series $\sum\limits_{n=1}^{\infty} \frac{\sin na}{(\ln 4)^n}$ convergent or divergent.

Hint. Investigate series made of absolute values of the terms of the given series.

Solution. $\left| \frac{\sin na}{(\ln 4)^n} \right| \leq \frac{1}{|(\ln 4)^n|} = \left(\frac{1}{\ln 4}\right)^n = q^n$, $q < 1$. Hence, the absolute values series converges by Comparison Criterion. It is absolutely convergent. Therefore, the given series converges.

Answer. Series converges.

59. Investigate if the series $\sum\limits_{n=1}^{\infty} b_n = \sum\limits_{n=1}^{\infty} (-1)^n \frac{n^3}{(n+1)!}$ convergent or divergent. Starting from what term, each consecutive term is smaller than the preceding term?

Hint. This is a Leibniz series.

Solution. Consider corresponding absolute value series, $\sum\limits_{n=1}^{\infty} a_n = \sum\limits_{n=1}^{\infty} |b_n|$, and using the Leibniz Theorem, we have $\lim\limits_{n\to\infty} a_n = \lim\limits_{n\to\infty} \frac{n^3}{(n+1)!} = 0$. Next, we find out starting from what term of the series, the value of each following term is less than the value of the previous term. Let us find the limit of the ratio of two consecutive terms: $\frac{a_{n+1}}{a_n} = \frac{(n+1)^3 (n+1)!}{(n+2)! n^3} = \left(\frac{n+1}{n}\right)^3 \cdot \frac{1}{n+2} < \left(\frac{2n}{n}\right)^3 \cdot \frac{1}{n+2} = \frac{8}{n+2}$. Because $\frac{8}{n+2} < 1$ if $n > 6$, then starting from the 7th term of the series the important condition is fulfilled and the given series converges as Leibniz series. Moreover, applying the D'Alembert Ratio Test (Corollary 3.3) and because $\lim\limits_{n\to\infty} \frac{a_{n+1}}{a_n} = 0 < 1$, then the alternating series is absolutely convergent.

Answer. Absolutely converges

60. Evaluate the sum of the series $\sum\limits_{n=1}^{\infty} (-1)^n \frac{n}{(1+n^3)^2}$ with accuracy of $\alpha = 0.001$.

Hint. See Problem 151.

Solution. Consider the series made of absolute values of the terms of the given series, such that $|a_n| = \frac{n}{(1+n^3)^2}$ and evaluate its third and fourth terms: $|a_3| = \frac{3}{(1+3^3)^2} = \frac{3}{784} > 0.001$ and $|a_4| = \frac{4}{(1+4^3)^2} = \frac{4}{4225} < 0.001$. So three terms are required: $\sum_{n=1}^{\infty} (-1)^n \frac{n}{(1+n^3)^2} \approx -\frac{1}{4} + \frac{2}{81} - \frac{3}{784} = -\frac{14551}{63504}$.

61. Find convergence radius of the series $\sum_{n=1}^{\infty} \frac{(n+1)^5 x^{2n}}{2n+1}$.

Hint. Use the Cauchy-Hadamard Formula.

Solution.

$$R^{-1} = \lim_{n \to \infty} \sqrt[n]{\frac{(n+1)^5}{2n+1}} = \lim_{n \to \infty} \frac{(n+1)^{\frac{5}{n}}}{(2n+1)^{\frac{1}{n}}} = \lim_{n \to \infty} \frac{\left((n+1)^{\frac{1}{n+1}}\right)^{\frac{5(n+1)}{n}}}{\left((2n+1)^{\frac{1}{2n+1}}\right)^{\frac{2n+1}{n}}} = 1 = R.$$

Here we used a standard fact that $\lim_{n \to \infty} n^{\frac{1}{n}} = 1$.

Answer. $R = 1$.

62. Find a Taylor expansion as powers of x for $f(x) = \ln\left(\frac{2+x}{1-x}\right)$.

Hint. First simplify $f(x)$ using the properties of logarithms, then use a power expansions of known functions.

Solution.

$$f(x) = \ln(2+x) - \ln(1-x)$$
$$= \ln 2 + \ln\left(1 + \frac{x}{2}\right) - \ln(1-x)$$
$$= \ln 2 + \sum_{n=1}^{\infty} \frac{(-1)^{n-1} \cdot x^n}{2^n n} - \sum_{n=1}^{\infty} \frac{(-1)^{n-1} \cdot (-1)^n \cdot x^n}{n}$$

which can be further simplified as $f(x) = \ln 2 + \sum_{n=1}^{\infty} (-1)^{n-1} \frac{x^n}{2^n n} + \sum_{n=1}^{\infty} \frac{x^n}{n}$. While the first series converges at $|x| < 2$, the second series converges at $|x| < 1$, so the obtained power series expansion for the given function will converge for $|x| < 1$.

Answer. $f(x) = \ln\left(\frac{2+x}{1-x}\right) = \ln 2 + \sum_{n=1}^{\infty} (-1)^{n-1} \frac{x^n}{2^n n} + \sum_{n=1}^{\infty} \frac{x^n}{n}$.

63. Find a Taylor series expansion as powers of x for $f(x) = \frac{3}{2-x-x^2}$.

Hint. Factor the denominator and rewrite the function as a sum of two other functions.

Solution. After factoring the denominator, the given function can be written as

$$f(x) = \frac{3}{2-x-x^2} = \frac{3}{(1-x)(2+x)} = \frac{1}{1-x} + \frac{1}{2+x}$$
$$= \frac{1}{1-x} + \frac{1}{2} \cdot \frac{1}{1-\left(-\dfrac{x}{2}\right)}$$
$$\sum_{n=0}^{\infty} x^n + \sum_{n=0}^{\infty} \frac{(-1)^n x^n}{2^{n+1}}.$$

The first series converges at $|x| < 1$ and the second at $|x| < 2$, hence this power series for the given function is valid only for $|x| < 1$.

Answer. $f(x) = \frac{3}{2-x-x^2} = \sum\limits_{n=0}^{\infty} x^n + \sum\limits_{n=0}^{\infty} \frac{(-1)^n x^n}{2^{n+1}}, \quad |x| < 1.$

64. Investigate the convergence of the series $\sum\limits_{n=1}^{\infty} \frac{2^n}{n \cdot 3^n}$.

Hint. Use the Cauchy Root Test (Theorem 3.6) or the D'Alembert Ratio Test (Corollary 3.3).

Solution.

1) Using   the   Cauchy   Root   Test,   we   have
$\lim\limits_{n \to \infty} \sqrt[n]{a_n} = \lim\limits_{n \to \infty} \sqrt[n]{\frac{2^n}{3^n n}} = \frac{2}{3} \cdot \lim\limits_{n \to \infty} \sqrt[n]{\frac{1}{n}} = \frac{2}{3} < 1$, hence the series converges.

2) Using the D'Alembert ratio test, we have $\lim\limits_{n \to \infty} \frac{a_{n+1}}{a_n} = \frac{2}{3} \cdot \lim\limits_{n \to \infty} \frac{n}{n+1} = \frac{2}{3} < 1$ therefore the series converges.

Answer. The series converges.

65. Is the series $\sum\limits_{n=1}^{\infty} \sin\frac{1}{n}$ convergent or divergent?

Solution. Because $\sin\frac{1}{n} \sim \frac{1}{n}$ then $\sum\limits_{n=1}^{\infty} \sin\frac{1}{n}$ behaves the same way as the harmonic series $\sum\limits_{n=1}^{\infty} \frac{1}{n}$, so it diverges.

Answer. It diverges.

66. Does the series $\sum\limits_{n=1}^{\infty} \frac{1}{n\sqrt{n}}$ converge or diverge?

Solution. The series can be written as $\sum\limits_{n=1}^{\infty} \frac{1}{n^{3/2}}$ which is the Dirichlet series for $p = \frac{3}{2} > 1$, hence it converges.

Answer. The series converges.

67. Using a Taylor series expansion as powers of x, evaluate the integral $\int_0^{0.1} \cos(100x^2)dx$ with accuracy of $\alpha = 0.001$.

Hint. See Problem 150.

Solution. Using the power series (Eq. 3.25),

$$\cos t = \sum_{n=0}^{\infty} \frac{(-1)^n t^{2n}}{(2n)!}, \quad t \in (-\infty, \infty), \quad \boxed{t = 100x^2} \Rightarrow \cos(100x^2)$$

$$= \sum_{n=0}^{\infty} \frac{(-1)^n 10^{4n} x^{4n}}{(2n)!}, \quad x \in R, \int_0^1 \cos(100x^2)dx$$

$$= \sum_{n=0}^{\infty} \int_0^1 \frac{(-1)^n 10^{4n} x^{4n}}{(2n)!}dx = \sum_{n=0}^{\infty} \frac{(-1)^n 10^{4n} x^{4n+1}}{(2n)!(4n+1)}\Big|_0^{0.1}$$

$$= \sum_{n=0}^{\infty} \frac{(-1)^n}{10 \cdot (2n)!(4n+1)}$$

Because we have an alternating series, if we take $(n-1)$ terms for the approximation of the integral the absolute value of the error of such estimation will be less than the following n^{th} term of the series. Thus the following inequality must be satisfied $|\text{error}| \leq |a_n| = \frac{1}{10(4n+1)(2n)!}$. Given the error of 0.001, we rewrite it as $\frac{1}{10(4n+1)(2n)!} < 0.001$. Instead of solving this inequality in general, we begin by evaluating the first several terms of the expansion,

$$a_0 = \frac{1}{10 \cdot 1 \cdot 1} = 0.1 > 0.001$$

$$a_1 = \frac{1}{10 \cdot (5) \cdot 2} = \frac{1}{100} > 0.001$$

$$a_2 = \frac{1}{10 \cdot (4 \cdot 2 + 1) \cdot 4!} = \frac{1}{2160} < 0.001$$

Because the value of the third term is smaller than 0.001, we need to approximate the integral by only the first two terms. Finally, $\int_0^1 \cos(100x^2)dx \approx$

$$\sum_0^1 \frac{(-1)^n}{10(4n+1)(2n)!} = 0.1 - 0.01 = 0.09.$$

Answer. 0.09

68. Find the first four terms of the Taylor's series expansion of $f(x) = x^x$ by powers of $x > 0$ centered at $x_0 = 1$.

Hint. Use the fact that $f(x) = x^x = e^{x \ln x}$, $x > 0$.

Solution. Let us find the first three derivatives of the function

$$f'(x) = e^{x \ln x}(1 + \ln x)$$

$$f''(x) = e^{x \ln x}\left(1 + 2\ln x + \ln^2 x + \frac{1}{x}\right)$$

$$f'''(x) = e^{x \ln x}\left(1 + 3\ln x + 3\ln^2 x + \ln^3 x + \frac{\ln x}{x} + \frac{2}{x} + \frac{2\ln x}{x} - \frac{1}{x^2}\right)$$

and evaluate their values at $x = 1$ obtaining that $f(1) = 1$, $f'(1) = 1$, $f''(1) = 2$, $f'''(1) = 2$. Substituting these into the Taylor's series, we get $f(x) = x^x = 1 + (x - 1) + \frac{2(x-1)^2}{2!} + \frac{2(x-1)^3}{3!} + \ldots$

Answer. $1 + (x - 1) + \frac{2(x-1)^2}{2!} + \frac{2(x-1)^3}{3!} + \ldots$

69. Find the nth term of the sequence 1, 6, 19, 44, 85, 146...

Answer. $a_n = \frac{n(2n^2+1)}{3}$

70. Find the nth term of the sequence 1, 5, 15, 35, 70, 126, 210, ...

Answer. $a_n = \frac{n(n+1)(n+2)(n+3)}{24}$.

71. Evaluate the sum of the first n terms of the series, $1 \cdot 4 + 2 \cdot 7 + 3 \cdot 10 + 4 \cdot 13 + \ldots$

Solution. It is easy to see that the nth term can be written as a product of the number of the term, n, and the corresponding number that divided by 3 gives a remainder of 1, $(3n + 1)$. You can visualize the second factor of each term as consecutive terms of an arithmetic progression with the first term of 4 and common difference of 3. Then the nth partial sum is $\sum_{n=1}^{n} n(3n + 1) = 3\sum_{n=1}^{n} n^2 + \sum_{n=1}^{n} n = \frac{n(n+1)(2n+1)}{2} + \frac{n(n+1)}{2} = n(n + 1)^2$.

Answer. $n(n + 1)^2$.

72. Using mathematical induction prove that $(1 + x)^n > 1 + nx$, $x > -1$, $n \in \mathbb{N}$.

Proof.

1. The statement is true for $n = 1$..
2. Assume that it is true for $n = k$, i.e.

$$(1 + x)^k > 1 + kx \tag{5.7}$$

3. Let us demonstrate that the statement is also true for $n = k + 1$ and that $(1 + x)^{k+1} > 1 + (k + 1)x$.

Multiplying both sides of Eq. 5.7 by $(1 + x) > 0$ we obtain the following true inequality that can be further simplified because $kx^2 > 0$.

$$(1 + x)^k (1 + x) \geq (1 + kx)(1 + x)$$
$$= 1 + kx + x + kx^2 > 1 + (k + 1)x.$$

The statement is proven.

73. Prove that for the Fibonacci sequence satisfying $a_1 = a_2 = 1$, $a_n = a_{n-1} + a_{n-2}$, $n > 2$, that for any natural n, the following statements are true:

a. $\sum_{k=1}^{n} a_k^2 = a_n \cdot a_{n-1}$

b. $\sum_{k=1}^{n} a_{2k-1} = a_{2n}$.

Hint. Use mathematical induction.

74. Prove that the recurrent sequence $\{a_n\} : a_1 = 3, a_2 = 5, a_{n+2} = 3a_{n+1} - 2a_n$ can be defined by $a_n = 2^n - 1$.

Hint. Evaluate several terms and then use mathematical induction as we did in Problem 74 of the book.

75. Prove that any term of the sequence $a_n = n^3 + 35$ is divisible by 6.

Hint. Use mathematical induction.

76. Prove that the number 6 divides each term of the sequence $a_n = n^3 + 17n$.

Hint. Use mathematical induction or a direct proof by rewriting the common term of the sequence.

Solution. $n^3 - n + 18n = (n - 1)n(n + 1) + 18n$. Now the common term consists of two terms, the first is always divisible by 6 as a product of three consecutive integers and the second term is also always divisible by 6.

77. Prove that any term of the sequence $a_n = 4^n + 15n - 1$ is divisible by 9.

Hint. Use mathematical induction.

78. Prove that $\sum_{n=1}^{\infty} \frac{6}{(n)(n+1)(n+2)} = \frac{3}{2}$.

Hint. See Problem 68.

79. Evaluate the sum of the first 100 triangular numbers.

Answer. 171,700.

80. For what real values of a and b will the sequence $x_0 = a$, $x_1 = 1 + b \cdot x_0$, ...,
$x_{n+1} = 1 + b \cdot x_n$ converge?
Solution. By induction we can establish that

$$x_n = \left(1 + b + b^2 + b^3 + \ldots + b^{n-1}\right) + a \cdot b^n = \frac{b^n - 1}{b - 1} + a \cdot b^n$$

$$= \frac{1}{1-b} + b^n \cdot \left(a - \frac{1}{1-b}\right).$$

This sequence converges to $\frac{1}{1-b}$.

There are two cases:

1. If $b \neq 1$, $a = \frac{1}{1-b}$, then the sequence is convergent for any real $b \neq 1$, $b \in \mathbb{R}$.
2. If $a \in \mathbb{R}$, $|b| < 1$.

81. Find the formula for the nth term of the Lucas sequence, 1, 3, 4, 7, 11, 18, 29, 47, 76,. . ..

Hint. Note that $L_n = L_{n-1} + L_{n-2}$.

Answer. $L_n = \left(\frac{1+\sqrt{5}}{2}\right)^n + \left(\frac{1-\sqrt{5}}{2}\right)^n$

82. Given $S_n = 1 + \frac{1}{\sqrt{2}} + \frac{1}{\sqrt{3}} + \frac{1}{\sqrt{4}} + \ldots + \frac{1}{\sqrt{n}}$, prove $2\left(\sqrt{n+1} - 1\right) < S_n < 2\sqrt{n}$.
Is the corresponding infinite series convergent or divergent?

Hint. Using a difference of squares formula to show that
$\sqrt{k+1} - \sqrt{k} = \frac{1}{\sqrt{k+1}+\sqrt{k}} < \frac{1}{2\sqrt{k}}$.

Proof. Applying the inequality to each term of the series, we have
$2\sqrt{k+1} - 2\sqrt{k} < \frac{1}{\sqrt{k}} < 2\sqrt{k} - 2\sqrt{k-1}$. Let us add these inequalities for all
n terms of the partial sum,
$$2\sum_{k=1}^{n} \sqrt{k+1} - 2\sum_{k=1}^{n} \sqrt{k} < \sum_{k=1}^{n} \frac{1}{\sqrt{k}} < 2\sum_{k=1}^{n} \sqrt{k} - 2\sum_{k=1}^{n} \sqrt{k-1}.$$ Because
$\sum_{k=1}^{n} \sqrt{k+1} = \sqrt{n+1} + \sum_{k=1}^{n} \sqrt{k} - 1 \Rightarrow \sum_{k=1}^{n} \sqrt{k+1} - \sum_{k=1}^{n} \sqrt{k} = \sqrt{n+1} - 1$. And
finally, $2\sqrt{n+1} - 2 < \sum_{k=1}^{n} \frac{1}{\sqrt{k}} = S_n < 2\sqrt{n}$. The statement is proven.

Next, let us decide if the infinite series is convergent of divergent. Clearly if
$\frac{1}{2} \sum_{k=1}^{\infty} \frac{1}{\sqrt{k}}$ diverges, then $1 + \frac{1}{\sqrt{2}} + \frac{1}{\sqrt{3}} + \ldots$ also diverges and vise versa. Denote
$\sum_{k=1}^{\infty} u_k = \sum_{k=1}^{\infty} \frac{1}{2\sqrt{k}}$ and $\sum_{k=1}^{\infty} v_k = \sum_{k=1}^{\infty} \frac{1}{\sqrt{k}+\sqrt{k+1}}$. Obviously $u_k = \frac{1}{2\sqrt{k}} >$
$\frac{1}{\sqrt{k}+\sqrt{k+1}} = v_k$, $\forall k \in N$.

As we demonstrated by solving Problem 60 of Chapter 2, for the partial sum of the second series $\sum_{k=1}^{n} v_k = \sum_{k=1}^{n} \frac{1}{\sqrt{k}+\sqrt{k+1}} = \sqrt{n+1} - 1$. This partial sum increases without bound so the corresponding infinite series is divergent. Therefore, using Theorem 3.5, because $\sum_{k=1}^{n} u_k > \sum_{k=1}^{n} v_k$, we can state that it is also divergent.

83. Prove that $\frac{\pi}{8} = \frac{1}{1\cdot3} + \frac{1}{5\cdot7} + \frac{1}{9\cdot11} + \frac{1}{13\cdot15} + \cdots$

Hint. Check Probs. 50 and 140.

Proof. The quantities inside each denominator differ by two, then the infinite sum on the right can be written as $\frac{1}{2} \cdot \left(1 - \frac{1}{3} + \frac{1}{5} - \frac{1}{7} + \frac{1}{9} - \frac{1}{11} + \cdots\right) = \frac{1}{2} \cdot \frac{\pi}{4} = \frac{\pi}{8}$. Inside the parentheses we recognize the infinite series representation for arctan 1.

84. Find approximate value of $\sqrt{3}$.

Solution. Let $\sqrt{3} = 2 \cdot \left(1 - \frac{1}{4}\right)^{\frac{1}{2}}$ and apply Eq. 3.27 for $x = 1/4$, $\sqrt{3} = 2\left(1 - \frac{1}{2}\cdot\frac{1}{4} - \frac{1}{2}\cdot\frac{1}{4}\cdot\frac{1}{16} - \frac{1\cdot1\cdot3}{2\cdot4\cdot6}\cdot\frac{1}{64} - \cdots\right) = 2 - \frac{1}{4} - \frac{1}{64} - \frac{1}{512} - \cdots$

85. Given a finite arithmetic progression of 100 terms with the first term 3 and common difference 4, how many terms are multiples of 11? Find them.

Solution. Because 11 is the 3rd term of the given progression, the mth multiple of 11 will have a position $n = 3 + (m-1) \cdot 11 = 11m - 8$, which must be less than 100, then

$$11m - 8 < 100$$
$$11m < 108$$
$$m \leq \left\lceil \frac{108}{11} \right\rceil = 9.$$

There are nine multiples of 11 among the 100 terms of the arithmetic progression. Each multiple of 11 is in a new arithmetic progression

$$b_m = 3 + (11m - 8 - 1) \cdot 4$$
$$b_m = 44m - 33 = 11(4m - 3)$$

The numbers are 11, 55, 99, 143, 187, 231, 275, 319, 363.

Answer. Nine multiples of 11 that are 11, 55, 99, 143, 187, 231, 275, 319, 363.

86. Give an example of an arithmetic progression with 100 integer terms such that any two selected terms are relatively prime?

Hint. See Problem 95.

Answer. $a_1 = 1 + 99! \, d = 99!$

87. Given a sequence of n consecutive natural numbers, prove that only one of its terms is divisible by n.

Proof. Let $a, a+1, a+2, \ldots, a+n-1$ is the required sequence. Any two numbers of this sequence cannot simultaneously be divisible by n because their difference is always less than n and hence cannot be divisible by n. Assume that $a = n \cdot q + r$, where $0 \le r < n$. If $r = 0$, the first term of such a sequence is divisible by n. If $r \ge 1$, then $a + n - r$ is a term of the given sequence and $a + n - r = n \cdot q + r + n - r = n(q+1)$. Therefore, $a + n - r$ is divisible by n.

88. Use mathematical induction to prove that the Fibonacci numbers satisfy $f_1 + f_2 + f_3 + \ldots + f_{2n-1} = f_{2n}$.

Proof.

1. If $n = 1 \Rightarrow f_1 = f_2$ $(1 = 1)$ is true.
2. Assume that the statement is true for $n = k$, i.e., $f_1 + f_2 + f_3 + \ldots + f_{2k-1} = f_{2k}$.
3. Let us demonstrate that the statement is true for $n = k + 1$ and that
$$f_1 + f_2 + f_3 + \ldots + f_{2(k+1)-1} = f_{2(k+1)}.$$

Indeed,

$$f_1 + f_2 + f_3 + \ldots + f_{2(k+1)-1} = (f_1 + f_2 + \ldots + f_{2k-1}) + f_{2k+1}$$
$$= \boxed{f_{2k} + f_{2k+1} = f_{2k+2}} = f_{2(k+1)}.$$

As shown in the box we used the property of Fibonacci recurrence. Therefore, the statement is proven.

89. Show that no term of an infinite arithmetic progression $a_n = 30n + 7$ can be written as a sum or difference of two prime numbers.

Solution. The given progression consists of only odd integers so if we assume that some of its terms can be written as the sum or difference of two primes, one of them must be 2. First, assume that a term can be represented by the sum of two primes,

$$a_n = 30n + 7 = 2 + p$$
$$30n + 5 = p$$
$$5(6n + 1) = p.$$

Clearly, p is not prime.
    Now assume that a term can be written as a difference of two primes:

$$a_n = 30n + 7 = p - 2$$
$$30n + 9 = p$$
$$3(10n + 3) = p.$$

Again we obtain the contradiction because this p is also not prime. Therefore, no term of the given infinite progression can be written as a sum of difference of two primes.

90. Prove that for every positive integer n, that $3(1^5 + 2^5 + 3^5 + \ldots + n^5)$ is divisible by $(1^3 + 2^3 + 3^3 + \ldots + n^3)$.

Hint. Look at Probs. 47 and 48.

Proof.

Method 1. Consider the difference of the sixth powers of two consecutive numbers. Next take the sum from 1 to n from the left and the right sides, using the fact that $\sum_{n=1}^{n} n^6 - \sum_{n=1}^{n} (n-1)^6 = n^6$.

$$n^6 - (n-1)^6 = 6n^5 - 15n^4 + 20n^3 - 15n^2 + 6n - 1$$

$$n^6 = 6\sum_{n=1}^{n} n^5 - 15\sum_{n=1}^{n} n^4 + 20\sum_{n=1}^{n} n^3 - 15\sum_{n=1}^{n} n^2 + 6\sum_{n=1}^{n} n - n$$

We solve this for $3\sum_{n=1}^{n} n^5$ which must be proven to be divisible by $\sum_{n=1}^{n} n^3$. Since all summations are from 1 to n, for simplicity we omit the indices of summation; also we do not expand the sum of the first n cubes:

$$3\sum n^5 = \frac{n^6 + n - 6\sum n + 15\left(\sum n^4 + \sum n^2\right)}{2} - 10\sum n^3$$

Using the summation formulas it can be shown that the following is true:

$$15\left(\sum n^4 + \sum n^2\right) = \frac{n(n+1)(2n+1)(3n^2 + 3n + 4)}{2}$$

$$n^6 + n - 3n(n+1) = n(n+1)(n^4 - n^3 + n^2 - n - 2)$$

$$3\sum n^5 = \frac{n(n+1)n(n+1) \cdot (2n^2 + 2n + 9)}{4} - 10\sum n^3$$

Factors of the first term can be recognized as the sum of the first n cubes, $\sum n^3 = \frac{n^2(n+1)^2}{4}$, so the following is valid:

$$3\sum n^5 = (2n^2 + 2n + 9)\sum n^3 - 10\sum n^3$$
$$3\sum n^5 = (2n^2 + 2n - 1) \cdot \sum n^3$$
$$= k \cdot \sum n^3.$$

The statement is proven.

Method 2. Use mathematical induction.

91. Evaluate the sum $4 - \frac{4}{3} + \frac{4}{5} - \frac{4}{7} + \frac{4}{9} + \ldots + \frac{(-1)^n \cdot 4}{2n+1} + \ldots$.

Solution. We know that this series is convergent. In order to find its sum, let us consider the following power series: $\displaystyle\sum_{n=0}^{\infty} \frac{(-1)^n x^{2n+1}}{2n+1}$. We know that for all $|x| \leq 1$, this series represents the Maclaurin series for $y = \arctan x$. Hence at $x = 1$, we obtain that

$$4\left(1 - \frac{1}{3} + \frac{1}{5} - \frac{1}{7} + \frac{1}{9} + \ldots + \frac{(-1)^n}{2n+1} + \ldots\right) = 4 \cdot \sum_{n=0}^{\infty} \frac{(-1)^n}{2n+1} = 4\arctan 1 = \pi.$$

Answer. π

92. Determine the nth term of the series and evaluate the infinite sum, $1 + \frac{2}{2} + \frac{3}{2^2} + \frac{4}{2^3} + \frac{5}{2^4} + \ldots$

Hint. Multiply or divide the series by 2. See also Probs. 59, 63 or 88.

Solution. This series can be written as $\displaystyle\sum_{n=1}^{\infty} \frac{n}{2^{n-1}}$, so the nth term is $a_n = \frac{n}{2^{n-1}}$. Assume that the requested sum of the infinite series is S, $S = 1 + \frac{2}{2} + \frac{3}{2^2} + \frac{4}{2^3} + \frac{5}{2^4} + \ldots$, and divide both sides by 2 to obtain half of the requested sum, $\frac{S}{2} = \frac{1}{2} + \frac{2}{2^2} + \frac{3}{2^3} + \frac{4}{2^4} + \frac{5}{2^5} + \ldots$. Subtracting the left and right sides of two sums and using the formula for the sum of infinite geometric progression, we obtain:

$$S - \frac{S}{2} = 1 + \left(\frac{2}{2} - \frac{1}{2}\right) + \left(\frac{3}{2^2} - \frac{2}{2^2}\right) + \left(\frac{4}{2^3} - \frac{3}{2^3}\right) + \left(\frac{5}{2^4} - \frac{4}{2^4}\right) + \ldots$$

$$\frac{S}{2} = 1 + \frac{1}{2} + \frac{1}{2^2} + \frac{1}{2^3} + \ldots = \frac{1}{1 - \frac{1}{2}} = 2$$

$$S = 4.$$

Answer. $S = 4$.

93. Evaluate the sum:

$$S = \frac{1}{1 \cdot 2} + \frac{2}{1 \cdot 3} + \frac{3}{2 \cdot 5} + \frac{5}{3 \cdot 8} + \frac{8}{5 \cdot 13} + \ldots + \frac{F_n}{F_{n-1} \cdot F_{n+1}},$$

where F_n the nth Fibonacci number.

Hint. Rewrite each fraction as difference of two other fractions.

Solution. Since $F_{n+1} = F_{n-1} + F_n$, then $\frac{F_n}{F_{n-1} \cdot F_{n+1}} = \frac{1}{F_{n-1}} - \frac{1}{F_{n-1}}$, and the given sum can be rewritten as follows

$$S = 1 - \frac{1}{2} + 1 - \frac{1}{3} + \frac{1}{2} - \frac{1}{5} + \frac{1}{3} - \frac{1}{8} + \frac{1}{5} - \frac{1}{13} + \ldots + \frac{1}{F_{n-4}} - \frac{1}{F_{n-2}} + \frac{1}{F_{n-3}}$$

$$- \frac{1}{F_{n-1}} + \frac{1}{F_{n-2}} - \frac{1}{F_n} + \frac{1}{F_{n-1}} - \frac{1}{F_{n+1}} = 2 - \frac{1}{F_n} - \frac{1}{F_{n+1}}.$$

Answer. $S = 2 - \frac{1}{F_n} - \frac{1}{F_{n+1}}$.

94. Given infinite series $\frac{1}{1} + \frac{1}{2} + \frac{2}{2^2} + \frac{3}{2^3} + \frac{5}{2^4} + \frac{8}{2^5} + \frac{13}{2^6} + \ldots$ Find the nth term of the series and evaluate its sum.

Hint. Note that the denominators of each fraction are powers of 2. Divide or multiply by 2.

Solution. The numerators are represented by Fibonacci numbers: 1, 1, 2, 3, 5, 8, 13, etc. and the denominators are powers of 2. Similarly to what we did in the previous problem, let us denote the requested sum by S,

$$S = \frac{1}{1} + \frac{1}{2} + \frac{2}{2^2} + \frac{3}{2^3} + \frac{5}{2^4} + \frac{8}{2^5} + \frac{13}{2^6} + \ldots$$

Divide both sides by 2,

$$\frac{S}{2} = \frac{1}{2} + \frac{1}{2^2} + \frac{2}{2^3} + \frac{3}{2^4} + \frac{5}{2^5} + \frac{8}{2^6} + \frac{13}{2^7} + \ldots$$

Next, subtract the first and the second sums :

$$S - \frac{S}{2} = 1 + \frac{1-1}{2} + \frac{2-1}{2^2} + \frac{3-2}{2^3} + \frac{5-3}{2^4} + \frac{8-5}{2^5} + \frac{13-8}{2^6} + \ldots$$

$$\frac{S}{2} = 1 + \underbrace{\frac{1}{2^2} + \frac{1}{2^3} + \frac{2}{2^4} + \frac{3}{2^5} + \frac{5}{2^6} + \ldots}_{\frac{S}{4}}$$

We notice that the series to the right of 1 is the series divided by 4, i.e., $S/4$ so $\frac{S}{2} = 1 + \frac{S}{4} \Rightarrow = 4$.

Answer. 4.

95. Find the sum, $S = 1^2 + 4^2 + 7^2 + 10^2 + \ldots + (3n+1)^2$.

Hint. The sum can be written as $1 + \sum_{n=1}^{n} (3n+1)^2 = 1 + \sum_{n=1}^{n} (9n^2 + 6n + 1) = 1$

$+9 \sum_{n=1}^{n} n^2 + 6 \sum_{n=1}^{n} n + n$ and then use Eqs. 1.29 and 1.30.

Answer. $S = \frac{(n+1)\left(6n^2+9n+2\right)}{2}$.

96. Evaluate $1 \cdot 1! + 2 \cdot 2! + 3 \cdot 3! + \ldots + 2017 \cdot 2017!$

Hint. Notice that $(n+1)n! - n! = n \cdot n! \Rightarrow (n+1)! - n! = n \cdot n!$. Replace each term of the finite series using this formula and cancel opposite terms.

Solution. $2! - 1! + 3! - 2! + 4! - 3! + 5! - 4! + \ldots + 2018! - 2017! = 2018! - 1$.

Answer. $2018! - 1$.

97. Evaluate the infinite sum: $\frac{1}{1 \cdot 3} + \frac{1}{5 \cdot 7} + \frac{1}{9 \cdot 11} + \frac{1}{13 \cdot 15} + \cdots$

Hint. Note that the numbers within each denominator differ by 2 and rewrite each fraction as a difference of two corresponding fractions.

Solution. We obtain $\frac{1}{2}\left(1 - \frac{1}{3} + \frac{1}{5} - \frac{1}{7} + \frac{1}{9} - \frac{1}{11} + \ldots\right) = \frac{1}{2} \cdot \frac{\pi}{4} = \frac{\pi}{8}$.

Answer. $\frac{\pi}{8}$.

98. Find the sum of the first 500 natural numbers with 3 as the last digit.

Solution. All natural numbers with last digit 3 can be written as $10n + 3$, so the sum of the first 500 of such numbers will be $3 + \sum\limits_{n=1}^{499}(10n+3) = 3 \cdot$
$500 + \frac{10 \cdot 499 \cdot 500}{2} = 1,249,000$.

99. Prove that the triangular numbers $T(k+n) = T(k) + T(n) + nk$ for all natural numbers n and k.

Hint. Use the formula for a triangular number.

Proof. Using the formula for a triangular number (Eq. 1.27), we rewrite the left side as

$$T(k+n) = \frac{(k+n)(k+n+1)}{2} = \frac{k(k+1)}{2} + \frac{n(n+1)}{2} + \frac{nk+kn}{2}$$
$$= T(k) + T(n) + kn.$$

100. Prove that $\int\limits_0^1 x^x dx = 1 - \frac{1}{2^2} + \frac{1}{3^3} - \frac{1}{4^4} + \cdots$

Hint. Use logarithms and Eq. 3.21, i.e., $x^x = \left(e^{\ln x}\right)^x = e^{x\ln x} = 1 + x\ln x + \frac{(x\ln x)^2}{2!} + \ldots + \frac{(x\ln x)^n}{n!} + \ldots$. Next, integrate the series from 0 to 1 and use the fact that $\int x^n(\ln x)^n dx = \frac{(-1)^n \cdot n!}{(n+1)^{n+1}} + C$

Proof.

$$\int_0^1 x^x dx = \int_0^1 \left(\sum_{n=1}^{\infty} \frac{(x \ln x)^n}{n!} \right) dx = 1 + \sum_{n=1}^{\infty} \frac{(-1)^n n!}{n!(n+1)^{n+1}} = 1 + \sum_{n=1}^{\infty} \frac{(-1)^n}{(n+1)^{n+1}}$$

$$= \sum_{n=1}^{\infty} \frac{(-1)^{n+1}}{n^n} = 1 - \frac{1}{2^2} + \frac{1}{3^3} - \frac{1}{4^4} + \dots$$

101. Evaluate the infinite sum $\sum_{n=2}^{\infty} \frac{1}{n^2+n-2}$.

Hint. Factor the denominator as $(n-1)(n+2)$.

Solution. Because the quantities within each denominator differ by 3, by shifting the summation index, we can rewrite the given series as

$$\frac{1}{3} \left(\sum_{n=2}^{\infty} \frac{1}{n-1} - \sum_{n=2}^{\infty} \frac{1}{n+2} \right)$$

$$= \frac{1}{3} \cdot \left(\sum_{n=1}^{\infty} \frac{1}{n} - \sum_{n=4}^{\infty} \frac{1}{n} \right) = \frac{1}{3} \cdot \left(1 + \frac{1}{2} + \frac{1}{3} \right) = \frac{11}{18}.$$

Answer. 11/18.

102. A debt of $2000 in one year is to be repaid by a payment due two years from now and a final payment of $1000 three years from now. If the interest rate is 4% compounded annually, what is the payment due in two years?

Solution. Let x be the unknown payment. Using the ideas presented in Chapter 4 and after sketching the diagram of the situation, we can equate the present values in both scenarios:

$$2000 \cdot 1.04^{-1} = x \cdot 1.04^{-2} + 1000 \cdot 1.04^{-3}$$
$$x = 1.04^2 \left(2000 \cdot 1.04^{-1} - 1000 \cdot 1.04^{-3} \right) \approx \$1118.46$$

Answer. The payment is $1118.46.

103. Given the sequence 1, 4, 10, 19, 31, 46, 64, ... find its nth term and the sum of the first n terms.

Hint. Consider the series made out of differences of the consecutive terms: 3, 6, 9, 12, 15, 18,... This is an arithmetic progression. Hence the solution is similar to Problem 32.

Solution. Because the first difference forms an arithmetic progression with first term 3 and common difference also 3, then in general it can be written as

$$a_2 - a_1 = d$$
$$a_3 - a_2 = 2d$$
$$a_4 - a_3 = 3d$$

$$\cdots$$

$$a_n - a_{n-1} = a_1 + d(1 + 2 + 3 + \ldots n - 1)$$
$$a_n = a_1 + d \cdot \frac{n(n-1)}{2}$$

So $S_n = \sum_{n=1}^{n} \left(a_1 + \frac{dn(n-1)}{2} \right) = na_1 + \frac{d(n-1)n(n+1)}{6}$. Substituting the values of the first term and common difference, we obtain the formulas for the given sequence, $a_n = a_1 + d \cdot \frac{n(n-1)}{2} = \frac{3n^2 - 3n + 2}{2}$ and $S_n = na_1 + \frac{d(n-1)n(n+1)}{6} = \frac{n^3+n}{2}$. In order to check the formulas, we can sum the first four terms: $1 + 4 + 10 + 19 = 34$ and verify that the sum is equal to that of our formula, $S_4 = \frac{4^3+4}{2} = 34$.

Answer. $a_n = \frac{3n^2-3n+2}{2}$; $S_n = \frac{n^3+n}{2}$.

104. Given an arithmetic progression with the first term a_1 and common difference d, denote by S_n the sum of the first n terms of the series. Evaluate $F(n) = S_{n+3} + 3S_{n+1} - S_n - 3S_{n+2}$.

Hint. Instead of expressing each partial sum using the formula for an arithmetic progression, think about what it means, i.e., "What is S_{n+3} ?" It is $S_n + a_{n+1} + a_{n+2} + a_{n+3}$. Replacing each sum properly, we obtain $S_{n+3} = S_n + a_{n+1} + a_{n+2} + a_{n+3}$; $3S_{n+1} = 3S_n + 3a_{n+1}$; $-S_n = S_n$; $-3S_{n+2} = -3S_n - 3a_{n+1} - 3a_{n+2}$. Adding the left sides of all equations and the right sides, we have $F(n) = a_{n+1} - 2a_{n+2} + a_{n+3} = 0$.

Answer. $F(n) = 0$.

Remark. The reason why the answer is 0 is because for any arithmetic progression a middle term is an arithmetic mean of its left and right neighbors.

105. Investigate the convergence of the series, $\frac{1}{10} + \frac{7}{10^2} - \frac{13}{10^3} + \frac{19}{10^4} + \frac{25}{10^5} - \frac{31}{10^6} + \ldots$.

Solution. The series terms change sign but not alternately ($+ - +$ or $- + -$) so we cannot apply Leibniz's Theorem for alternating series. Consider the corresponding absolute value series, $\sum_{n=1}^{\infty} |u_n| = \sum_{n=1}^{\infty} \left| \frac{a_n}{b_n} \right|$. By inspecting the numerators, $\{|a_n|\} = 1$, 7, 13, 19, 25, 31, $\ldots$ we can see that they represent an arithmetic progression with the first term and common difference of $a_1 = 1$, $d = 6$, respectively. On the other hand, the denominator sequence is a geometric progression with the first term and common ratio $b_1 = 10$, $r = 10$. The common term of the absolute value series can be written as $|u_n| = \frac{a_1 + (n-1)d}{b_1 r^{n-1}} = \frac{6n-5}{10^n}$. Applying the D'Alembert Ratio Test

(Corollary 3.3) to $\sum\limits_{n=1}^{\infty} |u_n| = \sum\limits_{n=1}^{\infty} \left|\frac{a_n}{b_n}\right|$, we obtain $\lim\limits_{n\to\infty} \left|\frac{u_{n+1}}{u_n}\right| = \lim\limits_{n\to\infty} \frac{6n+1}{6n-5} \cdot \frac{10^n}{10^{n+1}} = 0.1$
< 1. Therefore, the given series is absolutely convergent.

Answer. The series absolutely converges.

106. Investigate the convergence of the series $\sum\limits_{n=1}^{\infty} (-1)^{n+1} \arcsin\frac{1}{\sqrt[3]{n}}$.

Solution. This series is an alternating series that satisfies the Leibniz Theorem
(Theorem 3.9), $|u_{n+1}| < |u_n|$ because $\frac{1}{\sqrt[3]{n+1}} < \frac{1}{\sqrt[3]{n}}$ $\forall n \in \mathbb{N}$. Moreover,
$\lim\limits_{n\to\infty} |u_n| = \lim\limits_{n\to\infty} \left(\arcsin\frac{1}{\sqrt[3]{n}}\right) = 0$ and the necessary condition is satisfied. Because
$\arcsin\frac{1}{\sqrt[3]{n}} \sim \frac{1}{\sqrt[3]{n}}$, $n \to \infty$ (see formulas) and the series $\sum\limits_{n=1}^{\infty} \frac{1}{n^{\frac{1}{3}}}$ is divergent, the given
series $\sum\limits_{n=1}^{\infty} (-1)^{n+1} \arcsin\frac{1}{\sqrt[3]{n}}$ is conditionally (not absolutely) convergent.

Answer. The series is conditionally convergent.

107. (1985 USSR Mathematics Olympiad) Solve the equation $\cfrac{x}{2 + \cfrac{x}{2 + \cfrac{x}{\cfrac{...}{2 + \cfrac{x}{\sqrt{1+x}}}}}} = 1$.

Hint. See Problem 90.

Solution. Adding the number 2 to both sides of the equation, we be able to rewrite
its left hand side by the recursive formula, $a_{n+1} = 2 + \frac{x}{a_n}$ with the initial condition
$a_1 = 1 + \sqrt{1+x}$. The characteristic equation $r^2 - 2r - x = 0$ has two roots:
$r_1 = 1 + \sqrt{1+x}$, $r_2 = 1 - \sqrt{1+x}$. Because $r_1 = 1 + \sqrt{1+x}$, this sequence is
constant and the solution to the given equation can be found from solving
$1 + \sqrt{1+x} = 3$. The only solution to this equation is $x = 2$.

Answer. $x = 2$

108. Given an arithmetic progression, $\{a_n\}$, with S_k as the sum of the first k terms of
the progression, it is known that $S_n = m$, $S_m = n$. Evaluate the common difference
of the arithmetic progression.

Hint. Use the formula for a partial sum of an arithmetic progression and solve a
system of two equations.

Solution. Suppose that the first term and common difference of the progression are
a_1, d, respectively. We have the system,

$$\begin{cases} \dfrac{2a_1 + (m-1)d}{2} \cdot m = n \\[2ex] \dfrac{2a_2 + (n-1)d}{2} \cdot n = m \end{cases} \Leftrightarrow \begin{cases} 2a_1 + (m-1)d = \dfrac{2n}{m} \\[2ex] 2a_1 + (n-1)d = \dfrac{2m}{n} \end{cases}$$

Subtracting the left and the right sides of the equations of the system and using the difference of squares formula, we obtain $d = \frac{2(n+m)}{nm}$.

Answer. $d = \frac{2(n+m)}{nm}$

109. The functions $-\sin x$, $4\sin x \cdot \cot 2x$, $\cos x$ are the kth, $(k+1)$, and $(k+2)$ terms of an arithmetic progression, respectively. Find all values of x and k if the 7th term of this progression is $1/5$.

Solution. The middle term of the three consecutive terms of an arithmetic progression must be the arithmetic mean of two neighbors, i.e., it must satisfy the equation, $4\sin x \cot 2x = \frac{\cos x - \sin x}{2}$, simplifying which, we have $4\tan^2 x - \tan x - 3 = 0$ so there are two possibilities:

$$1)\ \tan x = 1 \qquad 2)\ \tan x = -\frac{3}{4}$$

$$x = \frac{\pi}{4} + \pi n \qquad \left[\begin{array}{l} x = -\arctan\frac{3}{4} + 2\pi k \\[2ex] x = \pi - \arctan\frac{3}{4} + 2\pi k \end{array}\right. \qquad n, k \in \mathbb{Z}.$$

Case 1. $\sin x = \cos x = \pm\frac{1}{\sqrt{2}}$, $\cot 2x = 0 \Rightarrow d = \mp\frac{1}{\sqrt{2}}$.

Using the value of the 7th term, we have

$$a_7 = \frac{1}{5} = a_1 + 6d = a_1 \mp \frac{6}{\sqrt{2}}$$

$$a_1 = \frac{1}{5} \pm \frac{6}{\sqrt{2}}.$$

Let us find the value of k:

$$a_k = a_1 + (k-1)d$$

$$a_k = -\sin x = \mp\frac{1}{\sqrt{2}}$$

From this we obtain $5(k-8) = \pm\sqrt{2}$, which has no solutions in the integers. Therefore, case 1 is not applicable.

Case 2. We evaluate that if

$$x = -\arctan\frac{3}{4} + 2\pi n \Rightarrow \sin x = -\frac{3}{5}, \cot 2x = -\frac{7}{24},$$
$$d = \frac{1}{10}, a_1 = -\frac{2}{5}, a_k = -\frac{2}{5} + \frac{k-1}{10} = \frac{3}{5},$$
$$k = 11.$$

Additionally, for the second possible value of $x = \pi - \arctan\frac{3}{4} + 2\pi n$ we obtain the following

$$\sin x = \frac{3}{5}, \cot 2x = -\frac{7}{24},$$
$$d = -\frac{1}{10}, a_1 = \frac{4}{5}, a_k = \frac{4}{5} - \frac{k-1}{10} = \frac{3}{5}$$
$$k = 15.$$

Answer.
$$\left[\begin{array}{l} x = \arctan\dfrac{3}{4} + 2\pi n, \ n \in \mathbb{Z}, \ k = 11 \ \left(a_1 = -\dfrac{2}{5}, \ d = \dfrac{1}{10}\right) \\ x = \pi - \arctan\dfrac{3}{4} + 2\pi n, \ n \in \mathbb{Z}, \ k = 15 \ \left(a_1 = \dfrac{4}{5}, \ d = -\dfrac{1}{10}\right) \end{array}\right.$$

110. Find Maclaurin series for $y = e^{\arctan x}$.

Hint. Use the method of undetermined coefficients (Problem 133).

Solution. Let us find the derivative of the given function and then express the function in terms of its derivative:

$$y' = e^{\arctan x} \cdot \frac{1}{1+x^2}$$
$$y = (1+x^2)y'$$

Next, we rewrite the functions as a power series and also in terms of its derivative:

$$y = a_0 + a_1 x + a_2 x^2 + \ldots + a_n x^n + \ldots$$
$$y = (1+x^2)(a_1 + 2a_2 x + 3a_3 x^2 + \ldots + na_n x^{n-1} + (n+1)a_{n+1}x^n + \ldots)$$

Equating the coefficients of x raised to the same power, we obtain the following true relationships:

$$a_1 = a_0$$
$$2a_2 = a_1$$
$$1 \cdot a_1 + 3a_3 = a_2$$
$$2a_2 + 4a_4 = a_3$$
$$\ldots$$
$$(n-1)a_{n-1} + (n+1)a_{n+1} = a_n$$

Using the initial condition and the formula above we can obtain coefficients of the power series:

$$x = 0 \cdot e^{\arctan 0} = 1 \; \Rightarrow a_0 = y(0) = 1,$$

$$a_2 = \frac{1}{2}, \; a_3 = -\frac{1}{6}, \; a_4 = \frac{7}{24}, \; a_5 = \frac{5}{120} = \frac{1}{24}, \; a_6 = \frac{29}{144}, \; \ldots$$

$$a_{n+1} = \frac{a_n - (n-1)a_{n-1}}{n+1}.$$

Finally, we have the requested Maclaurin series,

$$y = e^{\arctan x} = 1 + x + \frac{x^2}{2} - \frac{1}{6}x^3 - \frac{7}{24}x^4 + \frac{1}{24}x^5 + \frac{29}{144}x^6 + \ldots$$

Answer. $y = e^{\arctan x} = 1 + x + \frac{x^2}{2} - \frac{1}{6}x^3 - \frac{7}{24}x^4 + \frac{1}{24}x^5 + \frac{29}{144}x^6 + \ldots$

111.   Using   power   series,   solve   the   differential   equation $y'' = x \cos y'$, $y(1) = \frac{\pi}{2}$, $y'(1) = 0$.

Solution. We look for a solution in the form,

$$y = f(1) + \frac{f'(1)(x-1)}{1!} + \frac{f''(1)(x-1)^2}{2!} + \frac{f'''(1)(x-1)^3}{3!} + \ldots$$

Evaluating several derivatives, we have

$$y'' = x \cos y'$$
$$y''(1) = 1 \cdot \cos y'(1) = 1$$
$$y''' = \cos y' + x \sin y' \cdot y''$$
$$y'''(1) = 1$$
$$y^{(4)} = -\sin y' \cdot y'' - y'' \sin y' - xy''' \sin y' - xy'' \cos y' \cdot y''$$
$$y^{(4)}(1) = -1 \ldots$$

Finally, we obtain the Taylor series solution to the given differential equation,

$$y = \frac{\pi}{2} + 0 \cdot (x-1) + \frac{(x-1)^2}{2!} + \frac{(x-1)^3}{3!} - \frac{(x-1)^4}{4!} + \ldots$$

Answer. $y = \frac{\pi}{2} + \frac{(x-1)^2}{2!} + \frac{(x-1)^3}{3!} - \frac{(x-1)^4}{4!} + \ldots$.

112. How many terms of the infinite series $\sum\limits_{n=1}^{\infty} \frac{1}{n^4}$ are needed in order to evaluate it with accuracy of 10^{-3}?

Solution. This famous series converges to $\frac{\pi^4}{90}$ and was first evaluated by Euler. However, we want to approximate the sum of the infinite series so we use the Cauchy integral property for the remainder $r_n = S - S_n \leq \int\limits_{n-1}^{\infty} \frac{dx}{x^4} = \frac{1}{3(n-1)^3}$. We can see that the remainder is less than $1/1000$ if we approximate the sum by eight terms, $\frac{1}{3(8-1)^3} = \frac{1}{3 \cdot 343} = \frac{1}{1029} < \frac{1}{1000}$. Hence $n = 8$. $S_8 = 1 + \frac{1}{2^4} + \frac{1}{3^4} + \frac{1}{4^4} + \ldots + \frac{1}{8^4} \approx$ 1.08178. Note that $\frac{\pi^4}{90} \approx 1.082323234$.

Answer. We need 8 terms.

113. Evaluate the infinite sum $\sum\limits_{n=1}^{\infty} \frac{1}{n^2+3n}$.

Solution. If we factor the denominator, we can see that the two quantities differ by 3 and we can rewrite the common term of the series as the difference of two fractional expression multiplied by $1/3$,

$$a_n = \frac{1}{n(n+3)} = \frac{1}{3}\left(\frac{1}{n} - \frac{1}{n+3}\right).$$

The given sum now can be rewritten as $\sum\limits_{n=1}^{\infty} \frac{1}{n^2+3n} = \frac{1}{3}\left(\sum\limits_{n=1}^{\infty} \frac{1}{n} - \sum\limits_{n=1}^{\infty} \frac{1}{n+3}\right)$. Using the shifting summation index property, $\sum\limits_{n=1}^{\infty} \frac{1}{n+3} = \sum\limits_{n=4}^{\infty} \frac{1}{n}$, we can simplify this difference as $\frac{1}{3}\left(\sum\limits_{n=1}^{3} \frac{1}{n} + \sum\limits_{n=4}^{\infty} \frac{1}{n} - \sum\limits_{n=4}^{\infty} \frac{1}{n}\right) = \frac{1}{3}\left(1 + \frac{1}{2} + \frac{1}{3}\right) = \frac{11}{18}$.

Answer. 11/18.

114. Find integer solutions to $(x^2 + x + 1) + (x^2 + 2x + 3) + (x^2 + 3x + 5) + \ldots + (x^2 + 20x + 39) = 510$.

Solution. You can recognize this as the sum of an arithmetic progression where the first term, common difference and 20th term are: $a_1 = x^2 + x + 1$, $d = x + 2$, $a_{20} = x^2 + x + 1 + 19(x + 2) = x^2 + 20x + 39$. The formula for the sum of the first 20 terms of such a progression is a quadratic equation,

$$\frac{x^2 + x + 1 + x^2 + 20x + 39}{2} \cdot 20 = 510$$

$$2x^2 + 21x - 11 = 0$$

$$x = \frac{-21 \pm \sqrt{21^2 + 4 \cdot 2 \cdot 11}}{4} = \frac{-21 \pm 23}{4}$$

$$x = -11 \in Z, \quad x = \frac{1}{2}$$

$$a_1 = 111, \quad d = -9.$$

where we selected only the integer solution.

Answer. $x = -11$

115. A piece of charcoal from a tree killed by the volcanic eruption that formed the caldera of Crater Lake in the state of Oregon measured a radioactivity of 7.0 dpm/ g C. Assume that the steady state radioactivity of carbon-14 in the living tree is 16.0 dpm/g C (16 counts per second). How long ago did the eruption occur?

Solution. We calculate the age of the charcoal using the equation, $N(t) = N_0 \left(\frac{1}{2}\right)^{\frac{t}{t_{1/2}}}$, which can be rewritten also as

$$\ln\left(\frac{N_0}{N}\right) = \frac{t}{t_{1/2}} \ln 2$$

$$t = t_{1/2} \frac{\ln\left(\frac{N_0}{N}\right)}{\ln 2}.$$

Substituting $t_{1/2} = 5730$, $N_0 = 16$, $N = 7$, we obtain an approximate age, $t = 5730 \frac{\ln\frac{16}{7}}{\ln 2} \approx 6834$ years.

Answer. The volcanic eruption occurred approximately 6834 years ago.

116. Find the difference of two distinct positive numbers if their arithmetic mean is $3\sqrt{3}$ and geometric mean is $\sqrt{2}$.

Solution. Denote the numbers by x, y, $x > y > 0$. next, we have

$$\begin{cases} \dfrac{x+y}{2} = 3\sqrt{3} \\ xy = 2 \end{cases} \Rightarrow (x-y)^2 = (x+y)^2 - 4xy = \left(2 \cdot 3\sqrt{3}\right)^2 - 4 \cdot 2 = 100.$$

Because the numbers are positive, their difference is 10.

Answer. 10.

117. Estimate the value of irrational number e and estimate the error of the estimation.

Solution.   Consider   the   Maclaurin   series   for e,   i.e., $e^x = 1 + x + \frac{x^2}{2!} + \frac{x^3}{3!} + \cdots + \frac{x^n}{n!} + \ldots$, $x \in R$. Replacing x by 1 and cutting off the infinite   series   at   the nth term,   we   obtain   the   approximation   for $e, e \approx 1 + 1 + \frac{1}{2} + \frac{1}{6} + \ldots + \frac{1}{n!}$. For each value of n, the estimation will be different. Let us consider the remainder

$$R_n = \frac{1}{(n+1)!} + \frac{1}{(n+2)!} + \frac{1}{(n+3)!} + \cdots$$

$$= \frac{1}{(n+1)!}\left(1 + \frac{1}{n+2} + \frac{1}{(n+2)(n+3)} + \frac{1}{(n+2)(n+3)(n+4)} + \cdots\right)$$

$$< \frac{1}{(n+1)!}\left(1 + \frac{1}{n+1} + \frac{1}{(n+1)^2} + \frac{1}{(n+1)^3} + \cdots\right)$$

$$= \frac{1}{(n+1)!}\left(\frac{1}{1 - \frac{1}{n+1}}\right) = \frac{1}{(n+1)!} \cdot \frac{1}{n}$$

The quantity inside the parentheses is always less than the sum of the decreasing infinite geometric progression. Hence, the remainder of the approximation satisfies the inequality $R_n < \frac{1}{n!n}$. If we estimate the value of e by taking only four terms of the infinite series, the error of such estimation will be

$$R_4 < \frac{1}{4! \cdot 4} = \frac{1}{96} > \frac{1}{100},$$

$$\frac{1}{100} < R_5 < \frac{1}{5!5} = \frac{1}{600} < \frac{1}{1000}.$$

Approximation of e by four terms would give us 2.67 and by five terms 2.71.

118. Evaluate the infinite sum $S = -\frac{3}{2}x^2 + \frac{3}{8}x^4 - \frac{19}{720}x^6 + \frac{33}{8!}x^8 + \cdots$.

Hint. The given series can be written using sigma notation as $\sum\limits_{n=1}^{\infty} \frac{(-1)^n(2n^2+1)}{(2n)!} \cdot x^{2n}$.

Solution. We can rewrite the series as the sum of two series,

$$\sum_{n=1}^{\infty} \frac{(-1)^n(2n^2+1)}{(2n)!} \cdot x^{2n} = \sum_{n=1}^{\infty} \frac{(-1)^n x^{2n}}{(2n)!} + \sum_{n=1}^{\infty} \frac{(-1)^n 2n \cdot n}{(2n-1)!n} \cdot x^{2n}.$$

You can see that the first series is the Maclaurin series for $\cos x$. Next, we focus on the second series,

$$\sum_{n=1}^{\infty} \frac{(-1)^n}{(2n-1)!}nx^{2n} = x \cdot \sum_{n=1}^{\infty} \frac{(-1)^n x^{2n-1}n}{(2n-1)!} = x \cdot f(x).$$

Let us find this function! Notice that $f(0) = 0$ and consider the integral,

$$\int_0^x f(t)dt = \sum_{n=1}^{\infty} \frac{(-1)^n x^{2n} n}{2n \cdot (2n-1)!} = \frac{x}{2} \cdot \boxed{\sum_{n=1}^{\infty} \frac{(-1)^n x^{2n-1}}{(2n-1)!}} = -\frac{x}{2} \cdot \sin x.$$

Hence, after differentiation we obtain the function as $f(x) = -\frac{1}{2}\sin x - \frac{x}{2}\cos x$. Finally, the given sum is equal to $S = \cos x - \frac{1}{2}\sin x - \frac{x}{2}\cos x$.

Answer. $S = -\frac{1}{2}\sin x + \left(1 - \frac{x}{2}\right)\cos x$.

119. Find the nth term of the infinite series and evaluate the infinite sum,

$$\frac{0}{1} - \frac{1}{2} - \frac{1}{4} - \frac{1}{8} + \frac{0}{16} + \frac{2}{32} + \frac{6}{64} + \frac{13}{128} + \frac{25}{256} + \dots$$

Solution. It seems reasonable to look for the nth term in the form $u_n = \frac{a_n}{b_n}$, $b_n = 2^{n-1}$, $n \in \mathbb{N}$, $a_n = ?$ Consider the sequence of the numerators: $0, -1, -1, -1, 0, 2, 6, 13, 25, \dots$. Next, subtract two consecutive terms (find the first difference as we did in solving Prob. 32) $-1, 0, 0, 1, 2, 4, 7, 12, \dots$. This is the Fibonacci sequence minus 1! Let us again subtract pairs of consecutive terms and find the second sequence of differences: $1, 0, 1, 1, 2, 3, 5, \dots$ which is also a Fibonacci type sequence. Hence, the given series can be written as

$$\frac{1-1}{1} + \frac{1-2}{2} + \frac{2-3}{2^2} + \frac{3-4}{2^3} + \frac{5-5}{2^4} + \frac{8-6}{2^5} + \frac{13-7}{2^6} + \frac{21-8}{2^7}$$
$$+ \frac{34-9}{2^8} + \frac{55-10}{2^9} + \frac{89-11}{2^{10}} + \dots$$
$$= \left(\frac{1}{2^0} + \frac{1}{2^1} + \frac{2}{2^2} + \frac{3}{2^3} + \frac{5}{2^4} + \frac{8}{2^5} + \frac{13}{2^6} + \frac{21}{2^7} + \frac{34}{2^8} + \frac{55}{2^9} + \frac{89}{2^{10}} + \dots + \frac{F_n}{2^n} + \dots \right)$$
$$- \left(\frac{1}{1} + \frac{2}{2} + \frac{3}{2^2} + \frac{4}{2^3} + \frac{5}{2^4} + \frac{6}{2^5} + \frac{7}{2^6} + \frac{8}{2^7} + \frac{9}{2^8} + \frac{10}{2^9} + \frac{11}{2^{10}} + \dots + \frac{n}{2^{n-1}} + \dots \right)$$
$$= 0.$$

Here F_n is the nth Fibonacci number. Let us explain how we obtain the answer. Evaluating each sum inside parentheses separately, denote the first sum by $S = 1 + \frac{1}{2} + \frac{2}{4} + \frac{3}{8} + \frac{5}{16} + \frac{8}{32} + \frac{13}{64} + \frac{21}{128} + \frac{34}{256} + \frac{55}{512} + \frac{89}{1024} + \dots$. Dividing it by 2 we obtain $\frac{S}{2} = \frac{1}{2} + \frac{1}{4} + \frac{2}{8} + \frac{3}{16} + \frac{5}{32} + \frac{8}{64} + \frac{13}{128} + \frac{21}{256} + \frac{34}{512} + \frac{55}{1024} + \dots$. Subtracting S and $S/2$ we obtain a new expression for $S/2$: $\frac{S}{2} = S - \frac{S}{2} = 1 + \frac{1}{4} + \frac{1}{8} + \frac{2}{16} + \frac{3}{32} + \frac{5}{64} + \frac{8}{128} + \frac{13}{256} + \frac{21}{512} + \frac{34}{1024} + \dots$. On the other hand, dividing the first sum by four we obtain $\frac{S}{4} = \frac{1}{4} + \frac{1}{8} + \frac{2}{16} + \frac{3}{32} + \frac{5}{64} + \frac{8}{128} + \frac{13}{256} + \frac{21}{512} + \frac{34}{1024} + \dots$. We notice that this is the sum for $S/2$ minus one, which can be written as $1 + \frac{S}{4} = \frac{S}{2} \Rightarrow S = 4$.

Similarly, we can evaluate the sum inside second parentheses

$$\sigma = 1 + \frac{2}{2} + \frac{3}{2^2} + \frac{4}{2^3} + \frac{5}{2^4} + \frac{6}{2^5} + \frac{7}{2^6} + \frac{8}{2^7} + \frac{9}{2^8} + \frac{10}{2^9} + \frac{11}{2^{10}} + \cdots$$

$$2\sigma = 4 + \frac{3}{2} + \frac{4}{2^2} + \frac{5}{2^3} + \frac{6}{2^4} + \frac{7}{2^5} + \frac{8}{2^6} + \frac{9}{2^7} + \frac{10}{2^8} + \frac{11}{2^9} + \frac{12}{2^{10}} + \cdots$$

$$\sigma = 2\sigma - \sigma = 3 + \frac{1}{2} + \frac{1}{2^2} + \frac{1}{2^3} + \frac{1}{2^4} + \cdots = 3 + \frac{\frac{1}{2}}{1 - \frac{1}{2}} = 3 + 1 = 4.$$

Here we rewrote the second sum in a different form and applied the formula for the sum of infinite geometric progression. Finally, the difference of two identical quantities is zero ($S - \sigma = 4 - 4 = 0$).

120. Prove that $\frac{a+b}{a^2+b^2} + \frac{b+c}{b^2+c^2} + \frac{c+a}{c^2+a^2} \leq \frac{1}{a} + \frac{1}{b} + \frac{1}{c}$, $a, b, c > 0$.

Proof. Using the inequality between the arithmetic and geometric means, we obtain the following chain of true inequalities:

$$\left. \begin{array}{l} a^2 + b^2 \geq 2ab \ \Rightarrow \ \dfrac{1}{a^2 + b^2} \leq \dfrac{1}{2ab} \\[2mm] c^2 + b^2 \geq 2cb \ \Rightarrow \ \dfrac{1}{c^2 + b^2} \leq \dfrac{1}{2cb} \\[2mm] a^2 + c^2 \geq 2ac \ \Rightarrow \ \dfrac{1}{a^2 + c^2} \leq \dfrac{1}{2ac} \end{array} \right\}$$

and each of these inequalities can produce the following inequalities:

$$\frac{a+b}{a^2+b^2} \leq \frac{a+b}{2ab} = \frac{1}{2b} + \frac{1}{2a}$$

$$\frac{c+b}{c^2+b^2} \leq \frac{c+b}{2cb} = \frac{1}{2b} + \frac{1}{2c}$$

$$\frac{a+c}{a^2+c^2} \leq \frac{a+c}{2ac} = \frac{1}{2c} + \frac{1}{2a}$$

Adding the left and right sides,

$$\frac{a+b}{a^2+b^2} + \frac{b+c}{b^2+c^2} + \frac{c+a}{c^2+a^2} \leq \frac{2}{2a} + \frac{2}{2b} + \frac{2}{2c} = \frac{1}{a} + \frac{1}{b} + \frac{1}{c},$$

which completes the proof.

Appendix 1
MAPLE Program for Fibonacci Application

```
> restart:with(LREtools):
```

```
> eq:=R(n+2)=R(n+1)+R(n);
```

$$eq := R(n + 2) = R(n + 1) + R(n)$$

and solved for $R(n)$ subject to the two initial conditions $R(0) = 0$ and $R(1) = N$

```
> sol:=rsolve({eq,R(0)=0,R(1)=N},R(n));
```

$$sol := \frac{1}{5} \sqrt{5} \, N \left(\frac{1}{2} + \frac{1}{2} \sqrt{5} \right)^n - \frac{1}{5} N \sqrt{5} \left(\frac{1}{2} - \frac{1}{2} \sqrt{5} \right)^n$$

```
> number:=factor(sol);
```

$$number := \frac{1}{5} \, N \sqrt{5} \left(\left(\frac{1}{2} + \frac{1}{2} \sqrt{5} \right)^n - \left(\frac{1}{2} - \frac{1}{2} \sqrt{5} \right)^n \right)$$

```
> N:=1; n:=24;
```

$$N := 1$$

$$n := 24$$

```
> unassign('n'):
```

© Springer International Publishing Switzerland 2016
E. Grigorieva, *Methods of Solving Sequence and Series Problems*,
DOI 10.1007/978-3-319-45686-7

> `Rabbit_pairs:=radnormal(number);`

$$Rabbit\_pairs := \frac{1}{5}\sqrt{5}\left(\left(\frac{1}{2}+\frac{1}{2}\sqrt{5}\right)^{n}-\left(\frac{1}{2}-\frac{1}{2}\sqrt{5}\right)^{n}\right)$$

> `REplot(eq,R(n),{R(0)=0,R(1)=1},      1..24,      labels=["n","R"],`
`tickmarks=[3,3], color=green, thickness=3);`

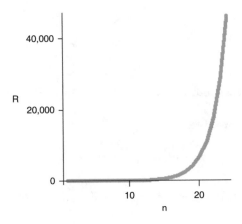

>

Appendix 2
Method of Differences

Consider a sequence $\{a_n\}$, a_1, a_2, a_3, a_4,... and another sequence, $\{b_n\}$, b_1, b_2, b_3, b_4,..., such that the following relationships are valid:

$$a_2 - a_1 = b_1$$
$$a_3 - a_2 = b_2$$
$$a_4 - a_3 = b_3$$
$$\cdots$$
$$a_n - a_{n-1} = b_{n-1}$$
$$a_{n+1} - a_n = b_n.$$

If we add the left and right sides of all equations and denote by S_n the nth partial sum of sequence $\{b_n\}$ we obtain the formula,

$$S_n = \sum_{k=1}^{n} b_k = a_{n+1} - a_1.$$

This formula gives us a straightforward method of finding the nth term of a sequence with integer terms when other methods do not work or perhaps demand too much creativity. Usually one has to subtract the consecutive terms of the obtained difference sequences until a difference sequence contains only the same numbers. Some ideas are demonstrated in the following problems.

Problem A1 Given the sequence 1, 5, 15, 35, 70, 126, 210, ... find its nth term and the sum of the first n terms.

Solution. Consider the series made out of differences of the consecutive terms: 4, 10, 20, 35, 56, 84,... , then a new sequence made out of the difference the consecutive terms of the new sequence 6, 10, 15, 21, 28, ... Finally, the last

© Springer International Publishing Switzerland 2016
E. Grigorieva, *Methods of Solving Sequence and Series Problems*,
DOI 10.1007/978-3-319-45686-7

sequence made of the difference of the consecutive terms of $4, 5, 6, 7, \ldots$ This is an arithmetic progression with the first term 4 and common difference 1. In general, the solution is similar to Problem 32 or to the homework problem. Let us consider the following sequences:

$$\{a_i\} \quad 1, \quad 5, \quad 15, \quad 35, \quad 70, \quad 126, \quad 210, \ldots$$
$$\{b_i\} \quad 4, \quad 10, \quad 20, \quad 35, \quad 56, \quad 84, \ldots$$
$$\{c_i\} \quad 6, \quad 10, \quad 15, \quad 21, \quad 28, \ldots.$$
$$\{d_i\} \quad 4, \quad 5, \quad 6, \quad 7, \ldots.$$
$$\{e_1\} \quad 1, \quad 1, \quad 1, \ldots.$$

in general it can be written as

$$c_2 - c_1 = d_1$$
$$c_3 - c_2 = d_2$$
$$c_4 - c_3 = d_3$$
$$\ldots$$
$$c_n - c_{n-1} = d_{n-1},$$

which is equivalent to $c_n = c_1 + \sum_{n=1}^{n-1} d_n$.

Because the last difference sequence $\{d_i\}$ forms an arithmetic progression with the first term 4 and common difference of 1, then its nth term can be written as $d_n = 4 + (n-1) \cdot 1 = n + 3$ and the sum of the first $n-1$ terms will be $S_{n-1} = \frac{2 \cdot 4 + (n-2) \cdot 1}{2} \cdot (n-1) = \frac{(n+6)(n-1)}{2} = \frac{n^2 + 5n - 6}{2}$.   Therefore, $c_n = 6 + \frac{n^2 + 5n - 6}{2} = \frac{n^2 + 5n + 6}{2} = \frac{(n+2)(n+3)}{2}$. This formula as we can see works just fine since $c_1 = \frac{(1+2)(1+3)}{2} = 6$, $c_2 = \frac{(2+2)(2+3)}{2} = 10$, $c_3 = \frac{(3+2)(3+3)}{2} = 15$, etc.

Going up we can now find the second sequence from the top (the first difference)

$$b_2 - b_1 = c_1$$
$$b_3 - b_2 = c_2$$
$$b_4 - b_3 = c_3$$
$$\ldots$$
$$b_n - b_{n-1} = c_{n-1}$$

which is equivalent to

$$
\begin{aligned}
b_n &= b_1 + \sum_{n=1}^{n-1} c_n = 4 + \sum_{n=1}^{n-1} \frac{n^2 + 5n + 6}{2} \\
&= 4 + \frac{1}{2}\left(\frac{(n-1)n(2n-1)}{6} + 5\frac{n(n-1)}{2} + 6(n-1) \right) \\
&= 4 + \frac{(n-1)(n^2 + 7n + 18)}{6} = \frac{(n+1)(n+2)(n+3)}{6}
\end{aligned}
$$

Finally,

$$
\begin{aligned}
a_n &= a_1 + \sum_{n=1}^{n-1} b_n = 1 + \sum_{n=1}^{n-1} \frac{(n+1)(n+2)(n+3)}{6} \\
&= 1 + \frac{1}{6}\sum_{n=1}^{n-1} n^3 + \sum_{n=1}^{n-1} n^2 + \frac{11}{6}\sum_{n=1}^{n-1} n + (n-1) \\
&= n + \frac{1}{6} \cdot \frac{(n-1)^2 n^2}{4} + \frac{(n-1)n(2n-1)}{6} + \frac{11(n-1)n}{12} = \frac{n^4 + 6n^3 + 11n^2 + 6n}{24} \\
&= \frac{n(n+1)(n+2)(n+3)}{24}
\end{aligned}
$$

We can see that each term of the given sequence can be evaluated using this formula

$$
a_1 = \frac{(1 \cdot 2 \cdot 3 \cdot 4)}{24} = 1, \; a_2 = \frac{2 \cdot 3 \cdot 4 \cdot 5}{24} = 5, \; \ldots, \; a_5 = \frac{5 \cdot 6 \cdot 7 \cdot 8}{24} = 70, \; \ldots
$$

Problem 37 Find the formula for the nth term of the sequence 1, 4, 10, 20, 35, 56, 84, 120, ... (*See Chapter 1 for geometric interpretation of this problem.*)

Solution. We can subtract consecutive terms until the difference of the consecutive terms become the same

$$
\begin{aligned}
&\{a_i\} \; 1, \; 4, \; 10, \; 20, \; 35, \; 56, \; 84, \; 120, \; \ldots \\
&\{b_i\} \; 3, \; 6, \; 10, \; 15, \; 21, \; 28, \; 36, \ldots \\
&\{c_i\} \quad 3, \; 4, \; 5, \; 6, \; 7, \; 8, \ldots \\
&\{d_i\} \qquad 1, \; 1, \; 1, \; 1, \; 1, \ldots
\end{aligned}
$$

Moving from the last row up, we obtain that

$$a_n = a_1 + \sum_{n=1}^{n-1} b_n$$

$$b_n = b_1 + \sum_{n=1}^{n-1} c_n$$

$$a_n = a_1 + b_1(n-1) + \sum_{n=1}^{n-1}\left(\sum_{k=1}^{n-1} c_k\right).$$

The sum inside the parentheses is the $(n-1)^{\text{st}}$ partial sum of the arithmetic progression $\{c_i\}$ with the first term 3 and common difference of 1.

$$S_{n-1} = \frac{2\cdot 3 + n - 2}{2}\cdot(n-1) \Rightarrow b_n = 3 + \frac{(n+4)(n-1)}{2} = \frac{n^2 + 3n + 2}{2}$$

$$a_n = 1 + \frac{1}{2}\cdot\sum_{n=1}^{n-1}(n^2 + 3n + 2) = 1 + \frac{1}{2}\cdot\left(\frac{(n-1)n(2n-1)}{6} + 3\frac{n(n-1)}{2} + 2(n-1)\right)$$

$$= 1 + \frac{n-1}{12}(n(2n-1) + 9n + 12) = \frac{n(n+1)(n+2)}{6}$$

Answer. $a_n = \frac{n(n+1)(n+2)}{6}$.

References

1. Williams, K.S., Hardy, K., The Red Book of Mathematics Problems (Undergraduate William Lowell Putnam competition). Dover, Mineola, NY (1996)
2. Grigorieva, E.V., Methods of Solving Complex Geometry Problems. Birkhäuser, Basel (2013)
3. Grigorieva, E.V., Methods of Solving Nonstandard Problems. Birkhäuser, Basel (2015)
4. Grigorieva, E.V., Complex Math Problems and How to Solve Them, vol. 1. TWU Press. Library of Congress, TX u 001007606/2001-07-11 (2001)
5. Dudley, U., Number Theory, 2nd edn. Dover, Mineola, NY (2008)
6. Barton, D., Elementary Number Theory, 6th edn. McGraw Hill, New York (2007)
7. Grigoriev, E. (ed.), Problems of the Moscow State University Entrance Exams, pp. 1–132. MAX-Press, Moscow (2002) (In Russian)
8. Grigoriev, E. (ed.), Problems of the Moscow State University Entrance Exams, pp. 1–92. MAX-Press, Moscow (2000) (in Russian)
9. Grigoriev, E. (ed.), Olympiads and Problems of the Moscow State University Entrance Exams. MAX-Press, Moscow (2008) (in Russian)
10. Wilf, H.S., Generating Functionology. Academic Press, New York (1994)
11. Lidsky, B., Ovsyannikov, L., Tulaikov, A., Shabunin, M., Problems in Elementary Mathematics. MIR Publisher, Moscow (1973)
12. Kaganov, E.D., 400 of the Most Interesting Problems with Solutions. MIR, Moscow (1997)
13. Rivkin, A.A., Problem Book for the Preparation to the Math Entrance Exam. MIR, Moscow (2003)
14. Vinogradova, Olehnik, and Sadovnichii, Mathematical Analysis, Vol. 2, Factorial (1996) (in Russian)
15. http://takayaiwamoto.com/Sums_and_Series/sumint_1.html
16. Dunham, W., Euler, The Master of Us All. The Mathematical Association of America, Washington, DC (1999)
17. Eves, J.H., An Introduction to the History of Mathematics with Cultural Connections, pp. 261–263. Harcourt College Publishers, San Diego, CA (1990)
18. Lander, L.J., Parkin, T.R., Consecutive primes in arithmetic progression. Math. Comput. **21**, 489 (1967)
19. Sierpinski, W., 250 Problems in Elementary Number Theory. Elsevier, New York (1970)
20. Alfutova, N., Ustinov, A., Algebra and Number Theory. MGU, Moscow (2009) (in Russian)
21. Currie, L.A., The remarkable metrological history of radiocarbon dating. J. Res. Natl. Inst. Stand. Technol. **109**, 185–217 (2004)

22. Anderson, E.C., Libby, W.F., Weinhouse, S., Held, A.F., Kirsohenbaum, A.D., Grosse, A.V., Radiocarbon from cosmic radiation. Science **105**, 576 (1947)
23. Grosse, A.F., Libby, W.F., Cosmic radiocarbon and natural radioactivity of living matter. Science **106**, 88 (1947)

Contest Problems for Further Reading

24. Shklarsky, D.O., Chentzov, N.N., Yaglom, I.M., The USSR Olympiad Problem Book: Selected Problems and Theorems of Elementary Mathematics. Dover, Mineola, NY (1993)
25. Past USAMO tests, The USA Mathematical Olympiads from the late 70s and early 80s can be found at. Art of Problem Solving website http://www.artofproblemsolving.com
26. Andreescu, T., Feng, Z., Olympiad books. USA and International Mathematical Olympiads (MAA Problem Books Series)
27. Andreescu, T., Feng, Z, Lee, G., Jr. (eds.), Mathematical Olympiads 2000-2001: Problems and Solutions from Around the World. MAA (2003)
28. Andreescu, T., Feng, Z. (eds.), Mathematical Olympiads 1999–2000: Problems and Solutions from Around the World. MAA (2001)

Index

© Springer International Publishing Switzerland 2016
E. Grigorieva, *Methods of Solving Sequence and Series Problems*,
DOI 10.1007/978-3-319-45686-7

Printed in the United States
By Bookmasters